The Rigger's Locker

The Good Luck Knot: elegant, graceful, useful.
See page 204 for an explanation.

The Rigger's Locker

Tools, Tips, and Techniques for Modern and Traditional Rigging

Brion Toss

Illustrated by Robert Shetterly

International Marine
Camden, Maine

This book is dedicated to the memory of
Nick Benton, Master Rigger.

Published by International Marine

10 9 8 7 6 5 4 3 2

Library of Congress Cataloging-in-Publication Data
Toss, Brion.
The rigger's locker: tools, tips, and techniques for modern and traditional rigging / Brion Toss; illustrated by Robert Shetterly.
p. cm.
Includes index.
ISBN 0-87742-961-8
1. Marlingspike seamanship. 2. Masts and rigging—Maintenance and repair. I. Title.
VM531.T67 1992
623.88'82—dc20 92-13028
CIP

TAB BOOKS offers software for sale. For information and a catalog, please contact TAB Software Department, Blue Ridge Summit, PA 17294-0850.

Questions regarding the content of this book should be addressed to:

International Marine
P.O. Box 220
Camden, ME 04843

Imageset by High Resolution, Camden, ME
Printed by Arcata Graphics, Fairfield, PA
Design by Faith Hague, Watermark Design, Rockport, ME
Edited by Jonathan Eaton, Tom McCarthy, Roger C. Taylor
Production by Janet Robbins

Contents

List of Miscellany: Useful Notes and Observations

Acknowledgments

Very little in this book is "mine." Instead, I have had the labor-intensive honor of recording, amplifying, and collating the ideas of others. If I knew who they were, they are credited in the text. So first of all, a warm thank you to all the unknown benefactors.

Thanks to *SAIL*, *WoodenBoat*, and *Classic Boat* magazines for supporting the book by printing excerpts from it. Special thanks to: David Jackson for support since before the beginning; Venita Robertson for a second wind; Robert Shetterly for letting me do his dialog; Jon Eaton for waiting; and all the gang at Point Hudson, for their inspiration. Thanks are also in order to the Artisans School of Rockport, Maine, for providing the material and setting for the cover photograph.

Foreword

I first met Brion Toss a dozen or so years ago. He was demonstrating traditional rigging techniques at a wooden boat revivalist encampment down by the shores of Lake Union in Seattle, Washington. He had a long length of rope stretched between two posts set in the ground, and he was worming, parcelling, and serving, which is to say he was up to his elbows in pine tar and enjoying every drop of it.

Brion was also lecturing the crowd about his belief that the word "marlinspike" was an unfortunate corruption of "marlingspike" and must therefore be rooted out of the American nautical lexicon, now and forevermore. He was introducing his rigging posts as Emily and Wiley ("Post!" he shouted to the bewildered audience, "Now do you get it?") and his rigging vises as Old Shep and his sister Edna St. Vincent Belay. His act, which was half carefully constructed and half anarchistic, was a cross between Popeye dancing the hornpipe and a street mime imitating Zsa Zsa Gabor. It is my favorite memory of the man.

My favorite piece of memorabilia from Brion is a leather jacket he gave me a few years ago. It's a tad out of date—a garment with a past—but I love it nevertheless. For one thing it smells of boatyards and rigging lofts, as it kept Brion warm while he learned the rigger's trade. For another, it has been customized with varnished star knots in place of the original buttons, a touch that reminds me of the work of Clifford Ashley, Cyrus Lawrence Day, Hervey Garrett Smith, and other writer-riggers of an earlier generation.

Writer-rigger. Now there's an interesting combination. Marlingspikes and manuscripts; pine tar and the pen. Who would ever imagine it? Yet within the broad nautical field there has always been a tradition of writing about rigging and a subgenre that could loosely be described as literature about the art of rigging.

No, I'm not talking about knot books. Those are a dime a dozen, and are usually written by people who discovered how to tie a bowline yesterday and are trying, unsuccessfully, to describe the process in print today. Lots of drawings with bights and frayed ends, and arrows pointing this way and that, and not enough text to make them understandable to a maritime savant. "Insert A in B, loop it around C, and pull" is about as literary as the run-of-the-mill knot book will ever get.

Yet there is art in rigging, and there is art in describing what rigging is all about. No wonder that, among the classics in the genre, there are titles such as *The Arts of the Sailor* by Hervey Garrett Smith and *The Art of Knotting and Splicing* by Cyrus Lawrence Day.

(Clifford Ashley was an artist, too, though not quite as highfalutin; his monumental work, superior to all others, was titled simply *The Ashley Book of Knots.*)

Smith, Day, Ashley—they were traditionalists who saw the understanding of rope and rigging as a continuum involving the drag of knowledge from the past into the present. Their additional responsibility, as they saw it, was to push that knowledge into the future, and in most respects, they were successful at it. But our times are different from their times. We have rope, of course, and they had rope, but ours is vastly different from theirs in both composition and manufacture, and therefore working with it is different. So, too, is describing how to work with it. The straight transfer of knowledge from the past to the present is not enough; with it must come an appreciation of the evolutionary, developmental nature of modern ropework and rigging. That is where Brion Toss, a traditionalist who can talk about the plusses and minusses of swaged terminals with the best of them, comes into the picture.

Brion Toss doesn't write knot books. Oh sure, he has his diagrams and his arrows, and he will go on about shoving this rope end through that loop, but this is just the beginning, the jumping-off point for discussions about where this knowledge all came from and where it might be leading. Brion is a true writer-rigger, a legatee of the gentlemen of the past who saw the art of rigging as part history, part scholarship, part mythical anecdote, and part down-and-dirty grunt work, complete with rope burns in the palms, wire cuts on the fingers, and leather jackets permeated with pine tar and bear grease. His, like that of his predecessors, is the layered approach, with the concept of the knot or splice or rigging element at the top of discussion and the meaning of it all at the bottom. In between is all manner of relevant, and sometimes entertainingly irrelevant, material on purposes, origins, and applications. In some ways Brion Toss's pitch is charming; in others it is profound. In no way is it boring.

Peter H. Spectre
Camden, Maine
April 1992

Introduction

This is a book about participation. It contains ways to get involved, mentally and physically, with sailing vessel rigging. It started out as a notebook, a series of scrawled observations on design and technique. I'd see something on a boat, or in a magazine, or in a shop, or someone would write, call, or visit and say, "Have you ever heard of *this* one?" and go on to describe some nifty wrinkle.

People have been coming up with such wrinkles for ages, but nowadays they are especially significant because they're so much harder to come by. Contemporary rigging is machine-made, and very technology-intensive; it tends by its nature to exclude sailors from participation. If we find it difficult to be involved, it's a short step to assuming that we *can't* be involved. I once took some yacht club members on a tour of the *Elizabeth II*, a working replica of a 15th-century British trading ship. The huge, heavy masts and yards of such vessels are held up by little bits of string—the carefully made marline seizings that secure the lower ends of the shrouds. The yacht club members had all seen old ships before, but they'd never looked closely at the details of the rig. When they understood what those seizings were doing, they were literally open-mouthed in astonishment. It had never really occurred to them that a human being, with nothing more than a stick, a piece of marline, and a little skill, could make something just as structurally significant as a modern swaged terminal.

What a contrast this makes with the days of the original *Elizabeth*. Because rig materials were so fragile and degraded so quickly, the crew's daily activities were intimately involved with the life of the vessel, so much so as to blur the distinction between the two. Modern rig materials require far less of our time for maintenance, but precisely because they are so evolved, so inaccessible to simple manual skills, they can require far more of our intelligence and resourcefulness. We can still be the life of our boats, albeit at an intellectual remove.

Of course, intelligence and resourcefulness are no guarantees of success. I once heard of a highly intelligent individual, a structural engineer, who was the skipper of a Great Lakes C-scow. Unlike some skippers, he liked to handle both the tiller and the mainsheet, instead of having another crewmember trim the main, but at 3:1, the sheet's mechanical advantage wasn't quite enough to work one-handed. But he was an engineer, and highly intelligent; it was easy for him to figure out that inverting the sheet tackle and adding a turning block would give him a 4:1 advantage, disregarding the slight extra friction from the extra block.

Just to be sure, he had his mate, who was also a highly intelligent engineer, check his figures. No problem, his mate said, and confirmed his skipper's 33 percent increase in advantage.

They did the work—it only took a few minutes—and shortly thereafter started a race. On the way to the first mark, the skipper was able to prove empirically what the calculations had predicted. No surprise: He was an engineer. But when they rounded the mark and let the sheet out for a run, they suddenly realized that when you increase a tackle's advantage by 33 percent, you also have to make the line 33 percent longer to span the same distance. They hadn't. The Figure-Eight Knot in the end of the sheet fetched up in the turning block with the boom half out, and the scow went over so hard it sheared the windvane off when the mast hit the water.

That's the trouble with intelligence: It doesn't always see far enough into the future. That's why technological fixes tend to spawn technological problems, which require more fixes, etc.

Nor is resourcefulness always enough, even when it's coupled with intelligence. I once heard of a sailor whose mast folded in half at the spreaders when the backstay parted. The top half didn't break completely off, but hung above deck, connected to the lower half by a thin strip of aluminum. It would have been suicide to climb up there to cut it away. Pulling it down from deck level didn't seem any safer. So the skipper sent the crew below and got out the assault rifle he had on board (he was cruising off Ethiopia and had felt it prudent to borrow the gun from a friend). Reclining in he cockpit, he set the rifle on single shot and proceeded to blow his mast away. And it worked.

Certainly the skipper gets points for resourcefulness—it takes a flexible mind to see a Kaloshnikov as a potential rigging tool. But this success masks a failure: The skipper had recently replaced the jibstay after finding broken wires in it. He had not replaced any other standing rigging, even though it was all the same age and had endured the same relative loads, because nothing was breaking yet. But prudence indicates that the failure of the jibstay should be considered, at least, as an indicator of the state of the rest of the gang.

That's the trouble with resourcefulness: It doesn't address what might have been. That's why it can easily degenerate into innovative-Band-Aids-as-a-way-of-life.

Successful innovation, in rigging or anything else, starts with understanding that a given question or problem is always part of a greater, more complex system. An appropriately context-based approach will automatically help weed out dysfunctional responses and encourage useful ones. Three steps I like to follow are:

1. Adapt the Old

Old rigging procedures tend to stick around, even in the face of radical changes in design and materials. They survive because they are effective, and adaptive, even in circumstances completely alien to their origins. Consider, for example, bungy cord. Remarkable, handy stuff, with no precedent in the history of rigging. Like all other cordage, it can only be put to work once you attach it to something. Unfortunately the stuff is so intractably slippery that old standby knots, like the Bowline, crawl right out of it. Hence the variety of clunky mechanical terminals and the corresponding lack of versatility—you can't readily adjust length or eye size as you can with a knot.

Enter the Angler's Loop (Figure 2-15), a simple old knot developed in the days of presynthetic fishing line. It was perfect for gut but became archaic with the introduction of extremely springy modern monofilaments. But the Angler's Loop did not die; it simply lay dormant until bungy came along.

If you're going to adapt the old, you have to know it. This can mean devoting yourself to the study of what might seem hopelessly outdated ideas. But think of this study as building a database, one which will reward you with maximum versatility. It's no coincidence that sailors well-steeped in traditional skills are the most valuable ones to have aboard in a modern boat emergency.

2. Invent the New

If you explore the old and come up dry, at least you'll know that you'll be inventing, not *re*inventing. Now you can go to work with an informed resourcefulness, free to find the truly new, and, with your tradition-bred prudence, the truly workable.

People have been innovating for a long time, so genuine novelties are rare. But there are still some doozies coming out. One of my favorites is Chapter 1's lead item and also involves bungy cord. It's called Tom Cook's Internal Bungy Snubber, and it's the kind of item that is so good it has no competition. So new materials can be opportunities as well as challenges.

It's always good to assume that any solution to a problem—whether new, old, or a combination of the two—will require some effort on your part to:

3. Work Out the Bugs

Continuing the bungy theme, there was once a sailor, sailing in choppy waters, who'd hung a kerosine-fueled anchor lamp from his topping lift. He'd secured it from swinging with a long piece of bungy cord, attached to the lamp at one end and the stern of the boat at the other. This involved less tension than a piece of rope would have generated, and was less likely to jar the lamp. Unfortunately it also let the lamp move around enough to loosen the bottom shackle on the topping lift. When the pin popped out, the lamp came zinging to deck, striking the sailor and spewing flaming kerosine all over the cockpit. With great presence of mind, the sailor picked up what was left of the lamp and threw it as hard as he could astern, neglecting, however, to detach the bungy cord. The fire-trailing lamp roared into the night, hung suspended for a moment against the stars, and then roared back to the cockpit, there to spew still more flaming kerosine. Fortunately, both sailor and boat escaped without serious damage, leaving us all with the lesson that a little novelty can be a dangerous thing.

Some of the elements of this book pretty much remained what they started out as—unrelated notes and observations. Some are related bits that congealed into essays. And some grew out of many hours of workshops, demonstrations, and magic shows. I've made no attempt at comprehensiveness; the aim here has been to assemble useful ideas, some profound and some trivial, in the hope they might stimulate you, the reader, into getting out on the docks, taking notes, and exploring possibilities. I've written this book to stand on its own, but you will find occasional references to my previous book, *The Rigger's Apprentice.* In that book I used relatively few procedures to illuminate basic rigging principles (friction, tension, rig loads). *The Rigger's Locker* is much more about nuts-and-bolts details, about how to go about applying those principles. Whether you read one or both books, remember that they're just books; it is people—their ideas, actions, and intentions—who are still the core of a vessel's life.

Chapter 1
Sheer Ingenuity

"I'd rather make it up than look it up!"
– Lavinia Jordan

Nothing brings out creative drive like a length of rope, a bit of hardware, and a pressing need either to move something or keep it from moving. People have been facing the endless permutations of this challenge for thousands of years, and they're still coming up with innovative responses.

TOM COOK'S INTERNAL BUNGY SNUBBER

For example, prudent sailors know the importance of a gentle "snubbing" action on mooring lines; a boat moving with the waves can fetch up with a jerk against its tether. It's hard on deck fittings and on the comfort of anyone aboard. You can buy bulky, expensive rubber sausages that the mooring line wraps around to solve this problem, but light-to-medium- displacement craft are better off with an Internal Bungy Snubber (Figure 1-1).

The brainchild of Seattle-ite Tom Cook, this snubber is simple, durable, won't foul on chocks and hawses, and can be coiled down with the mooring line. To make it, you just lay a length of bungy cord into the middle of a three-strand line, and stitch the ends down with the cord under some tension. When there's no load on, the bungie'd section accordions up neatly. As the boat

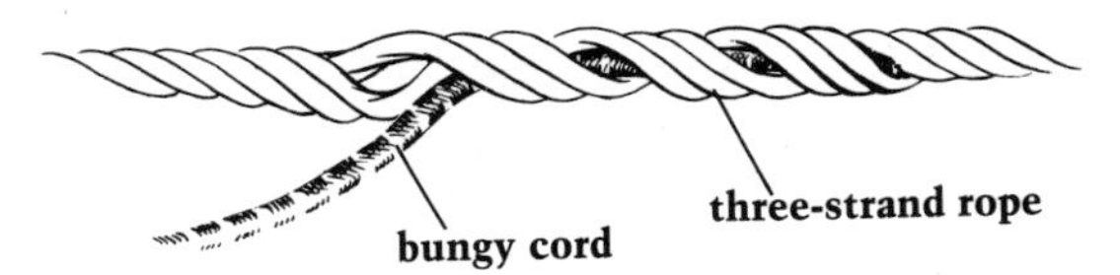

Figure 1-1. *The Bungy Snubber. Use a 2- to 3-foot length of bungy cord that is half the diameter of the three-strand rope. The idea is to insert the bungy cord in the rope at any convenient point along its length. Unlay the strands of the rope enough to work one end of the cord into the middle, and stitch with four to six passes of sail twine through first one strand, then the cord, then a second strand. Then lay the cord into the rope by twisting the rope strands open as you go. Fair it periodically by stretching the cord out and milking the line down over it. The cord should be completely covered except where it exits the rope. Work in as much of the cord as you can, then stitch it where it exits the line, using the same stitching as before. Trim the excess.*

moves, the cord stretches to take up the loads.

GOODYEAR EXTERNAL SNUBBER (Figure 1-2)

Not nearly as elegant as Cook's inspiration, this is still an admirably effective snubber for heavier vessels. A short line or chain connects the tire to cleat, rail, piling, etc. The mooring line connects to the other side of the tire. Using another short line with a thimble in its end on the mooring line side will keep the mooring line clean and chafe-free.

NEW AGE SELVAGEE (Figure 1-3)

Many "new" ideas are updated classics. The Selvagee was originally a marline doughnut made by passing twine repeatedly around a pair of pins, then hitching around the circumference of the doughnut with one of the ends to bundle all the strands together. A Selvagee is stronger than rope of the same diameter because in use it lies flat against what it's hitched to, so that all its strands take an even strain. Today's high-tech parallel-fiber lines are Selvagee evolutions.

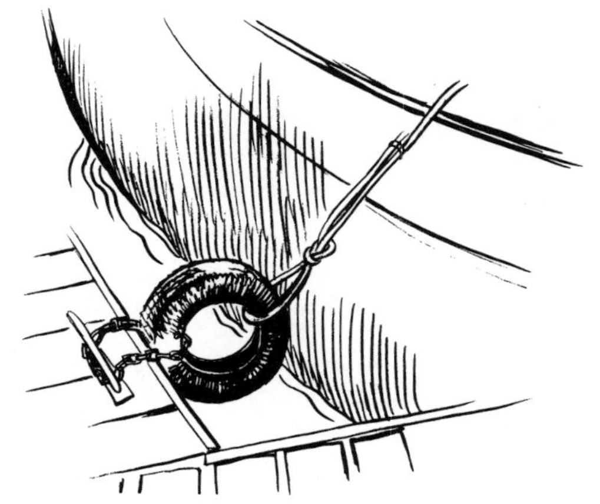

Figure 1-2. *The Goodyear Snubber—ingeniously simple, if not elegant.*

A length of double-braid rope with its core removed and ends knotted together makes a New Age Selvagee. It's neater and much quicker to make than its predecessor, and like it will lie flat to grip the hitchee better than whole rope. Note: The cover of a double-braid rope provides about 50 percent of the strength of that rope, so make your Selvagees and strops (see below) from

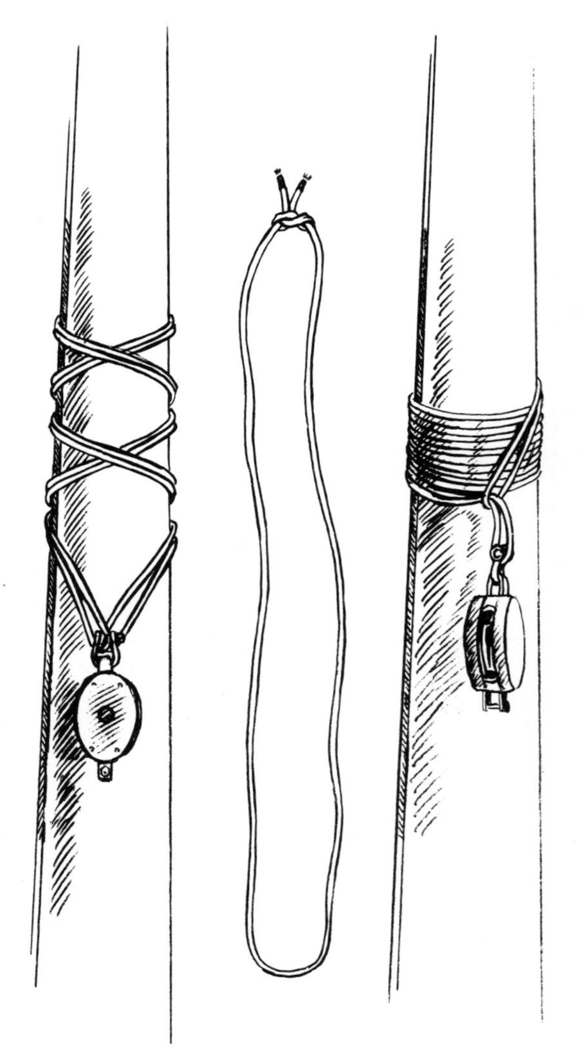

Figure 1-3. *A New Age Selvagee in repose (center) and in action (left and right).*

INSTANT DOUBLE SNUBBER

Whether you have a rope or chain anchor rode, it's a good idea to attach a snubber to the rode after the anchor is set. The snubber acts as a shock absorber for chain rodes and as a sacrificial chafe piece for nylon ones. The snubber itself can be a length of nylon double-braid, of strength equal to the rode, and long enough to reach from a belay at the bitts or foredeck cleat to outside the bow.

Ordinarily the snubber is attached with a Rolling Hitch or similar knot, or with a chain claw. In heavy winds, two snubbers are a prudent idea, with one to take up initial slack and another, slightly longer one to handle heavier surges and as backup in case the first slips or chafes through. The two separate lines get complicated, but you can make two snubbers instantly from a single piece of line.

Extrude a length of core from the outboard end, and whip it and the cover end. Use this two-tailed line as a single snubber by spiraling the two ends around the rode, selvagee-style, or hitch each tail on separately. For the latter procedure, use line about twice as strong as the rode. Make the core tail, which is sometimes weaker than the cover, about 2 feet (0.6 m) shorter than the cover, and hitch it on first. Hitch the cover on beyond it. As the core stretches under load, the cover takes the strain.

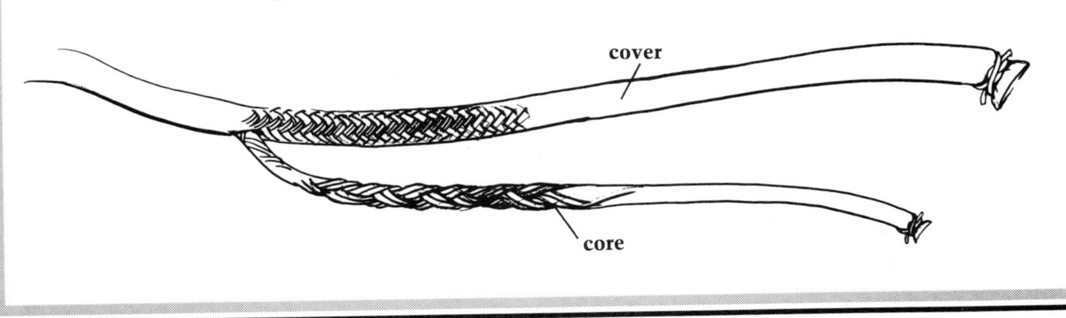

oversize line to compensate for the removal of the core. Either Selvagee will work for traditional uses such as setting up shrouds with deadeyes and lanyards (the Selvagee is wrapped around the shroud, a come-along is hung from it and pulls tight on the lanyard), and for more modern rig jobs, too. For example, say you have a "wrap" (fouled turns) on a winch. Clap a Selvagee on ahead of the winch, tie a line to the Selvagee, and lead the line to another winch. Take up the strain, putting slack into the original line so you can clear the wrap.

You can also use the Selvagee for non-essential jobs like hanging a hammock from mast and rigging, or for very essential jobs like supporting a damaged shroud or stay, or for anything in between.

NEW AGE DECK STOPPER (FIGURE 1-4)

A related item is the New Age Deck Stopper or Deck Strop. For this, a piece of double-braid is bent around a thimble, with the standing parts either Brummel-spliced or seized together. The core is then removed from the ends of the rope, leaving the rope whole in the middle only. Shackle this strop to the deck next to a halyard turn-

HASSE'S OCTOPUS MAGAZINE

Next to a downhaul, the single greatest low-tech jib-taming tool is an item called a "magazine." It's a piece of wire rope with an eye in either end, and with wooden or plastic disks fixed in place near each eye. After lowering the sail, you clip one end of the magazine to a deck eye and the other end either to an eye on the pulpit or to the sail's halyard. You then take the sail hanks off the stay one at a time and hank them to the magazine. No fumbling, no wrestling, no getting two-thirds of the hanks off the stay and losing the whole works to leeward in a malicious gust or wave.

During a recent Pacific cruise, sailmaker extraordinaire Carol Hasse came up with a magazine refinement called an "Octopus." It's a strip of 1-inch (25 mm) webbing the length of the sail's foot, one end of which is sewn to the top loop of the magazine. Three-quarter-inch (19 mm) strips of webbing, long enough to function as sail ties, are sewn to the long piece of about 2-foot (0.6 m) intervals. You can vary this spacing, making the crosspieces closer together where the sail is bulkiest. If the sail will typically be left on deck for some time, you can also space the ties so that they coincide with stanchions, chocks, or other deck gear you can tie to.

Make a different-colored magazine for each sail, to make sorting easier, and either leave them on deck or in the sailbag while the sail is up. If you always leave the sail on deck, it's a simple step to make an Octopus Magazine House of acrylic cloth.

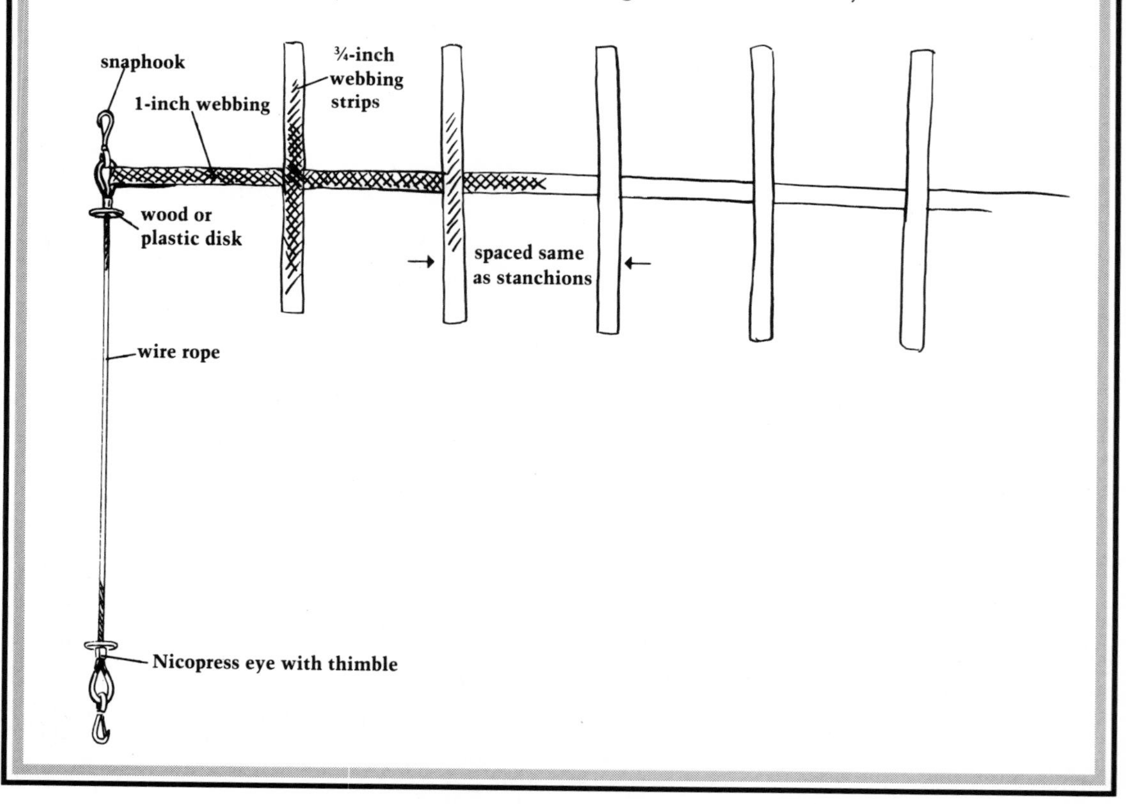

Figure 1-4a. To make a New Age Deck Stopper or S.H.A.T. strop, Brummel-splice or seize a double-braid eye around a thimble, then remove the core from the legs of the strop. (See Figures 4-54 and 4-55 for Brummel Splice.)

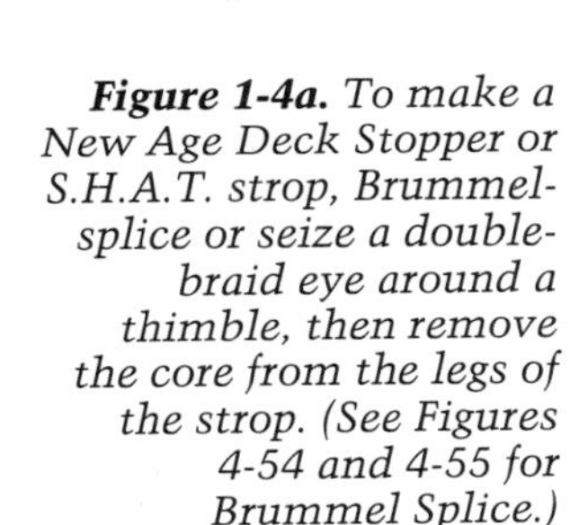

Figure 1-4b. The strop in use, with ends spiraled around a length of wire, in this case to rig a temporary stay.

ing block and wrap it around the standing part of a taut halyard. You can now release tension on the hauling part of the halyard; the strop will hold on while you belay the line. This is precisely the same job that modern mechanical "clutch stoppers" do, and you might want to keep a spare strop around for a modern boat, just in case one of those mechanical marvels malfunctions.

Because this strop is Solid, Hollow, and Ambi-Tailed, it is also known as a S.H.A.T. strop.

'BINERS (Figure 1-5)

The Carabiner, or 'Biner, is a roughly oval, spring-loaded shackle originally developed for mountaineers. But in recent times it has found its way aboard boats, primarily for safety tethers. The best models lock shut, and the best of the locking models are easy to operate one-handed. That way you avoid the potentially fatal irony of going overboard because you let go of a secure handhold to deal with your safety tether.

If you have no locking carabiners, try the mountaineer's trick of hooking two of them in from opposite directions. That way, at least one of them will stay closed, no matter how you're washed, dropped, lurched, or bumped around.

As long as you have two carabiners, though, have two locking ones, and put them on separate tethers, one about 18 inches long, the other about 3 feet long. Now you can secure yourself with the short tether when you're standing at the helm and don't need to move around much. When you need to go forward, switch to the longer tether for greater mobility. Anytime you need to move past an obstruction, clip the short tether in before you detach and reattach the long tether on the far side of the obstruction.

When there's nothing that the carabiner will clip around near at hand, pass the tether around any available, round-cornered object—a stanchion, handrail, whatever—and clip the 'biner to its own standing part, forming a slip knot. To clip around a mast, pass the long tether around, then clip it to the short tether.

If you have an extra-large foredeck, you can expand your mobility by making the long tether about 5 feet long and securing a third 'biner halfway down its length. Leave this one clipped to your safety harness, and you'll have a short and a moderately long

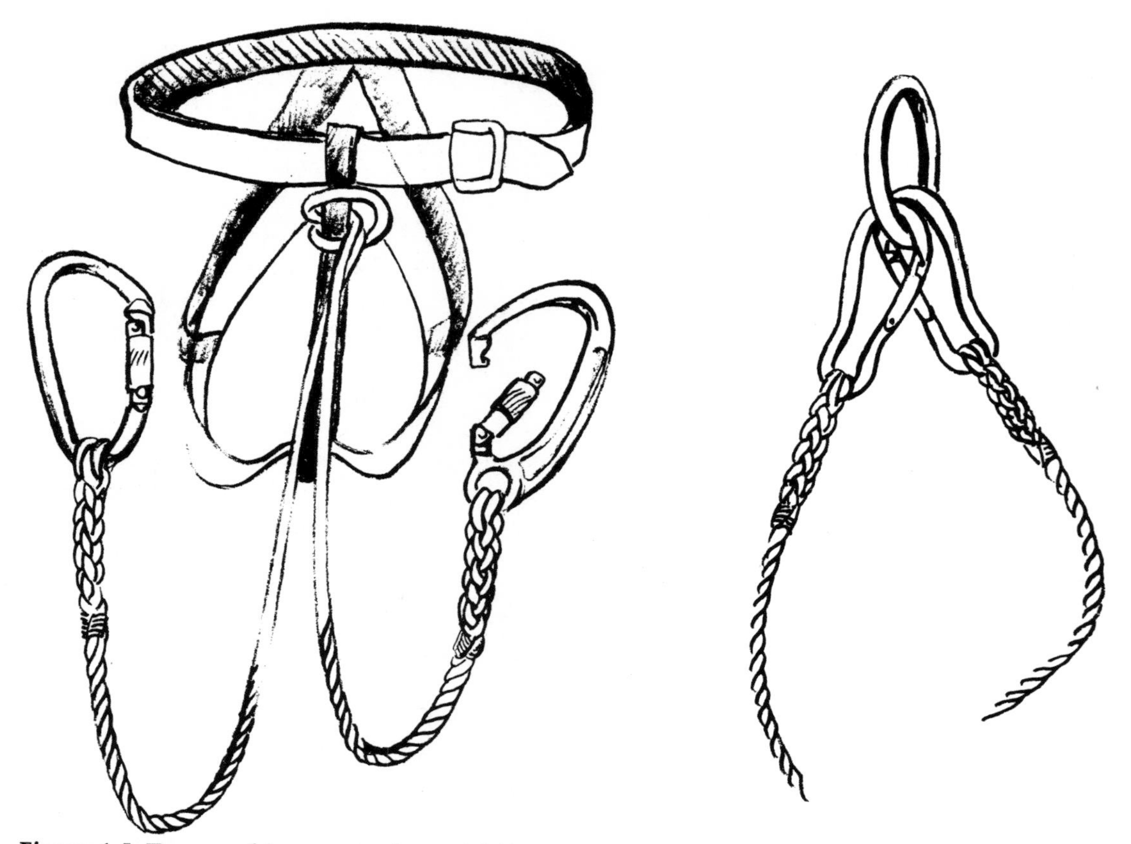

Figure 1-5. *Two carabiners on tethers of different lengths fastened to your safety harness give you handy mobility without loss of safety (left). If you have no locking carabiners, try the mountaineer's trick of hooking two of them in from opposite directions (right).*

tether. Unclip it and you'll have a short and an extra-long tether.

For more carabiner techniques, see the "Living Aloft" section of Chapter 4.

THE PARDEY TIE-ROD (Figure 1-6)

Rigging extends past the tangs and chainplates; the entire hull is a member of the rigging system, absorbing the power of the sails and transforming that power into vessel motion. So the rigging strains the hull—the weather side of the hull is in tension right down to the keel from the upward pull of the shrouds. The shrouds also pull in, because they connect to the hull at an angle. Even if the vessel's deck is heavily reinforced, it can slowly buckle upward from the pressure of the shrouds. That's why a vertical tie-rod in front of or behind the mast, to hold the deck down in the middle, is a good idea.

Veteran cruisers Lin and Larry Pardey planned to install such a rod on their 30-foot cutter *Taleisin*, but found that it was going to be in the way whether they put it ahead of or behind the mast. They wanted storage locker doors by the foot of the mast. Their solution was a mini-rod attached to a tang on the mast face and running diagonally up to a deck beam. The rod is tensioned, like a turnbuckle, by a barrel in its middle. It's com-

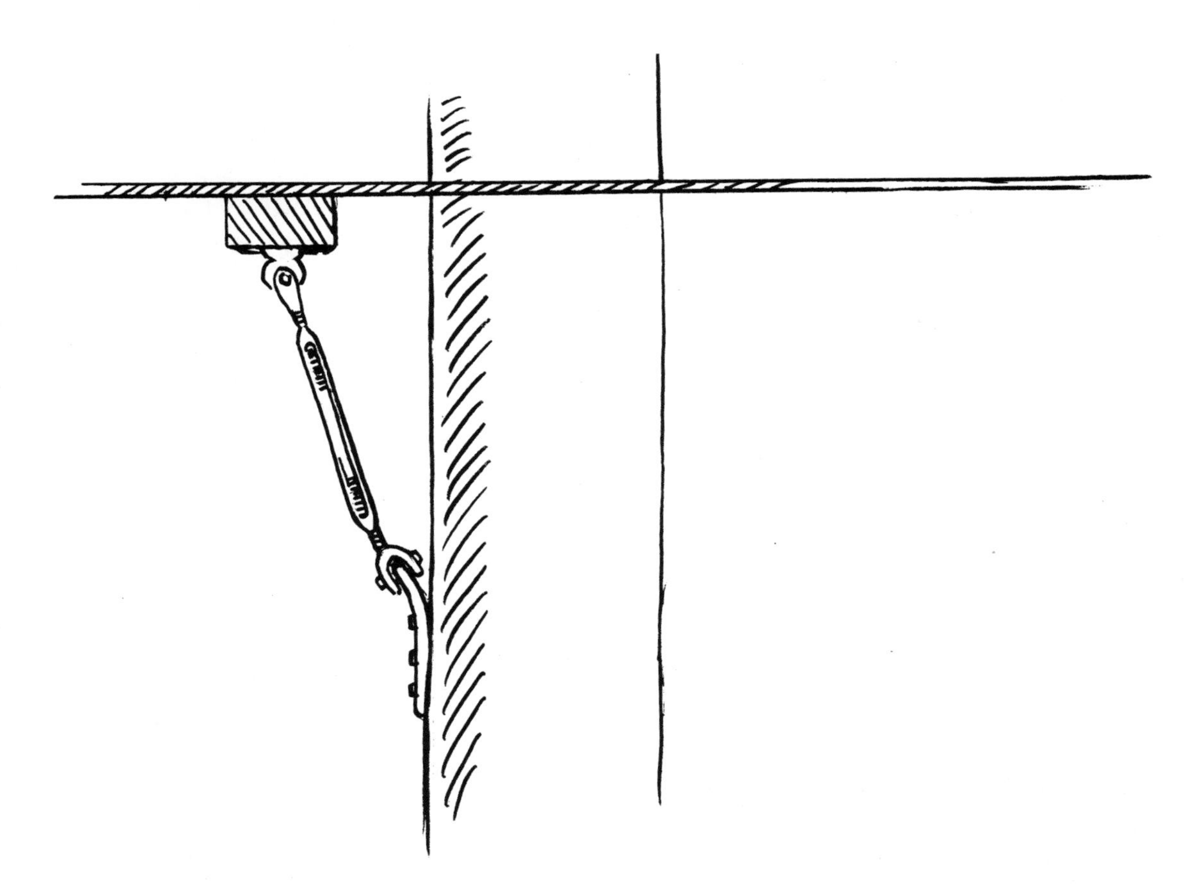

Figure 1-6. *The Pardey Tie-Rod.*

pletely out of traffic, provides a convenient handhold or towel hanger, and is easily detached if the mast needs to come out.

SONIA STAY (Figure 1-7)

The builders of the 34-foot double-ended yawl *Sonia* were equally creative. The main backstay splits partway down to miss the mizzenmast, and reaches the deck well aft on this fine-sterned vessel. Ordinarily, this would put a lot of upward strain on a delicate part of the hull.

But instead of making the stern heavier (and more expensive), *Sonia*'s builders carried the backstay legs right through the deck, then angled them down and in to attach to the sternpost. Now the only load at deck-level is a slight compression between the legs of the backstay. And the sternpost is itself supported from sagging.

CHART-RIGGING

Once you start extending the purview of rigging into the hull itself, it's hard to stop. Here's a new idea that brings sailboat hardware into a completely unrelated area—chart stowage.

You start by screwing lengths of boltrope track to the underside of the chart table or the top of a designated chart locker. Then you get your friendly local sailmaker to sew lengths of the appropriate-sized rope to the

QUICK CHAFE GEAR

A strip of leather with a mooring line-size hole in either end makes tough, adjustable chafing gear. Sewing the leather on is better for staying put, but is not adjustable.

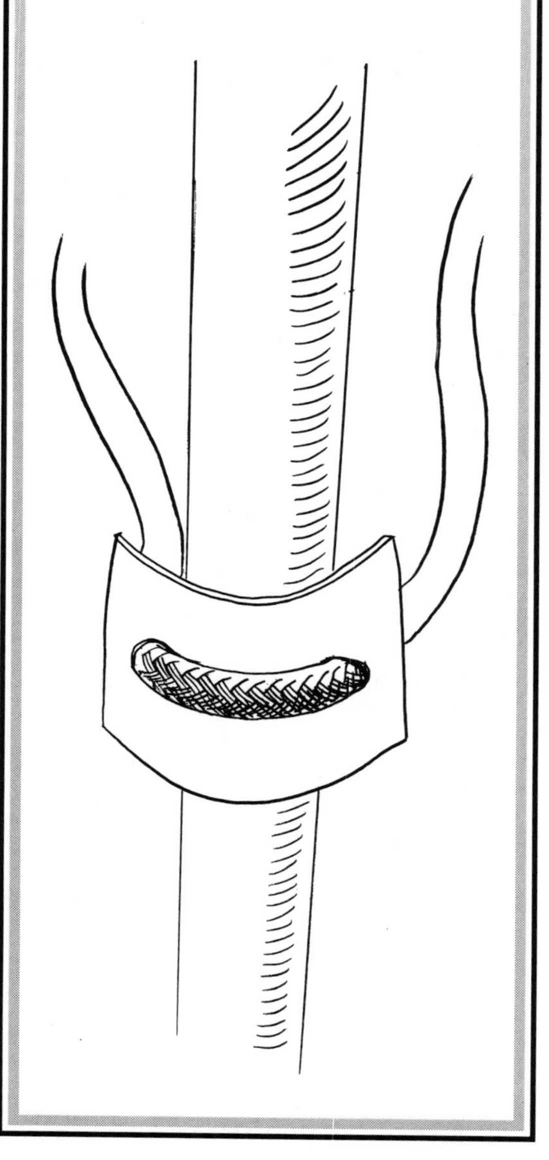

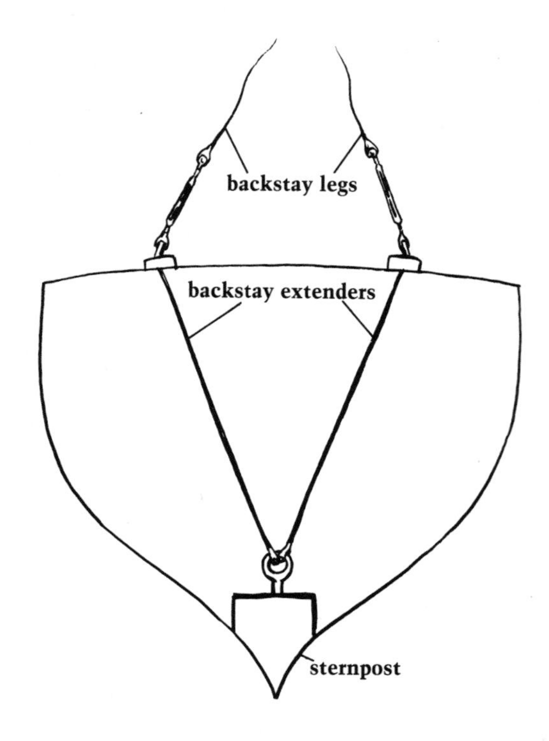

Figure 1-7. *Sonia's through-the-deck backstay configuration.*

top edges of zipper-opening clear-plastic chart pouches. You now have the most compact, easily-accessed, tidiest method of stowing charts ever known. Keep at least two charts in each pouch (one facing out on either side, others in the middle) and arrange the pouches according to your sailing territory. The pouches can be laid on the chart table or taken right out on deck, where the plastic cover will keep the wet out. You can even install a strip of track in the cockpit to keep the pouch in place there.

PRETTIFICATION

Moving on to matters aesthetic, we come to the Ugly Crimp-On Wire Fitting Dilemma. These fittings, generally known under the (proprietary?) term Nicopress, are commonly used to form eyes on wire rigging,

TED'S "ELIMINATE NICOPRESS WRESTLING" TIP

Nicopress fittings are wonderfully convenient, but can involve some wrestling during setup. The end is threaded through one side of the fitting, bent to form an eye, then threaded back through on the other side of the fitting. The wrestling comes when you're trying to hold the end so that it projects just slightly beyond the end of the fitting, while you pull on the standing part to snug the eye down around the thimble. Further wrestling ensues when you try to hold the entire spring-load assembly still while you crimp the fitting. To eliminate both problems, lightly crimp the corner of the fitting where the end projects, while the eye is still loose. Use a vise, pliers, Vise-Grips, etc. With the end locked in place it's easy to adjust and hold eye size while pressing.

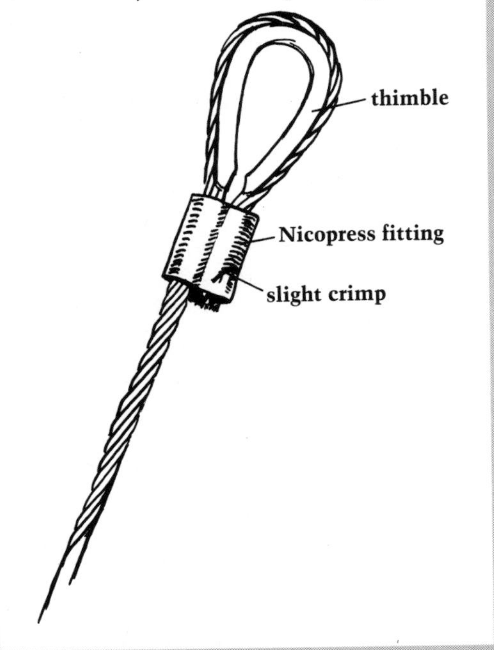

NICOPRESS NOTES

When setting up, allow for sleeve lengthening by positioning the sleeve slightly below the thimble, and with the end projecting slightly beyond the sleeve. If this is not done, the sleeve will be distorted as it mashes into the thimble, and the end will slip inside the sleeve, losing bearing surface and thus strength and security.

particularly for small (¼ inch and less) wire. The fittings, properly applied, are inexpensive, strong, and quick to make. But they are also ugly and have hard corners and a little bit of bristly wire end sticking out. To correct these drawbacks, try wrapping some heat-shrink tape over the finished fitting, or thread on some heavy-duty heat-shrink tubing before making the eye, then slide the tubing down over the finished fitting. Apply a little hot air and you have a suddenly smoother, nearly attractive terminal. Industrial-cute. It's also good practice to slather the fitting, and any wire that will be shrink-wrapped, with anhydrous lanolin.

TRANSFORMATION

No need to cover up the loop-and-button becket in Figure 1-8; it's inherently lovely as well as practical. The strands that form the eye splice are transformed into the button that secures the eye. And the extra bulk of the splice is chafing gear for whatever you're buttoning around. This novel becket, the brainchild of James McGrew, is just the thing for hanging fenders and for tack pendants, small-craft halyards, or, in small stuff, as a belt lanyard for tools and keys.

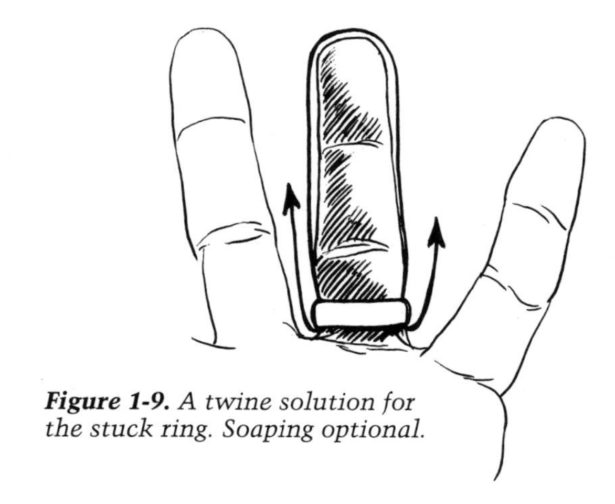

Figure 1-9. *A twine solution for the stuck ring. Soaping optional.*

SHEER INGENUITY

Rigging isn't just for boats. Take, for example, the elegant twine solution to the ancient ring-stuck-on-finger dilemma, illustrated in Figure 1-9. Or how about tightening a loose fan belt by running a turnbuckle out?

If your problem is a broken fan belt and no spare, try panty hose (Figure 1-10). All of the procedures in this section are products of ingenuity; this is just a particularly off-the-wall-but-by-golly-it'll-get-us-home example. Now, if only one could time an engine with a pair of cuff links . . . .

Figure 1-8. *The Loop-and-Button Becket.*

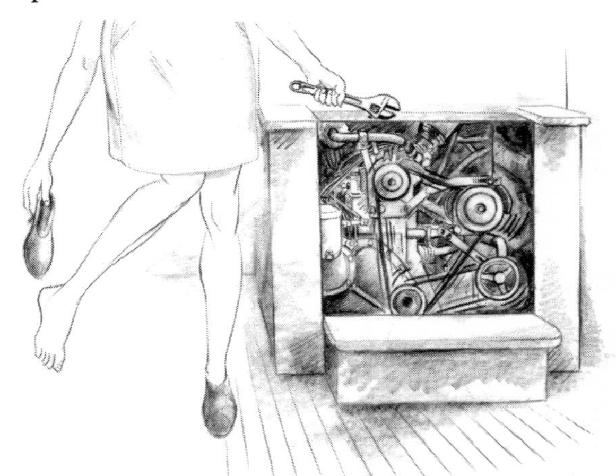

Figure 1-10. *The pantyhose fan belt.*

Chapter 2

Knots

". . . the simple act of tying a knot is an adventure in unlimited space."

– Clifford Ashley

Ropework is a sine qua non of sailing, even aboard the most gee-whiz mechanized boat. But owing to radical developments in materials and design in this century, ropework has had to undergo some changes to escape being archaic. Specifically, knots have had to become more secure to hold in slick synthetics, and more jam-proof because of higher loads. In addition, people are still coming up with innovations that would be valuable in any material in any century.

By and large, the most successful developments have been refinements on or revivals of old, old knots. New knots are often flashes in the pan; they founder on circumstances that tradition long ago anticipated.

CONSTRICTOR KNOTS

Old knots never die; they just wait for us to come to our senses. For example, hose clamps are the emergency recourse of choice for binding cracked tillers, spars, boathooks, etc. Once I even saw one on the end of a ravelling line. The prevailing attitude about them is that though they are expensive, time-consuming to apply, snag on everything, and look awful, they're better than anything else for temporary repairs, right?

Wrong. For all the above jobs, and for hundreds of others besides, hose clamps can do little that Constrictor Knots can't—including clamping hoses. A Single or Double Constrictor made with a piece of job-scaled nylon or polyester twine is a quick, easy, unobtrusive, durable, and essentially free way

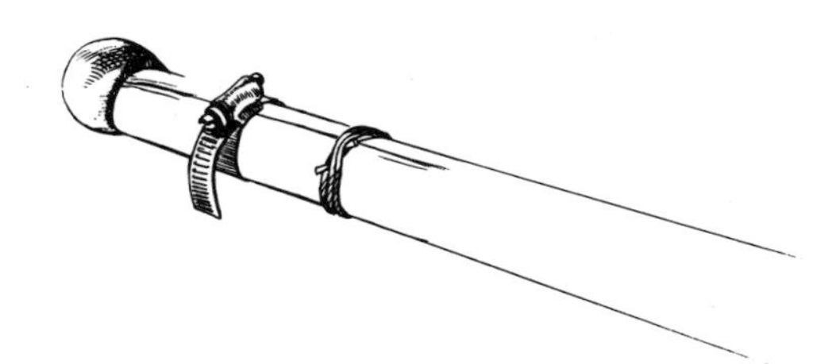

Figure 2-1. *The old and the new side by side on a sprung tiller: John Henry versus the steam hammer, the Constrictor versus the hose clamp.*

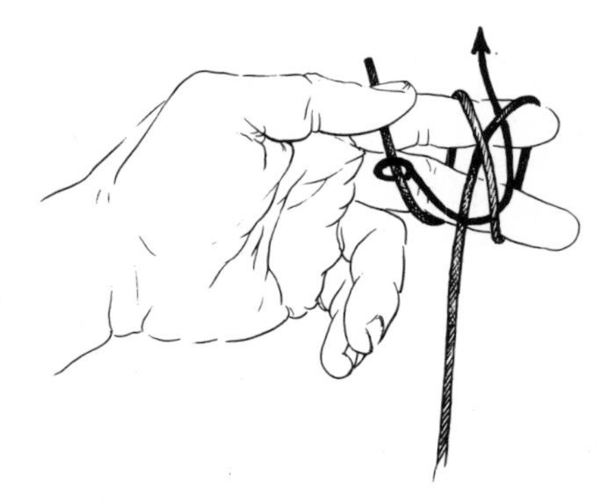

Figure 2-2a. *To tie a Single Constrictor with the end, make a crossed round turn, crossing from right to left. Bring the end up on the left side of the standing part, then lead it over to the right and under the crossing point, away from you.*

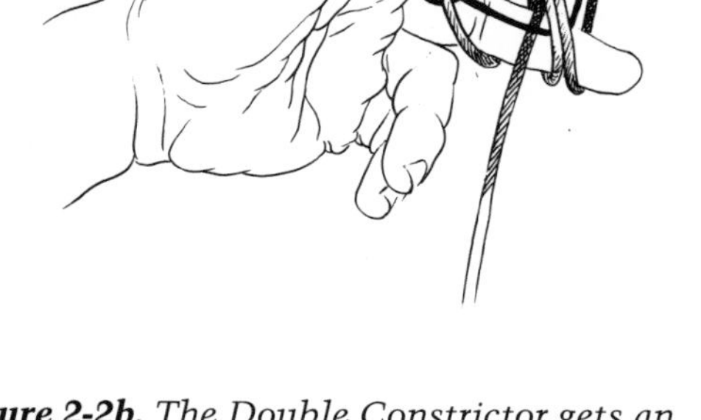

Figure 2-2b. *The Double Constrictor gets an extra crossing turn, parallel with and to the left of the first. The end goes under three parts as it passes under the crossing point.*

to bind things together. If the Bowline is the King of Knots, surely the Constrictor is the Queen.

In recent years, sailors and landspeople alike have been coming to their senses in sufficient numbers that a Single Constrictor Knot is no longer a rarity. But the Double Constrictor still is. The Double is for those situations where extra strength and security is a must, as for semi-permanent lashings, whippings, or for large gluing jobs where a hose, "C," or other kind of clamp might be unavailable or too bulky.

Either the Single or Double can be tied with the end (Figure 2-2) when you need to fasten it around a ring, stanchion, spoke, etc.

But whenever possible—whenever you can make the knot and then drop it over the constrictee—tie Constrictors in the bight (Figures 2-3 and 2-4), a faster method.

Bear in mind that the Double does not draw up as easily as the single; work out as much slack as you can with the tip of a spike before pulling on the ends. The best way to tighten a Constrictor is to hitch a spike,

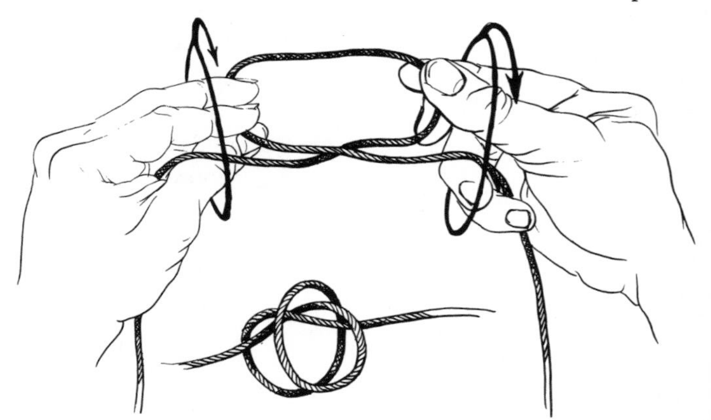

Figure 2-3. *To tie the Single Constrictor "in the bight" (without using the ends), pick up the line with your hands about a foot apart, palms away from you. Holding the line with your ring and little fingers, use your other fingers to make a loop, right over left. Arrange your hands exactly as in the drawing, right palm facing you, left palm away. To complete the knot, just turn your hands over. Once you get it figured out, the whole process takes about four seconds.*

Figure 2-4. *To tie the Double Constrictor in the bight, make a Clove Hitch and arrange it on your left hand as shown, with the upper end on the left. Cross the upper end over the right, then pull slack into the right-hand turn, and twist it 180 degrees, counterclockwise (the part nearest you moves to the right, toward your fingertips). Place the twisted loop over your fingertips to complete the knot.*

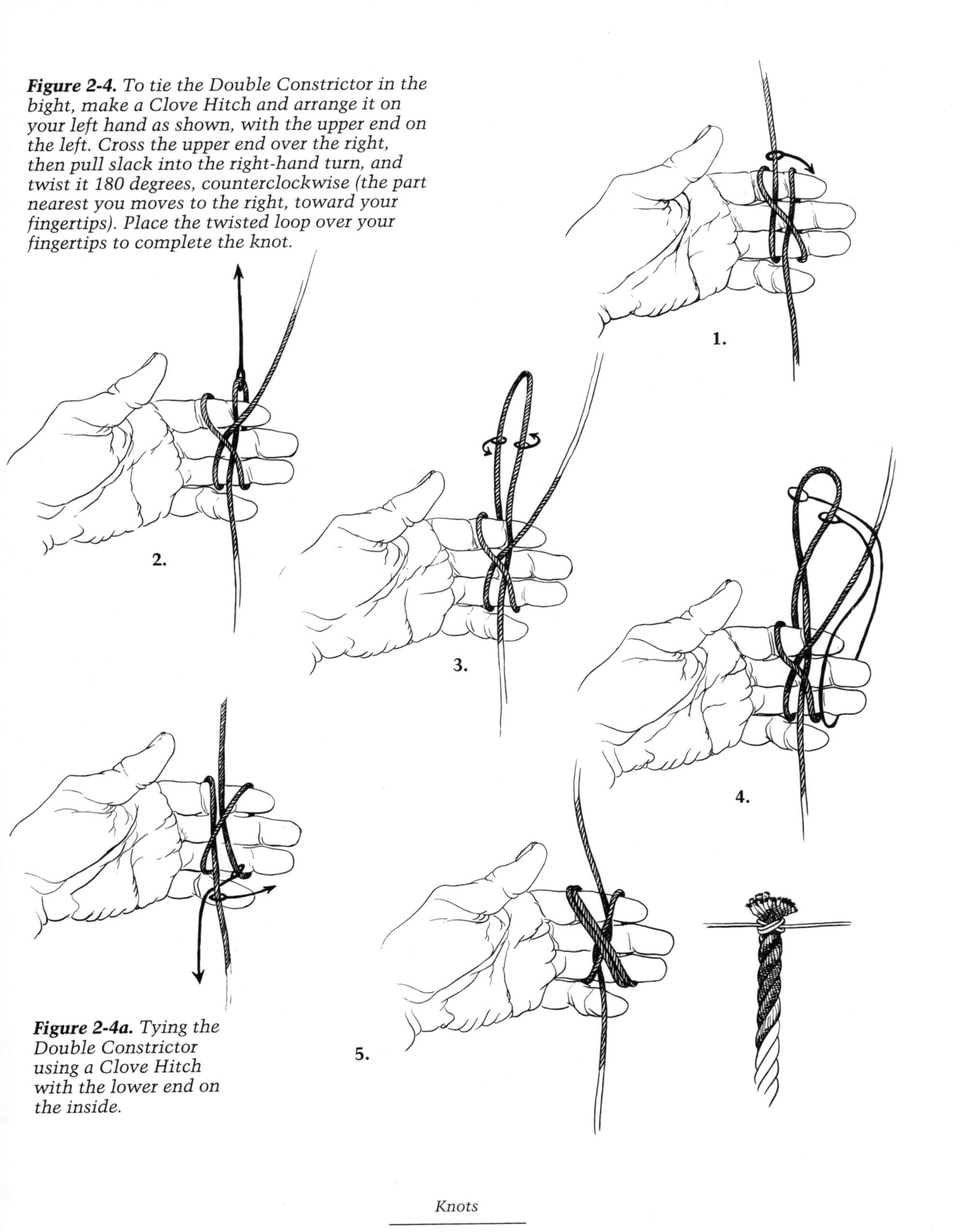

Figure 2-4a. *Tying the Double Constrictor using a Clove Hitch with the lower end on the inside.*

stick, or the like to each end. Pull. With heavy nylon twine you can exert even more force by bracing one stick between your feet and holding the other with your hands (Figure 2-5). For extremely tight Constrictors made with rope for large jobs (splinting a broken boom, for instance), position the Constrictee between two sheet winches and crank away. No matter what the scale or tension, always arrange the knot so that its Overhand Knot portion lies over a convex surface.

TURK'S HEADS

Moving aside from strict utility for the moment, try making the circular braid known as a Turk's Head from a Single Constrictor (Figure 2-6). When doubled or tripled (see Rigger's Apprentice), this knot makes a decorative ring for ditty bags, bell-ropes, bottles, wrists, oars, etc.

TRANSOM KNOT

A close relative of the Constrictor is the Transom Knot (Figure 2-7a). Its ends come out at right angles to the turns, and it's particularly handy when you need to fasten two things together at right angles to one another. Originally developed by Clifford

Figure 2-5. *For very tight Constrictors, seat yourself on deck and hitch a spike (or functional substitute) onto each end of the twine. Brace one spike between your feet and hold the other in your hands. Pull.*

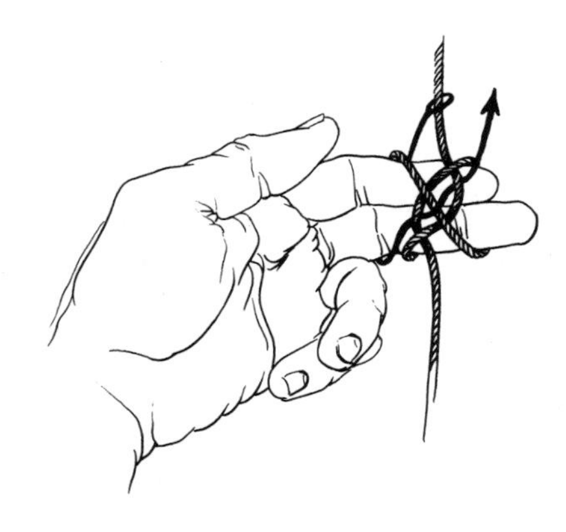

Figure 2-6. *To turn a Single Constrictor into a Turk's Head, arrange the knot around your left-hand fingers and open it up as shown. Pass the upper end down behind your fingers, up on the left side of the standing part, then pass it under, over, and under as shown, tucking up and to the right. Double and triple the knot by leading the end back into the knot, parallel with the standing part. A four-lead, three-bight knot results.*

Ashley to hold the crosspieces of a kite together, this knot is perfect for lashing winter boat-cover frames, trellises, electrical wiring, and as a quick repair for ladders, ratlines, and fences. For extra security and strength, a second knot can be made opposite and at right angles to the first.

DOUBLE TRANSOM

When the aim is to bind a number of long thin objects such as sticks or wire bundles use a series of transom knots (Figure 2-7b) and draw each knot firmly before moving on to the next. Unlike most hitches, one transom knot can be chafed apart and the others will remain.

SHOES

Moving along to bound feet, we come to what might be the best-known knot of all,

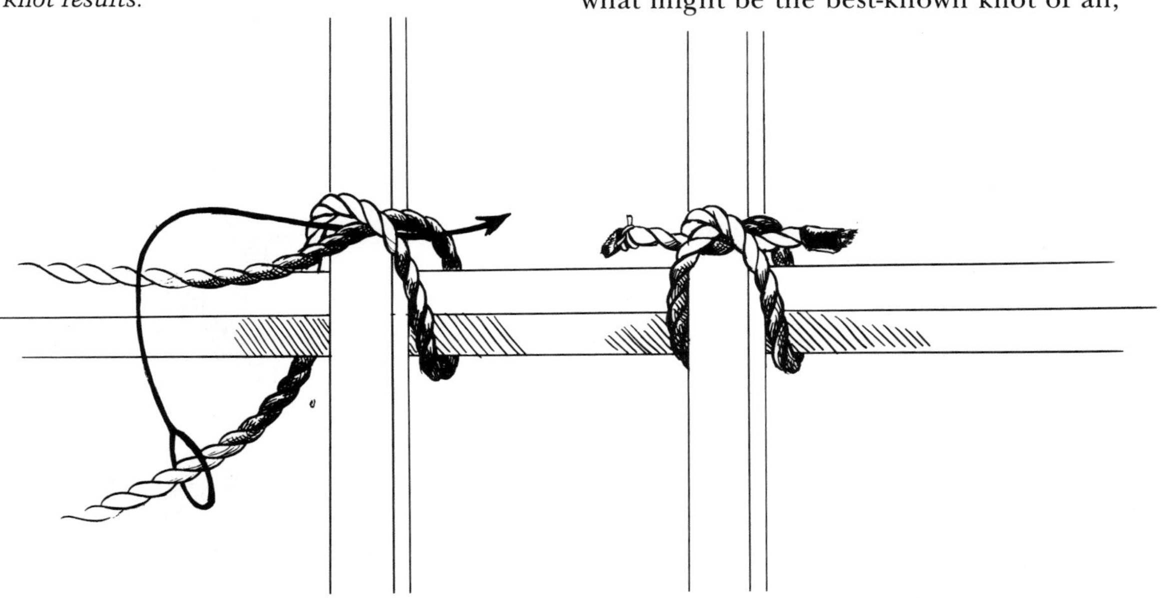

Figure 2-7a. *A Transom Knot can be used to lash a framework. Lay the vertical piece over the horizontal one. Lead the line across from left to right, down behind the horizontal piece, and up to the left diagonally across the vertical one. Then pass it down behind the left side of the horizontal piece, and up and under the crossing, left to right. For extra-firm lashings, add a second Transom Knot opposite and at right angles to the first. "Frap" by tying a Constrictor Knot between the two crosspieces to further tighten one or two Transom Knots.*

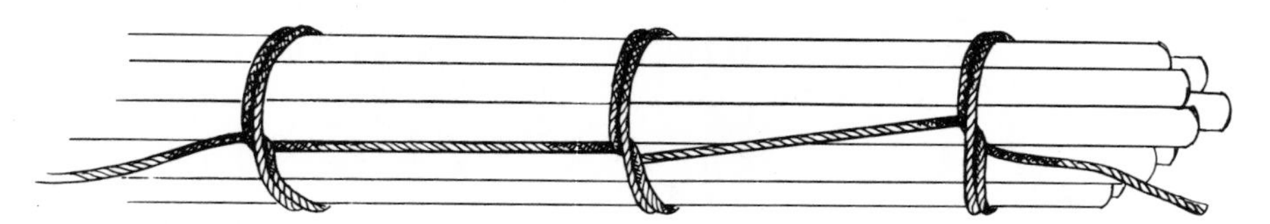

Figure 2-7b. *The Double Transom series in action, taming a bunch of spare battens.*

the Bowknot (Figure 2-8a), which most people tie their shoelaces with. I say "most" because this relative of the Square Knot is often mistied into a relative of the (shudder) Granny Knot (Figure 2-8b). If your shoes have laces, look at them now. If the bows sit athwartshoe, they're Square; if fore-and-aft, Granny. Aside from superior appearance, the Square version offers superior security. If you've been tying the Granny, the easiest way to switch is to reverse the way you tie the Overhand Knot that is the first half of the Bowknot, then make the bow part as you always have. The new method will be second nature in no time.

Even a properly made Bowknot is no paragon of security, especially in slick modern laces, or even in new leather ones. The usual way to improve security is to make an extra Overhand Knot with the bights (Figure 2-8c). The result is secure but can be hard to untie when wet and is about as attractive as, say, a hose clamp. The Turquoise Turtle Knot (Figure 2-9) is a handsome alternative—simple, always easy to untie with a pull on the ends, and very easy to remember.

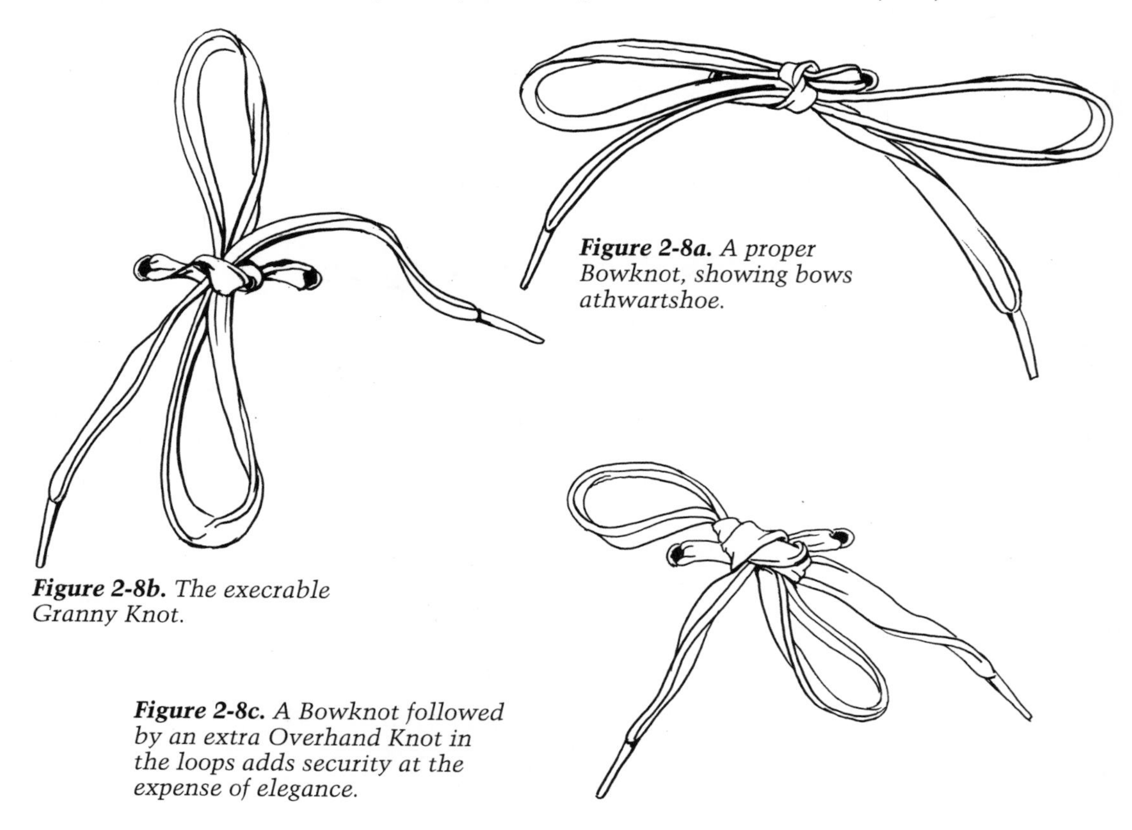

Figure 2-8a. *A proper Bowknot, showing bows athwartshoe.*

Figure 2-8b. *The execrable Granny Knot.*

Figure 2-8c. *A Bowknot followed by an extra Overhand Knot in the loops adds security at the expense of elegance.*

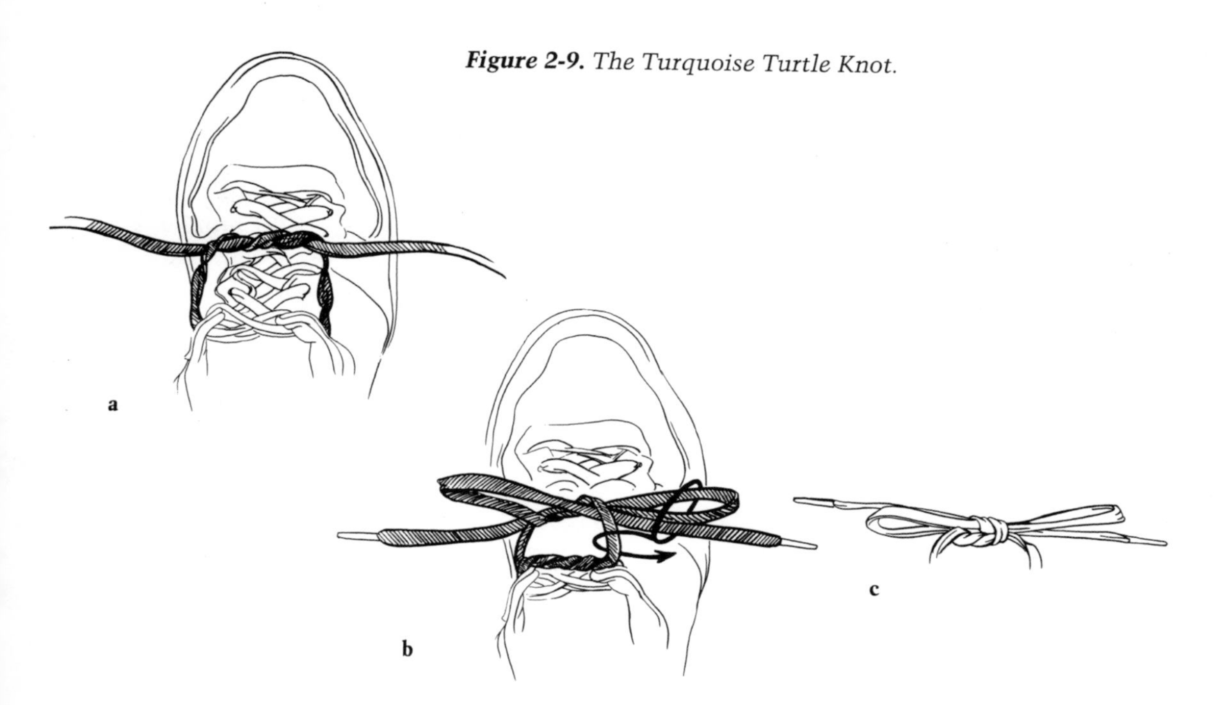

Figure 2-9. *The Turquoise Turtle Knot.*

Start with the usual Overhand Knot, but pass an end around a second time. Make the loops as usual, but leave a little space between them and the Overhand Knot. Pass the end and loop on one side through this space. Draw up as you usually would, by pulling on the ends.

THE NICKNOT

And now for a shoelace knot in which the ends can't tangle or be pulled loose, because with this knot there are no ends. Introducing the Nicknot (Figure 2-10), a copyrighted, patent-applied-for brainstorm of one Nicholas Hyduke.

Best-suited to thick laces, the Nicknot is a compact delight for shoes, as well as for the drawstrings on sweatpants, pajamas, duffle bags, etc. (No more drawstring ends disappearing inside their tunnel.)

The tying steps are simple variations on the standard Bowknot. The bead or Figure-Eight Knot ends up on top of the finished knot, giving you a little "handle" to pick the knot quickly apart to untie. To prepare for the knot, just lace your shoes toward the toe, then tie the ends together at the bottom.

SOME KNOTS FOR SLIPPERY ROPE

A pause here, before we go on to other knots, to consider the material we tie them with. Modern synthetic rope is much stronger and longer-lived than the natural-fiber ropes of yore. But these virtues have not been obtained without cost: Modern rope also tends to be much springier and slicker than its ancestors—so much so that knots which used to be paragons of security are now liable to come undone all by themselves.

All knots hold together because of friction, so it would seem that slick-rope knot adaptations would simply be a matter of taking extra turns and tucks to generate more friction and thus more security. And this is a

TARP TIE-DOWN

If you haul your boat for the winter, it's prudent to drape a tarp over it to keep out the elements. But how do you keep the tarp from flapping in the breeze, eventually shredding itself to pieces after beating your topsides to death? Simply cinching the tiedown lines supertight is usually not enough; the rope stretches over time, and can overstrain the tarp grommets.

But if you interject strips of inner tube between the tiedowns and the plank they belay to, you'll have a resilient configuration that can compensate for rope stretch and wind.

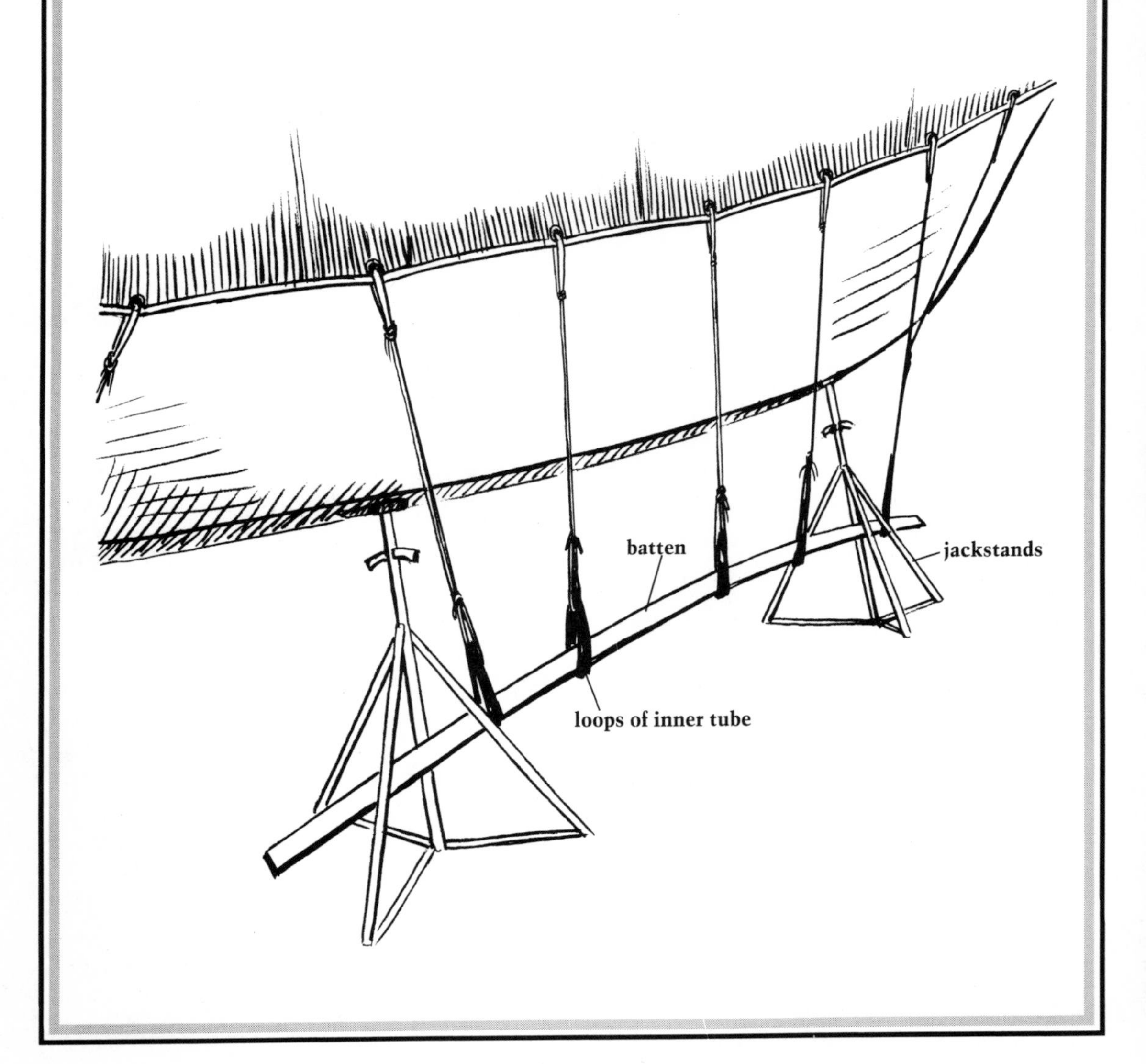

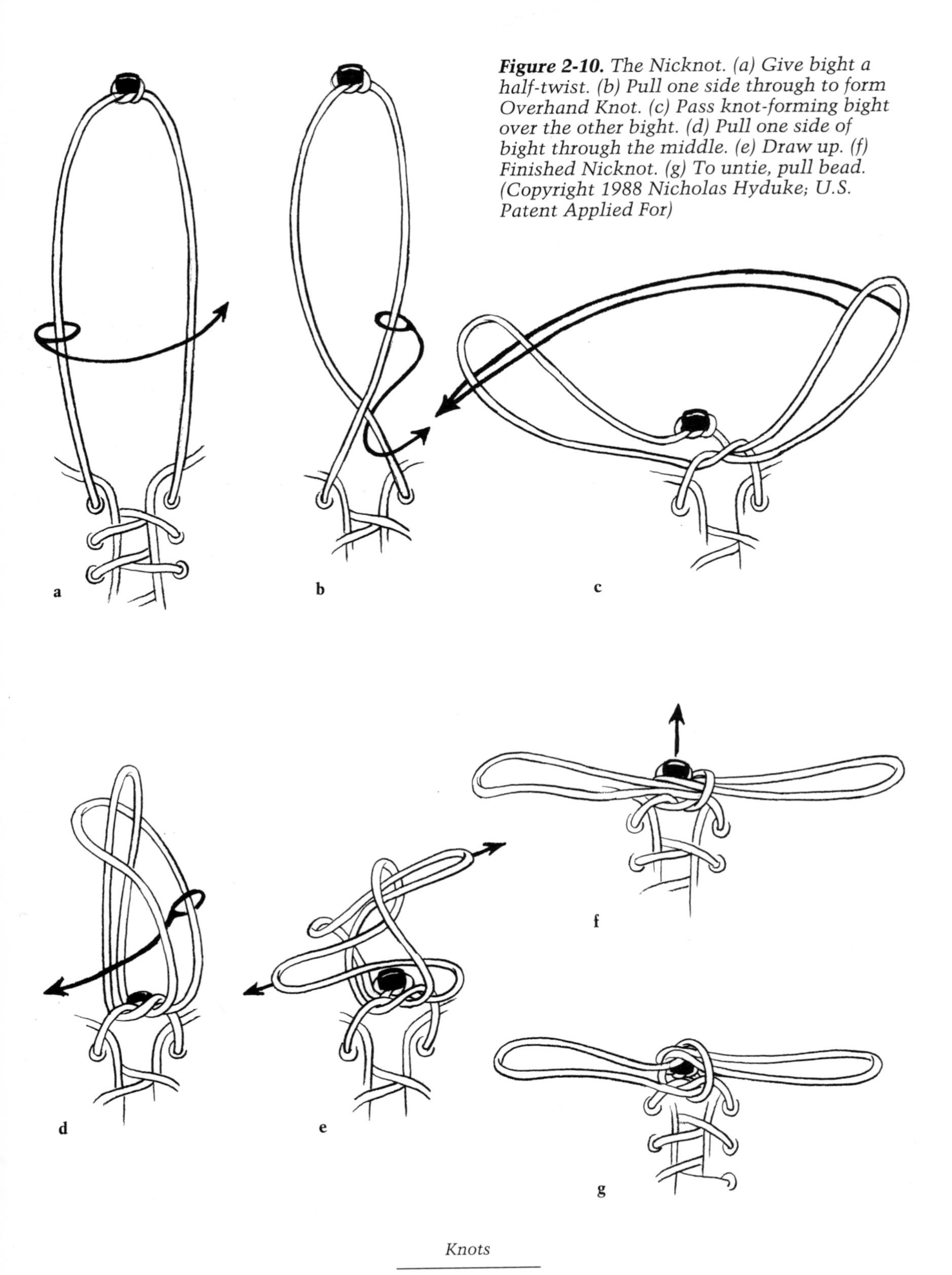

Figure 2-10. *The Nicknot. (a) Give bight a half-twist. (b) Pull one side through to form Overhand Knot. (c) Pass knot-forming bight over the other bight. (d) Pull one side of bight through the middle. (e) Draw up. (f) Finished Nicknot. (g) To untie, pull bead. (Copyright 1988 Nicholas Hyduke; U.S. Patent Applied For)*

TIDAL AND STORM MOORING

In strong tidal areas or in storms, set Bowlines around pilings for your mooring lines. The loops can ride up and down, needing no adjustment.

BELAY AIDS

Whether you're pulling a stump, reinforcing mooring lines to prepare for a hurricane, or dragging a boat up the beach on its cradle, this setup will give you a firm anchor when there's no convenient tree and you're fresh out of piledrivers.

To make, drive two long stakes, crowbars, or what-have-you's into the ground at a slight angle away from the load. Set a stout log or beam behind them. If the latter, round the corners so the line won't chafe. The stakes must be at right angles to the pull.

Drive two more stakes in right behind, then two more a little farther back.

Lash the pairs of stakes together, *starting with the last two.* This will help assure that the load will be more evenly shared among the stakes.

If you have spare stakes, no harm in driving four or five sets. But beyond that you're probably better off setting an entirely separate additional anchor.

good approach for splices, at least; three tucks with each strand suffices for Manila and hemp, while slippery synthetics should get five or six tucks.

But most modern line adaptations are not so simple, partly because completely different rope constructions as well as materials have created a need for completely new knots, and partly because, with most knots, security is just one of several important considerations. Is the knot quick and simple to tie? Is it compact? Is it versatile? Can it be tied under strain? Can it be untied easily, even after a heavy load? This last consideration is particularly important, since modern rigs tend to load lines very heavily. The more heavily loaded a line is, the more likely the knot it is attached with will jam. This brings up a nice irony, in that modern lines under low or flogging loads will tend to untie themselves, but after high loads can't be untied at all. Considering all this, it's clear that an "If you can't tie a knot, tie a lot" philosophy simply won't do.

The Bowline

The Bowline is so universally revered that people feel they've done something wrong if it slips or jams. So it's with a sense of bewilderment and guilt that we try to reinforce it with everything from shackles to duct tape. But a simple extra tuck, or an extra turn before tucking, is all it takes to restore the King of Knots to its accustomed regal security

Figure 2-11a. A Bowline for slick line. Merely tuck the end back through as shown.

Figure 2-11b. Another way to "lock" a Bowline for greater security. Start as usual, but before threading the end back through the loop, take a Round Turn around the standing part.

(Figure 2-11). And this also seems to lessen the Bowline's inclination to jam under extremes of loading.

Hitches

The Round Turn and Two Half-Hitches (Figure 2-12) is another old workhorse that doesn't get along well with modern rope; even under load, the half-hitches can slowly untie themselves, one at a time. On the other hand, this is one of the few hitches you can make under load. So if you want it to stay put, seize the end to the standing part with a Double Constrictor. This is easily cut away when you want to untie the knot. If you can tie while the line is relaxed, the Stunsail Hal-

Figure 2-12. The Round Turn and Two Half-Hitches can be tied under load. Here it's Double Constrictored to the standing part for greater security. Though shown here in side view for clarity, the overhand part of the Constrictor should in practice fall on one of the two round surfaces.

yard Hitch is a fine knot and will hold without seizing (Figure 2-13. See also "Stays'ls," in Chapter 4).

The Buntline Hitch, another excellent old knot, has to be the most underused hitch around. It is extremely simple in structure, utterly secure, and so compact that it's almost transistorized (Figure 2-14). Its one drawback is that it can jam, so you need to be selective about its use, employing it on lines that don't have to be untied often. It's perfect at the standing end of the mainsheet, for example, or for a halyard shackle. And the Buntline Hitch is easy to untie when it hasn't been loaded heavily, which is why it's perfect to attach a halyard to a bosun's chair or climbing harness, or to hang fenders, or to hang a hammock.

SPECIALISTS

The knots considered so far have possessed, in addition to their other virtues, tremendous versatility. Constrictors are for everything from kite frames to handcuffs, the Bowknot wraps packages, and of course the word "Bowline" is derived from a Middle English word that means "use it dang near any-

KENNETH BATES'S BINDING ADJUSTABLE HITCH

This knot performs the same function as the venerable Rolling Hitch: It holds under lengthwise pull but can be slid along to adjust tension. Bates's knot has the added feature of jamming the end to prevent accidental untying once the knot is slid up firmly against the hitchee.

Make a Round Turn around the standing part, working away from the eye. Then cross back toward the eye, make a half-hitch around the standing part, and lead the end around and under the half-hitch once more. All the turns are made with the same direction of rotation. Draw up well.

Figure 2-13. The Stunsail Halyard Hitch, an excellent choice if you can tie when the line is not under strain.

where." Versatility is a hallmark of great knots, and a good thing it is; if you learn all the great ones, you only have to learn a few.

But just as mechanics have a few specialized gizmos around for jobs their sockets and screwdrivers can't handle, so a skillful tyer will know some specialized knots.

Angler's Loop

For example, there's the problem of tying a loop in the cloth-covered rubber rope called bungy cord. Because it is by definition so springy, even a locked Bowline will crawl right out of it. That's why bungy ends usually

> **PAINTER HEIGHT**
>
> Painters are often located quite high up on a dinghy's stem. But this can result in the bow's being pulled down when the boat is towed. Locate the painter above the waterline, but low enough that it will lift the bow a bit under tow. If the boat is sometimes trailered, make sure the painter attachment is high enough that the painter leads fair to the trailer winch.

Figure 2-14. *The Buntline Hitch. Make a Round Turn around the standing part, then reenter with the end for a Half Hitch around the standing part. Draw up.*

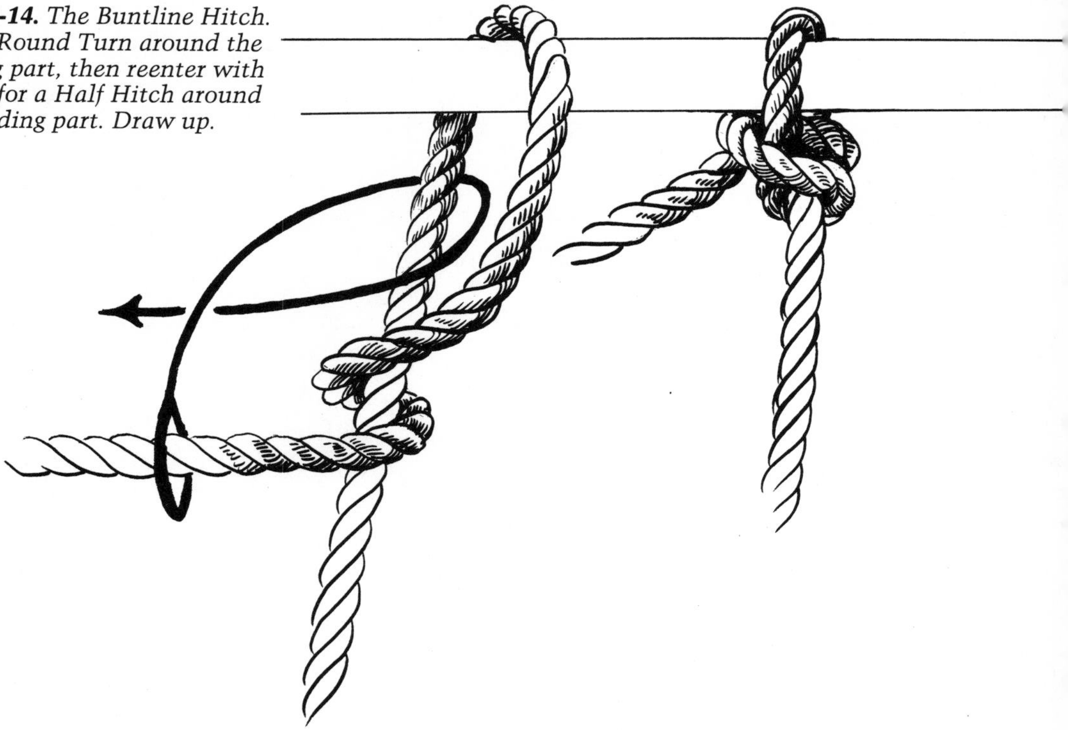

have a plastic-covered hook into which the end has been threaded, folded, and crimped (Figure 2-15). An expedient method, but one lacking in flexibility; the cord length can't be adjusted, and the bulky, snag-prone hook can be a nuisance.

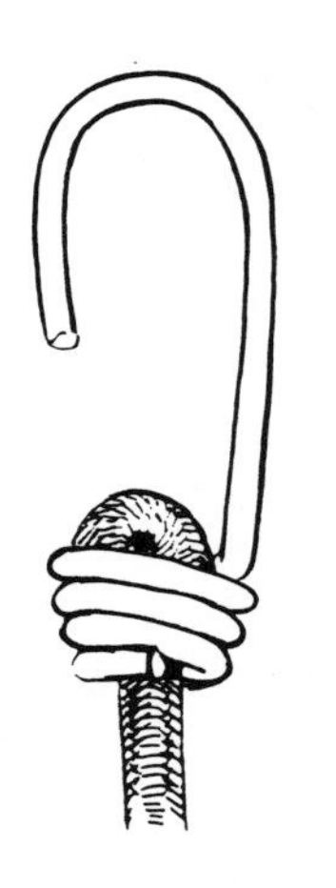

Figure 2-15. *Typical bungy hook: clunky, no adjustment.*

Geoffrey Budworth, a past Secretary of the International Guild of Knot Tyers, faced these bungy drawbacks when he was trying to secure gear on the deck of his kayak. "There are no knots that will hold in this stuff," other kayakers said. "Nonsense," Mr. Budworth said, and went home determined to come up with a bungy loop. And he soon found an old knot, the Angler's Loop (Figure 2-16), which neither slips nor jams in bungy. With it one can adjust cord length by making the loop larger or smaller, with or without a hook in the loop.

By way of showing the permutations old knots go through, here is an Angler's Loop history: It started out as the name implies, as a knot for fishing line, back when the line was made of gut. Both line and knot were rendered obsolete by the introduction of synthetic line. Then someone started using it as a knot for large line, because although it jammed, it was very quick and easy to tie, whether by the method shown here or the showy "Flying Bowline" technique (see *Rigger's Apprentice*). So the Angler's Loop survived, used by a few as a speedily produced but little-known alternative to the Bowline. Then along comes bungy and Geoffrey Budworth, and suddenly this antique knot has a new, mainstream life.

Two from Switzerland

Sam Rogers sails singlehanded on Swiss lakes. When he leaves the dock, there's no one there to help him cast off mooring lines. When he returns, he can usually hail some-

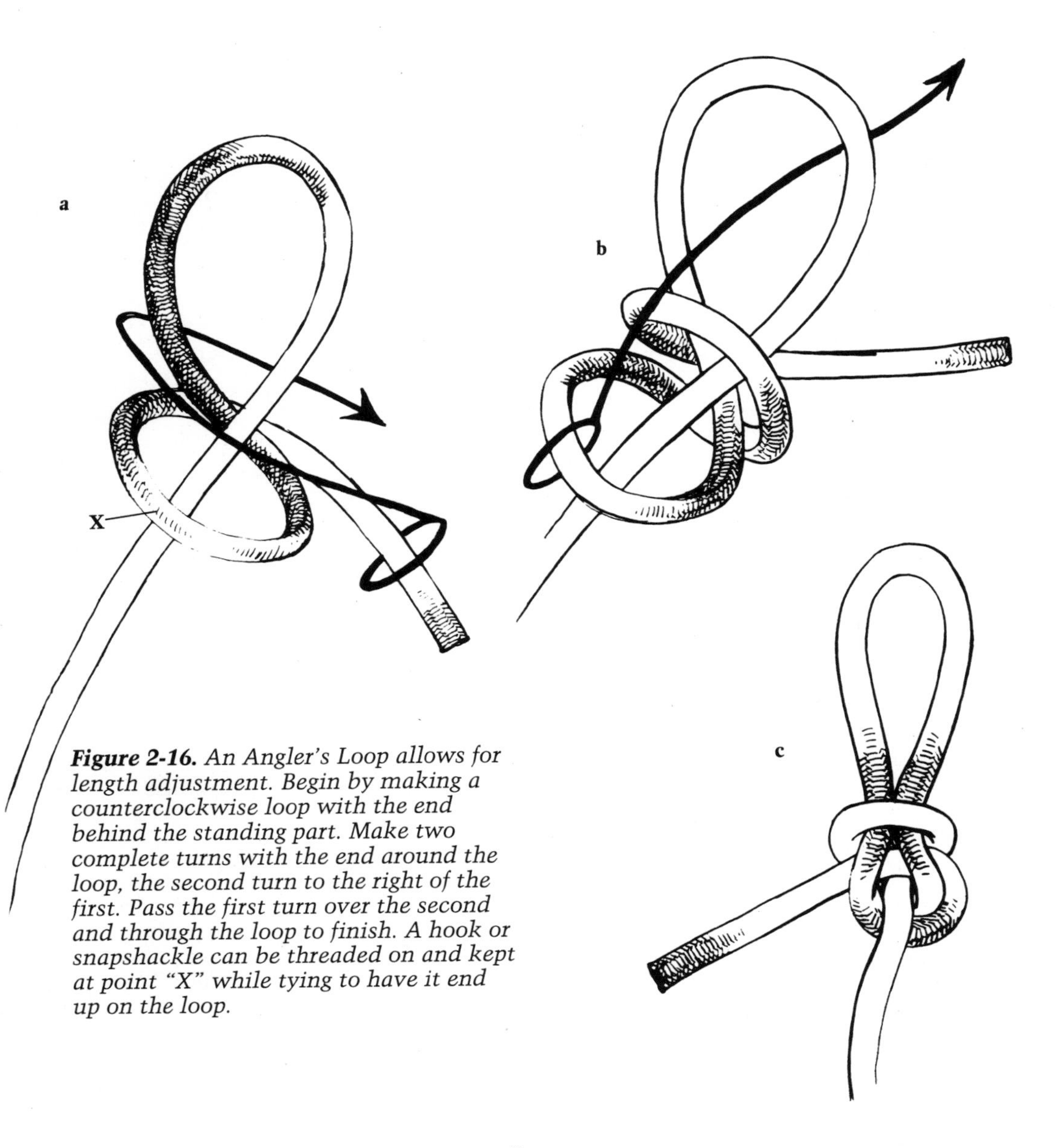

Figure 2-16. *An Angler's Loop allows for length adjustment. Begin by making a counterclockwise loop with the end behind the standing part. Make two complete turns with the end around the loop, the second turn to the right of the first. Pass the first turn over the second and through the loop to finish. A hook or snapshackle can be threaded on and kept at point "X" while tying to have it end up on the loop.*

one to take a mooring line to a cleat, but how to get the line to them? These problems have been thornier for singlehanders than the actual sailing is, and Sam has addressed them better than anyone else I know of.

Let's start with leaving the dock. Traditionally, this has been dealt with by tying some form of slip knot in the mooring line, then leading the end back to the vessel. Trouble is, most slip knots aren't very secure, and so can come undone before you want them to. And the ones that are secure don't clear readily and completely enough, so you can get hung up by a snagged mooring line in mid-departure. Sam's knot of choice is the Pseudo Bowline (Figure 2-17). Because it is based on the Bowline, learning it is not nearly so involved as learning a whole new knot—so is this really a specialized knot, or just another example of the Bowline's versatility?

As for returning to the dock, all you need

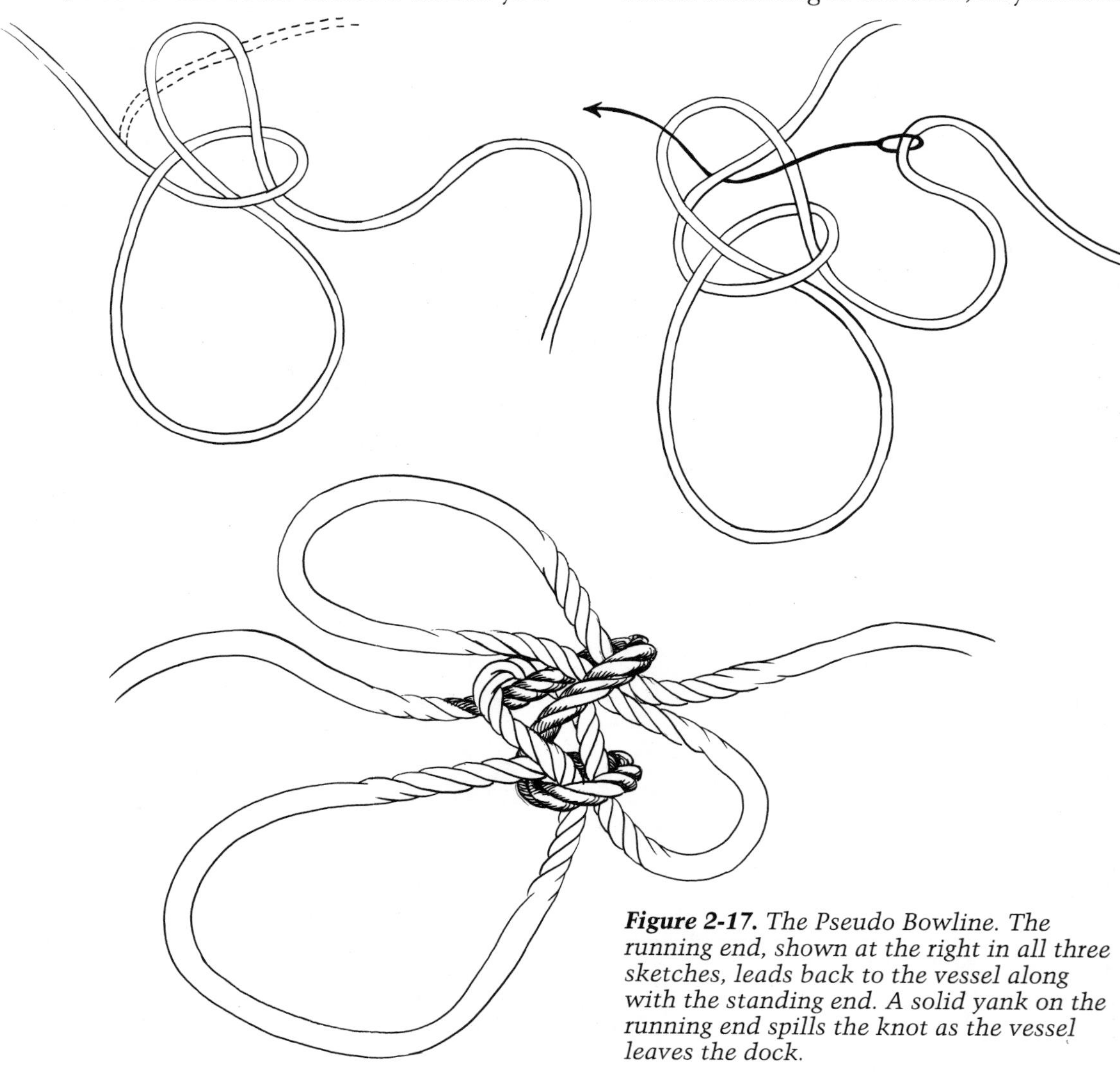

Figure 2-17. *The Pseudo Bowline. The running end, shown at the right in all three sketches, leads back to the vessel along with the standing end. A solid yank on the running end spills the knot as the vessel leaves the dock.*

TOWING BRIDLE

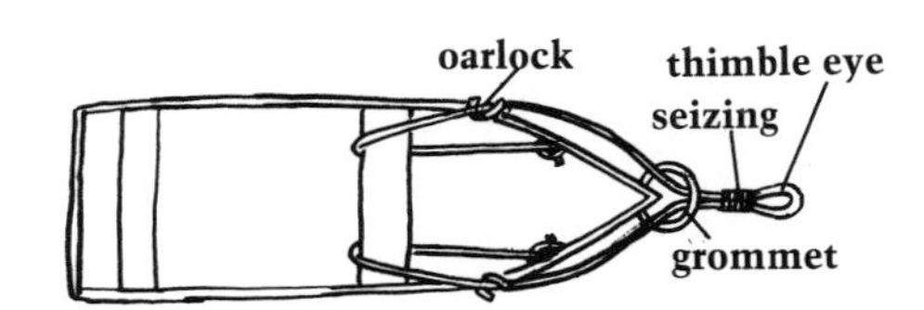

When towing large rowing craft, particularly in a chop, the painter can put an excessive strain on the stem. For this situation, make a bridle that runs from a forward oarlock, out through the painter grommet in a bight, then back to an oarlock on the other side. This distributes the strain over more of the hull. In addition, the painter will hold even if one side of the bridle pulls out or chafes through.

is a heaving line, which can be any light line tied to the mooring line at one end and weighted at the other. Throwing this is a lot easier than throwing the mooring line itself.

But the traditional heaving line has two serious drawbacks, which I believe have limited its use aboard yachts. The first is that the traditional weight has been a metal-cored knot called a Monkey's Fist. When this thing hits something, be it another boat or the skull of someone trying to catch the heaving line, it can very likely break it. As if this actionable quality weren't enough, Monkey's Fists have a sadistic habit of landing at the edge of a dock, then bouncing merrily into the water. This necessitates recoiling and reheaving as your boat drifts toward disaster.

The only thing worse than disaster for most sailors is embarrassment. And the heaving line also offers plenty of opportunities for that because it's devilishly tricky to throw accurately. The usual procedure is to coil the line, hold it in one hand, and heave it underhand or sidearm, and it's very easy to snag the coil with a finger or mistime the release. The result is a lovely arc of rope that goes nowhere near the dock, usually in full view of scornful onlookers.

Well, friends, if you'll examine Figure 2-18 you'll see a tamed heaving line. The weight is a canvas bag filled with sand. Any canvas shop can whip one up for you in minutes. The bag won't cripple, doesn't bounce, and has a certain utilitarian grace to it. The coil is ready for throwing, held by a "trigger" made from the line itself. To release, you just lift your thumb. Thanks, Sam.

DIAGONAL TOWING

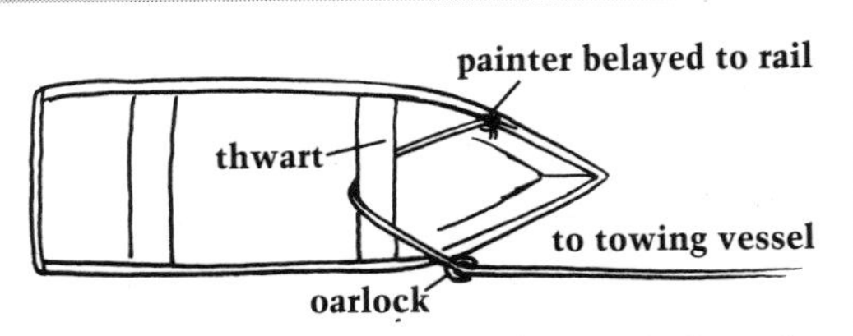

In light airs, a dinghy towed astern will often "run up" on its tow and smack it in the transom, due to wave action. To prevent this, lead the painter to a forward thwart. This will cause the dinghy to angle out to the side, away from the transom. Adjust the lead forward or aft for the ideal angle.

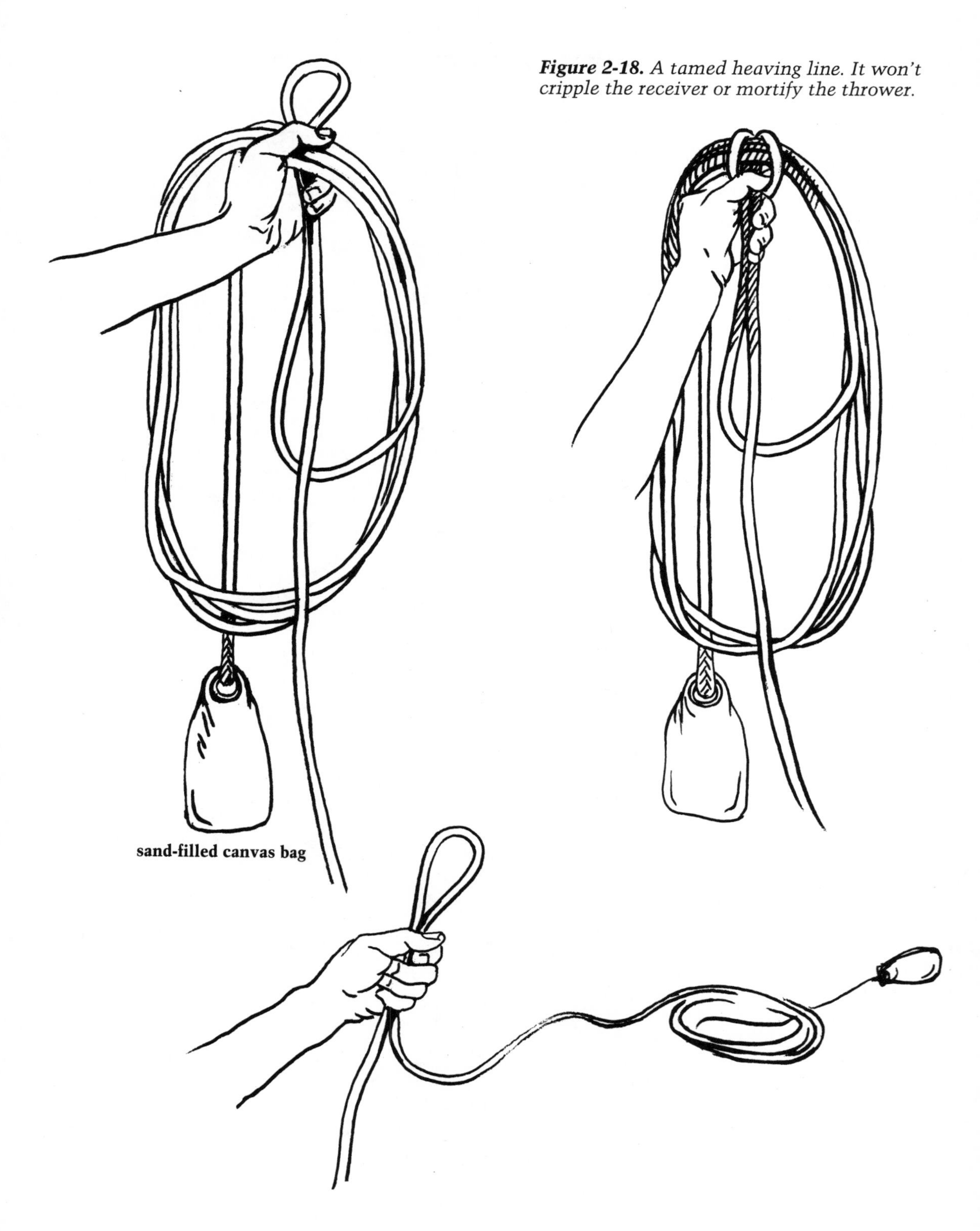

Figure 2-18. *A tamed heaving line. It won't cripple the receiver or mortify the thrower.*

Chapter 3
Splices

"Nothin' don't seem impossible once you've clapped eyes on a whale."
– Elizabeth Goudge

As with basic knots, splices have undergone evolutionary changes in this century due to stronger slicker materials and more concentrated rig loads. But whereas contemporary knots are mostly evolved from old knots, the introduction of double-braid rope has necessitated from-scratch invention of splices completely different from those for three-strand. So contemporary splicing involves upgraded traditional skills for three-strand and a whole new vocabulary for double-braid.

Many sailors' response is to say, "Forget that braided stuff; I'll stick with something I can understand". Others say, "I don't understand braid, but I can hire someone else to splice it." I believe that double-braid is worth using where minimal stretch is important, and that the concept of splicing it, once grasped, is extraordinarily simple. So I hope you'll approach it in this chapter with an open mind. I'd also like to stress that three-strand is cheaper for its size and perfectly suitable where moderate stretch is acceptable, and is my preference where you want stretch—mooring and anchor lines.

In writing this chapter, I've assumed you already know the basic three-strand eyesplice. If not, see *The Rigger's Apprentice* or *Knowing the Ropes* (Roger Taylor. International Marine, 1989) for instructions. The splices that are included here have been selected for their value. Many are little-known. All deserve to be better-known.

THE DOUBLE-BRAID SPLICE MADE HUMAN

Materials needed:

- 20 feet (6 meters) of ½-inch (or 13-mm) braided nylon
- light sail twine
- electrical or Scotch tape
- felt-tip pen
- heavy-duty scissors or knife
- tape measure
- splicing tools

ROUND EYE SPLICE ENTRY

The teardrop shape of a regular spliced eye is fine for thimbled eyes, mooring lines, and most anything else. But it won't set fair when put on tight on round thimbles, masts, and yards. The entry shown here results in a fairer-leading round eye.

To make:

(1) Set the eye size, then select the end strand that lies closest to the standing part. Tuck it under the nearest standing part strand once, against the lay (from right to left when the eye is toward you).

(2) Select the middle end strand and tuck it under the first *two* standing part strands, against the lay.

(3) Turn the eye over. There's one untucked end strand, and one standing part strand waiting for it. Pass the end over, from left to right (with the lay) then tuck back under, from right to left (against the lay).

Finish by tucking as for a regular splice: two more full tucks for Manila and hemp; three more for spun Dacron; and four more minimum for nylon and other slick fibers.

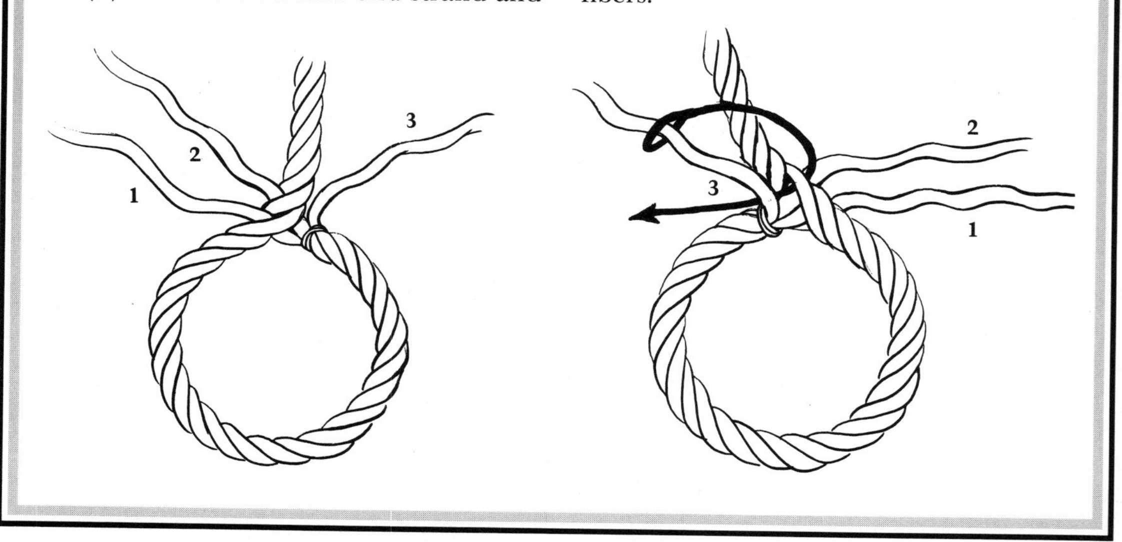

Double-braid rope is an alien construct. It might be in the collective unconscious of E.T.'s people, but it sure ain't in ours. That's why many sailors who can instinctively grasp the weaving principle behind splicing three-strand are hopelessly disoriented by braid.

For most of us, splicing double-braid is something only undertaken armed with manufacturers' instruction booklets and an arsenal of splicing tools. The long and short fid lengths, number- and letter-tagged marks, and all the other stage business serves to distract us from the knowledge that we have absolutely no idea what's going on. That's a pity, for once you get the concept, eliminating the air of superstitious ritual, you'll work more smoothly, turning out good splices quickly.

The Concept

To de-alienize the procedure, first picture the rope. It's made of two braided tubes, one inside the other. Why make rope this way? Because you can get more yarns into the same diameter than with three-strand, so the rope is stronger for its size. And because of the angle of the braids, the yarns travel a shorter distance between the ends, so there's less stretch. The two tubes are not attached to one another, so if the ends are not whipped or (shudder) melted, you can slide back the outer tube (the "cover") to expose the inner tube (the "core"). It's just like sliding the wrapper back to expose a straw. This independence of core and cover is the key to splicing double-braid.

So. Here's the concept: The core sneaks out through the side of the cover, a little way from the end; core and cover are joined end-to-end; then the core is drawn back into the cover, and it takes the end of the cover with it. This forms an eye.

There has to be more to it, right? Sure, lots, but it's all just a series of details and refinements that help the splice come out strong and pretty.

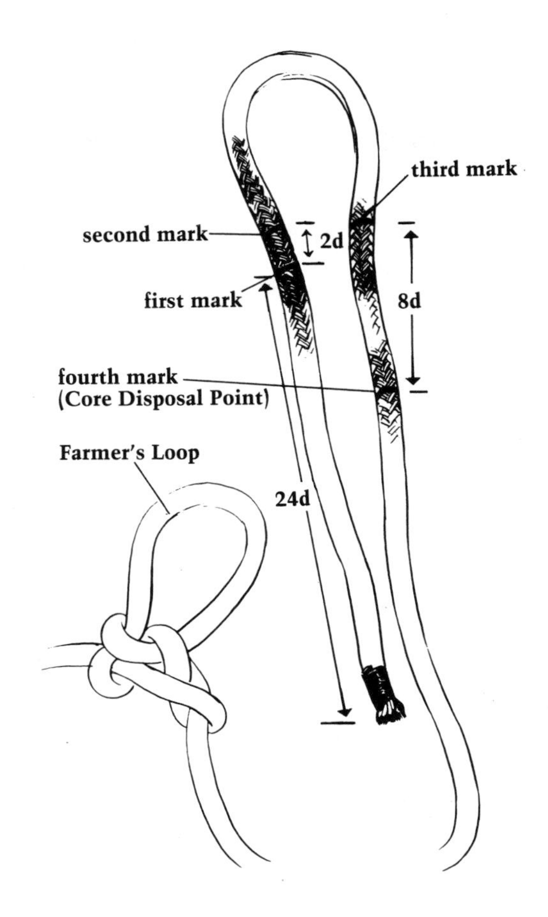

Figure 3-1. *Marking the cover for the Double-Braid Eyesplice.*

Measuring the Cover The first and most important detail is measuring. This splice relies on specific measurements that vary with rope diameter (d). We'll dispense with the usual, bizarre "fid-length" code and base all our measurements on one magic number: 24. What's the number? 24. Right.

For your first splice, use a piece of new ½-inch nylon rope. Get about 20 feet (6 meters), enough for a few practice pieces.

With a felt-tip pen, make a mark 24 times the diameter (24d) from one end. With ½-inch line (13 mm), that means 12 inches (305 mm). If this were ⅜-inch (9.5 mm) line, the length would be 9 inches (229 mm), ⅝-inch (16 mm) line gets 15 inches (381 mm), and so on. The core will join the cover at this point.

Make a second mark 2d away from the first one, working away from the end. Just remember the 2 in 24. This mark is where the eye will start. Everything from this mark to the end will be buried inside the rope.

To make the burial easier, mark now for a taper, which you'll cut later. Count five sets of strands toward the end from the first mark. The ones you count can angle left or right, it doesn't matter as long as you're consistent. Mark the fifth set of strands and the set that intersects it, making a little chevron (Figure 3-1). Count four more sets and make another chevron. Then five sets and mark, then four, etc., right to the end.

Now measure the circumference of the

eye you want. For this splice, let's call it 10 inches (254 mm). Make a third mark 10 inches (254 mm)from the second one, away from the end. When the splice is done, this length of rope will form the eye.

Make a final mark on the cover 8d farther on. Remember this by multiplying the 2 and 4 in 24: 2 x 4 = 8d. In our example, that's 4 inches (102 mm). Allow me a little mystery here by referring to this mark for now as the Core Disposal Point (CDP). You'll see its significance soon.

To review, so far we've made marks at 24d from the end (where the core joins the cover), at 2d (where the eye starts), at 10 inches (254 mm) beyond the second mark (a sample eye circumference), and at 8d beyond that (the mysterious CDP).

Excising the Core Put down your pen and make a loop knot, like a Farmer's Loop (shown) or Figure-Eight made with a bight. Just something that will stay tied. Put it at 6 to 8 feet (2 meters or so) from the last mark. In just a moment, when you slide the cover back, the slack in the cover will bunch up against the knot instead of wandering on down the standing part.

Pick up your awl. Bend the rope sharply at the third mark and use the point of the awl to nudge the cover yarns apart, clearing a little hole at the mark. You'll see the core inside (Figure 3-2). Use a light touch to avoid excessive yarn distortion and to avoid clearing little holes in your fingers.

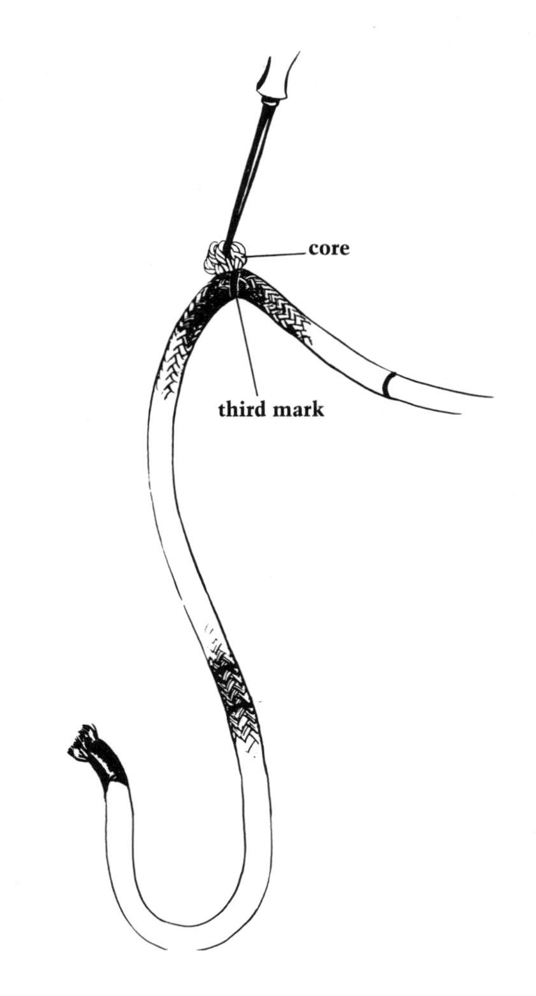

Figure 3-2. *Extracting the core.*

When the hole is as wide as it will readily go, use the point of the awl to pry a bight of the core out of that hole. Just herniate it right into daylight. Pull on this bight to pull the core end right out of the cover (Figure 3-3a). Throw a little tape on the core end to keep it from ravelling.

In taking the core out, you've disturbed the relationship between it and the cover. You need to restore that relationship so that the reference points you're about to make on the core will complement those on the cover. So take a moment to anchor the standing part, on the far side of the loop knot, to something solid. Then grasp the rope firmly on the splice side of the loop knot and slide your hands toward the end, milking any slack out of the cover, so there's even tension on core and cover. Detach the standing part from its belay.

Measuring the Core Now you can mark the core at the point it exits the cover (Fig-

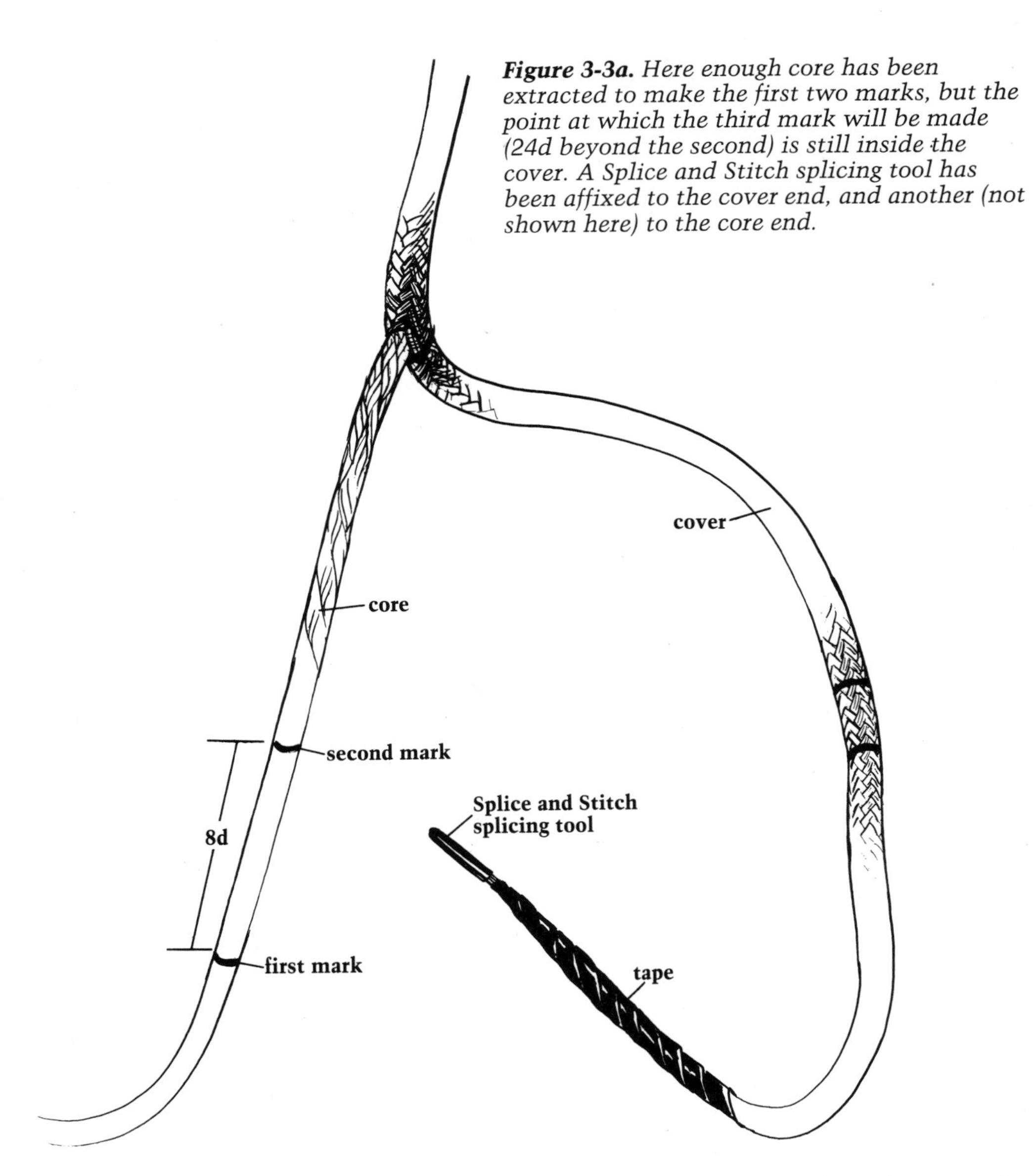

Figure 3-3a. Here enough core has been extracted to make the first two marks, but the point at which the third mark will be made (24d beyond the second) is still inside the cover. A Splice and Stitch splicing tool has been affixed to the cover end, and another (not shown here) to the core end.

ure 3-3a). That done, slide the cover firmly back toward the knot; we've got our reference point and can now mark the core some more.

Measure 8d from the mark you just made on the core, away from the end, and make a second mark. In our example that's 4 inches (104 mm). It's the same as the distance between the third and fourth mark on the cover, right? This is where the cover will join the core. This corresponds to the mysterious CDP. Make a final mark 24d farther on.

God Bless Mr. McGrew Your rope is all marked and ready to splice. Time for a brief aside to discuss tools.

Double-braid requires special tools, since instead of weaving, as in three-strand, you'll be threading the tubes into one another. This is where it gets alien.

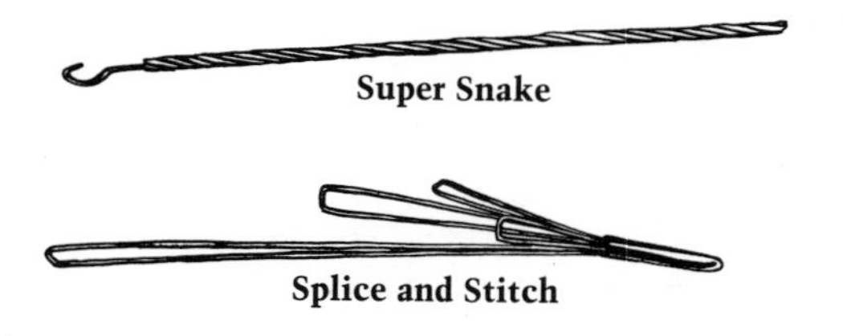

Figure 3-3b. *Two useful purpose-built splicing tools. Substitute as necessary.*

The original and still dominant tools for the job are the "fids and pushers" available at any chandlery. Don't use them! They're expensive, bulky, slow, and clumsy to use. And you have to buy a separate fid for each size of rope. They'll work with these instructions, but you're much better off with a "Super Snake" or "Splice and Stitch," both made by Fid-O-McGrew, 8120 Rio Linda Blvd., Elverta, CA 95626. I have no financial interest in these tools—they're just far and away the cheapest, most compact, fastest, and easiest-to-use gizmos you can get (Figure 3-3b). One size fits all yacht-diameter rope.

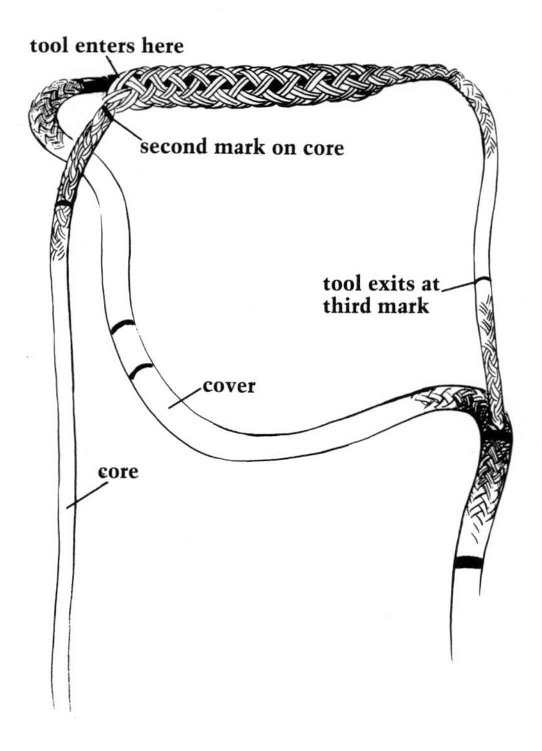

Figure 3-4. *The cover end enters the core at the second mark and travels along the standing part of the core to exit at the third mark.*

The Splice Enough promo. Using the instructions that come with the tools, put a Snake or Stitch on the core and cover ends (Figure 3-3a). Work the cover end into the core at the second mark, pointing down the standing part (Figure 3-4).

Thread the tool along inside the core, being careful to avoid snagging any yarns. The best way to do this is to push some slack into the section you're threading through, as shown, to expand the diameter of the tube, and to open the braid so you can see what you're doing.

Exit the tool at the third mark on the core. Draw the cover out and remove the tool.

It's time to cut the taper. Pull on the cover end until the first two marks are exposed (Figure 3-5). With your scissors, cut the chevron-marked strands farthest from the end. Pry these strands up a little before cutting so the ends will protrude. After cutting, take hold of the ends and pull the yarns right out of the end. These first ones might be reluctant; be gentle but firm.

Continue cutting and removing all marked yarns. When you're done you should have a slightly dishevelled but smoothly tapering end. If instead you have a choppy, scraggly mess, you miscounted your strands when marking. Start over.

Half-hitch the very end of the cover around the core, right where it exits (Figure 3-6). Bunch the core down against that hitch and pull on the bight of the cover until the first mark on the cover emerges at the second mark on the core.

Insert the core into the cover at this

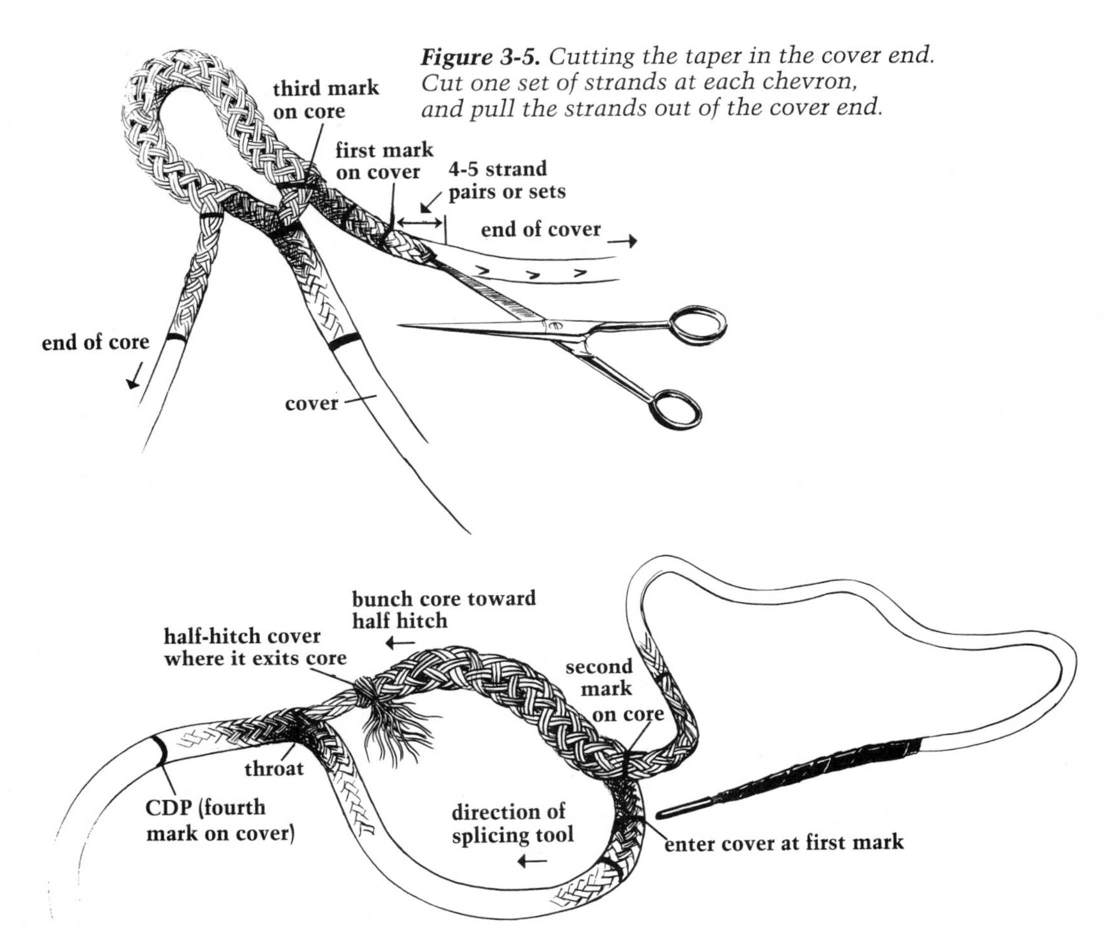

Figure 3-5. *Cutting the taper in the cover end. Cut one set of strands at each chevron, and pull the strands out of the cover end.*

Figure 3-6. *Half-hitch the tapered cover end to the core, right where it exits. Bunch the core against the hitch until the first mark on the cover emerges. The core enters the cover at this point.*

mark, again pointing away from the end (Figure 3-7). Thread along the inside of the cover, and by and by you'll come to the throat of the eye. Keep going; you want to emerge at the CDP, the spot where the core will be buried sufficiently for security. Working along between throat and CDP, be extra careful to avoid snagging yarns in this constricted area.

Exit and remove the tool. You've now buried core and cover, trapping them inside each other à la Chinese Handcuffs with a Martian twist (Figure 3-8). We're almost done.

Tightening the Crossover Look at the area where core and cover dive into one another (Figure 3-8). Before you finish the splice, you want to make this "crossover" point compact and secure. So unhitch the tapered cover end and pull on it, bunching the core back toward the crossover. Then pull on the core end, bunching the cover back toward the crossover. Core and cover should jam up against each other where they interlock.

Repeat pulls until the crossover is really firm. Then smooth the bunched-up line away a little bit on either side, so it's not dis-

Figure 3-7. *Snake the core through the cover, past the throat, to reemerge at the CDP.*

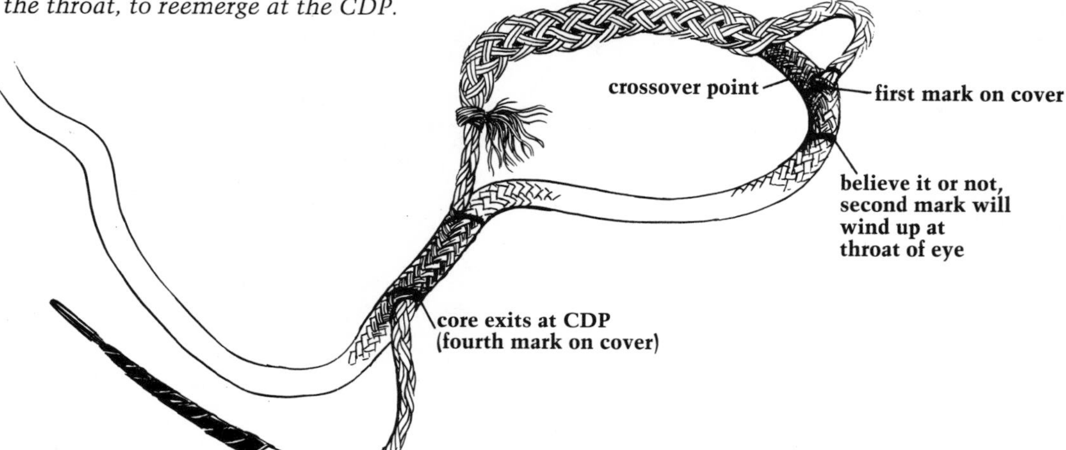

Figure 3-8a. *Letting the cover disappear inside the core after stitching the crossover point.*

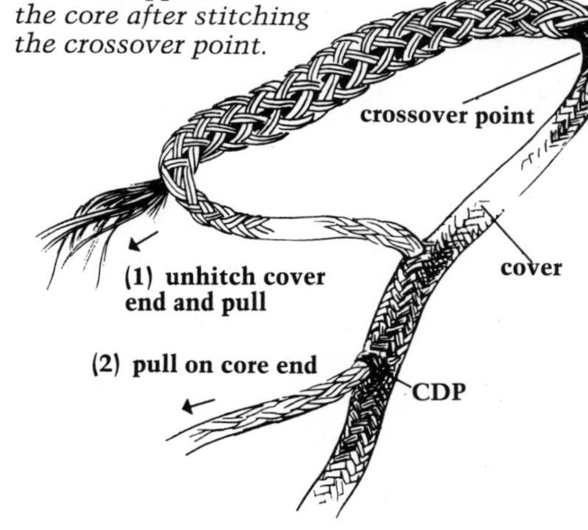

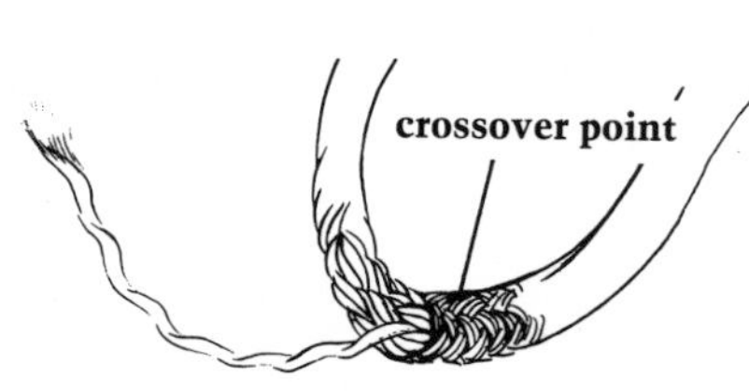

Figure 3-8b. *Pull one yarn of the core out at the crossover point and stitch through the crossover with it to lock it and make it more compact. Use the same thread later to stitch the throat of the finished splice.*

torted. With the tip of a spike, pry up a yarn of the core at the crossover, then pull its end out (Figure 3-8). Use this yarn to stitch three or four times through the crossover so it cannot pull apart. Do this neatly so the crossover stays firm and compact. When you're done stitching, don't trim the yarn. You can use it later to stitch through the throat of the completed splice. This will keep it from working loose when it's not under load.

The Finish Milk the core firmly away from the crossover. As you smooth the core out, the tapered cover end will disappear inside. That's one end taken care of.

Next, milk the cover away from the crossover. The core end will draw back but is too long to disappear. Just get all the slack out of the cover, then mark the core where it exits the cover. You'll be glad to know that taper-

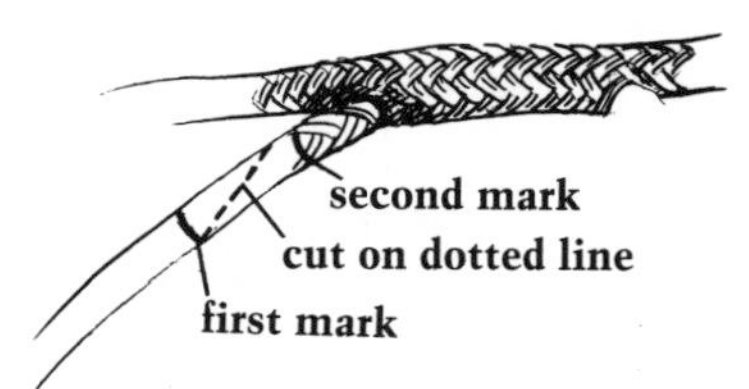

Figure 3-9. *Trimming the core end.*

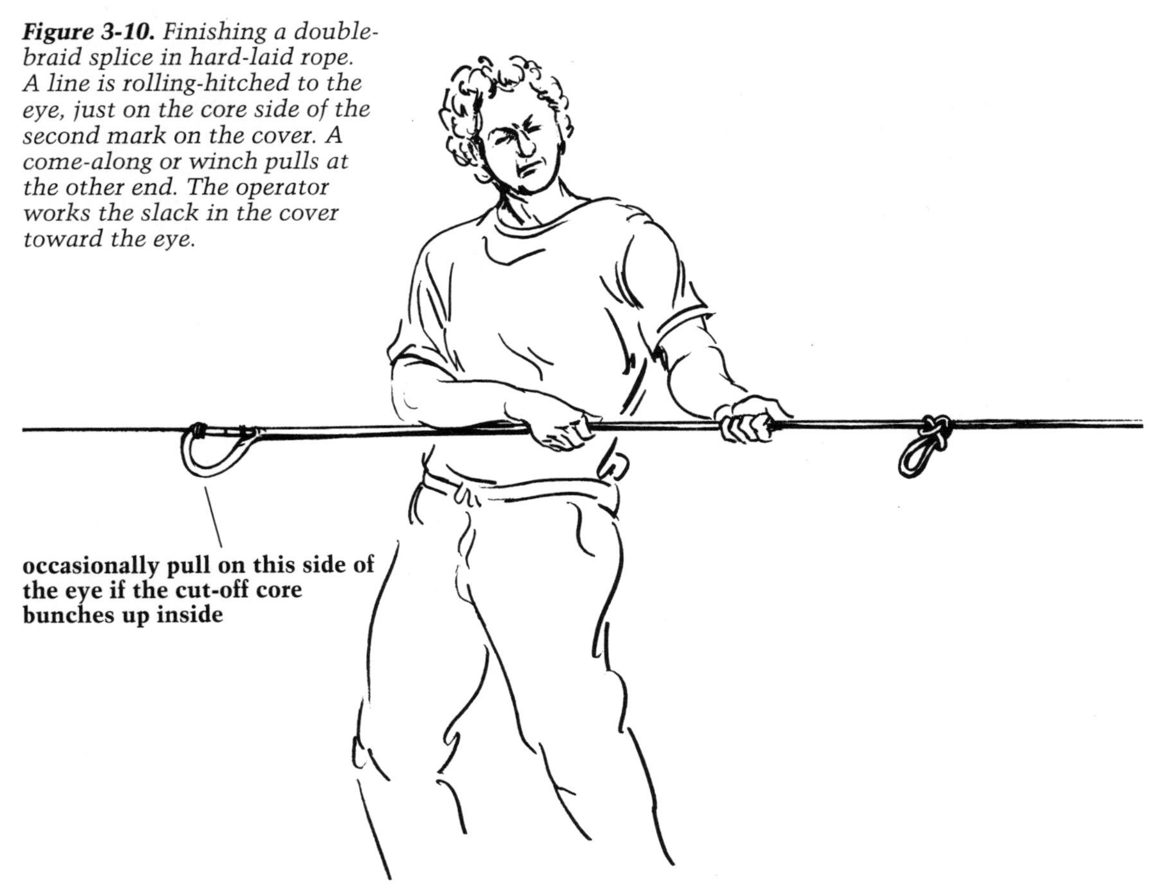

Figure 3-10. *Finishing a double-braid splice in hard-laid rope. A line is rolling-hitched to the eye, just on the core side of the second mark on the cover. A come-along or winch pulls at the other end. The operator works the slack in the cover toward the eye.*

ing the core is much easier than tapering the cover.

Just pull the core out and cut the core on a taper above the mark (Figure 3-9). You can do this with a very sharp knife on a block of wood or with a good pair of pruning shears, or you can comb the core apart and cut the yarns with scissors. When you're done, milk away from the crossover once more. This time the core, too, will disappear inside.

The Grand Finale: Retie the standing part to something solid. Rolling-hitch a light line to the eye, right next to the second mark on the cover. Take this line to something equally solid and haul tight (Figure 3-10).

Milk the cover firmly away from the loop knot, using both hands. The core and crossover will be drawn inside. This is a wonderfully gratifying conclusion to all that has gone before.

You should run out of slack just as the cover reaches the second mark. More likely, there'll be a little slack left in the cover. Untie the loop knot, anchor the eye to something solid, and milk the cover very firmly for the entire length of the rope. This will remove any slack left from splicing, plus any that might be there from uneven tensioning in the ropemaking machinery. Milk twice, and you're done.

Options and Variations

A. Rendering The orneriest and most frequently encountered obstacle to double-braid splicing comes at the very end, when

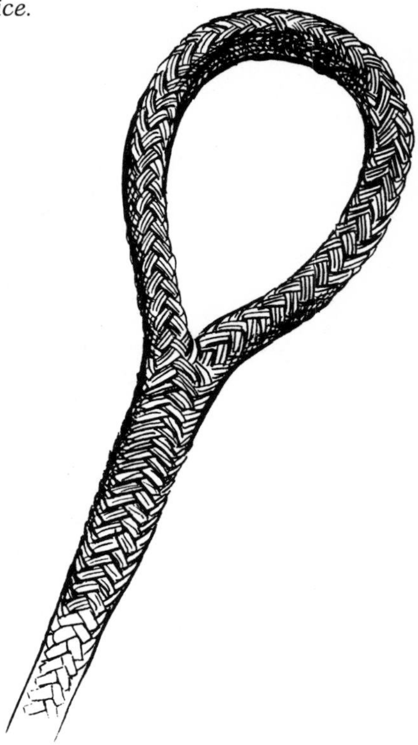

Figure 3-11. *A finished Double-Braid Eyesplice.*

1. More tension on the eye. The small line you hitched to the eye in Figure 3-11 made rendering easier, because anytime you tension a line its diameter shrinks. Physics. For extra-tough rope, hook the other end of the small piece to a block-and-tackle, come-along, or winch. Tighten up, then milk. Be sure the small piece pulls only on the crossover side of the eye. Otherwise, it will also tension the cover, making rendering more difficult.

2. Lubricate the core. Go to an electrical supply store and get some of the gel that electricians use to pull wire through conduit. Ask for it in those terms; it comes in a variety of brand names. The stuff is water soluble, so you can (and must!) rinse it out after the splice is done.

3. Put more slack into the cover. When you lessen the tension on a line, it expands. More physics. Undo the loop knot, then go to the other end of the line, cast off any tape or seizings, and pull a couple of feet of core out. This will put slack into the cover. Milk

you're trying to work the cover up to that second mark. Sometimes this is a result of mismeasurement, and sometimes the seizing or stitching at the crossover was made clumsily, so the crossover stands up and snags the cover. More often, unfortunately, the problem is caused by the structure of the rope itself; some rope is much harder-laid or otherwise "sticky" than others, and unwilling to render. If all you ever splice is slick, soft-laid rope—user-friendly stuff that makes splicing a holiday but is increasingly rare—then what you've read so far is all you need. What follows is an escalating series of tactics to defeat less docile rope. The stiffer the material, the more involved combination of tactics you'll need. Learn them all and no rope can stand against you.

PAINTERS

For a pram painter, bore a hole in the forward transom just large enough to accept the rope. Bore a second hole in the knee that supports the forward transom. Lead the painter through both holes, after fairing the edges so the wood won't cut the rope, then make a Figure-Eight or button knot in the end of the rope. You now have a painter that is ultimately strong and secure, without buying and installing any hardware. And the painter can be instantly removed, in case you need to use the line for something else.

this slack up to the splice and retie the loop knot behind it. Reanchor the standing part and put tension on the eye. Render away, and be careful to avoid going too far with the newly capacious cover. When the splice is done, undo the loop knot one last time and very thoroughly milk all possible cover slack back toward the end.

4. Build a splice shuttle. If tension, lubrication, and a baggy cover can't close up the eye, and if you measured right and the crossover is smooth and compact, you have one lean, mean piece of rope on your hands. Your only hope now is a bit of devious technique, courtesy of the ingenious Mr. McGrew.

The first step is to form the splice slightly differently: Bring the core out at the throat of the splice, rather than at the fourth mark on the cover (Figure 3-12). With no extra core taking up space inside the rope, it's much easier to get the eye home.

And because it is so easy, most splicing instructions you'll see recommend this method as standard—they'll tell you to trim the core off here. Trouble is, a core end thus treated will inevitably pull back under load, leaving a hollow space at the throat of the eye. For years, manufacturers have assured rope users that this hollow spot "does not materially affect" splice strength. The argument for this is that since the two halves of the eye share the load evenly, only 50 percent of the load should come on either half. Since the cover alone constitutes 50 percent

Figure 3-12. *Building a Splice Shuttle, an alternative approach for stubborn rope. This drawing picks up where Figure 3-6 leaves off.*

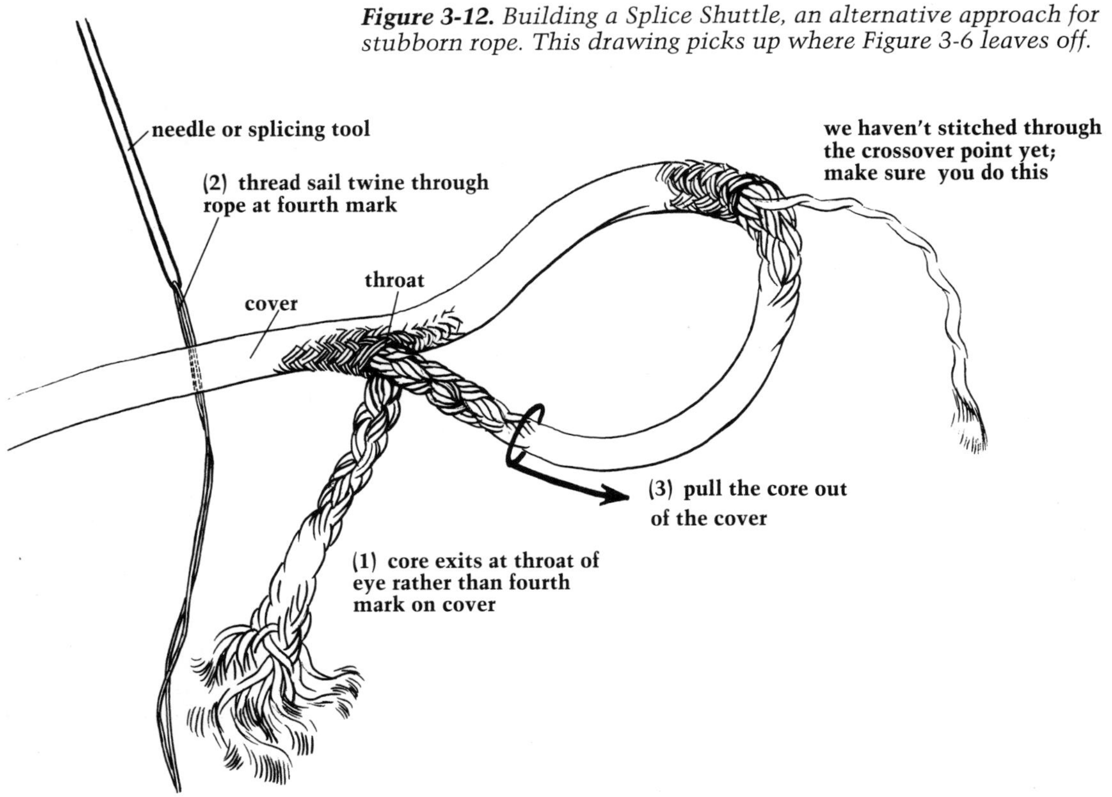

of rope strength, the throat should be the strongest part of the splice.

However, the two legs often do not share the load evenly. For example, picture a mooring eye around a bollard. The other end is made fast to a vessel alongside. As the vessel shifts in the wind, the angle of line to bollard shifts. But due to friction around the bollard, the eye cannot shift, especially if the line is wet or the surface of the bollard less than perfectly smooth. The vessel only has to shift a little for most or all of the strain to fall on one leg of the eye. If this happens to be the leg with the hollow spot, the strength of the line has been reduced by about 50 percent. The same thing can happen with an eye looped around a bitt, or on a halyard with an even slightly cocked thimble. Burying the core well into the standing part is crucial for ultimate splice security.

So how do you get this done in hard, tight rope? Finish the splice with the core exiting at the throat, but leave it uncut. Then stick a #9 sail needle, or the longest loop on the Splice and Stitch, or even a piece of double seizing wire through the rope at the fourth mark on the cover. Thread the tool with about 3 feet of sail twine, preferably unwaxed. Pull the tool out, leaving the twine in the rope. Then pull the core out of the cover, drawing the twine up and out at the throat of the eye (Figure 3-13). Pull the twine out clear of the rope, but leave it threaded on the tool. You've just built a Splice Shuttle.

Render the splice home, then pull the eye up out of the throat with the string. Unbraid the core and stick a few yarns into the eye of the tool. Pull it down to shuttle the yarns down inside the standing part and out at the mark (Figure 3-14). Then use the twine to haul the tool back up to the throat. Reload with yarns and pull down again. Continue shuttling until you can't get any more yarns through. Half of them are for security. Cut off flush any that remain at the throat. Trim likewise the ones at the mark.

B. Untapered Cover The step of tapering the cover is the most time-consuming part of

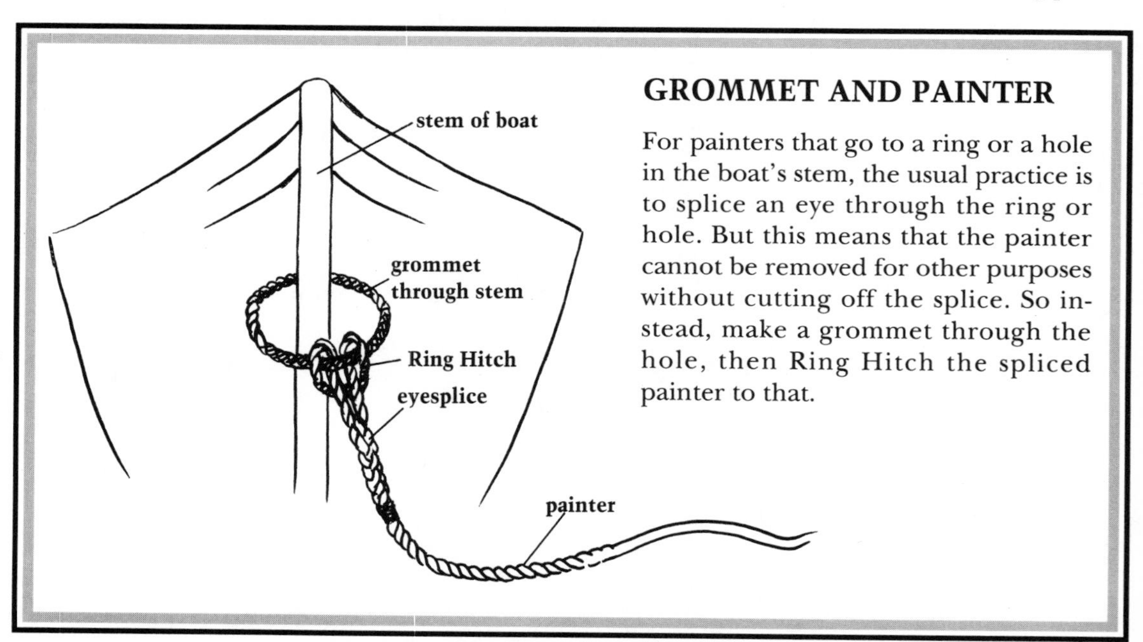

GROMMET AND PAINTER

For painters that go to a ring or a hole in the boat's stem, the usual practice is to splice an eye through the ring or hole. But this means that the painter cannot be removed for other purposes without cutting off the splice. So instead, make a grommet through the hole, then Ring Hitch the spliced painter to that.

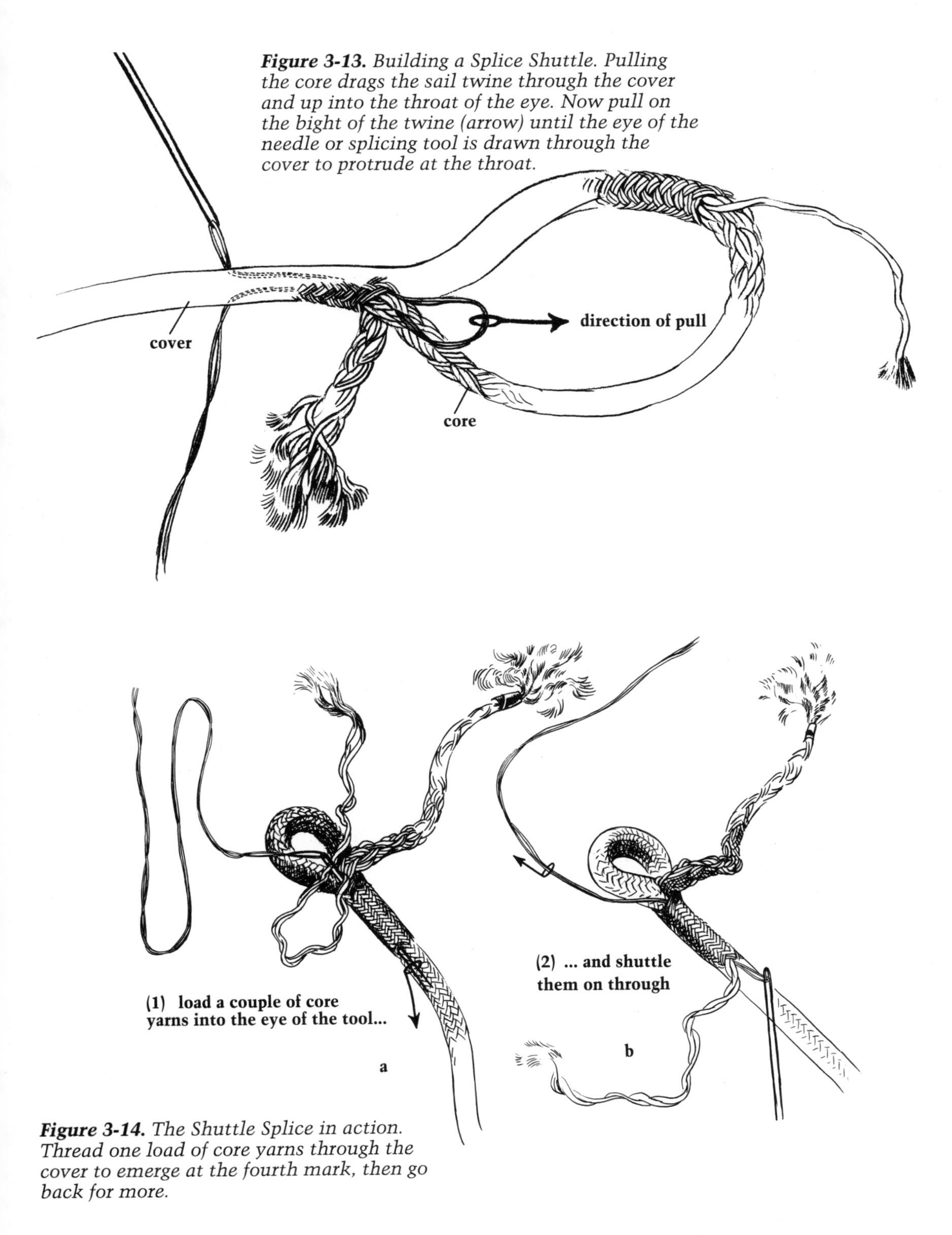

Figure 3-13. *Building a Splice Shuttle. Pulling the core drags the sail twine through the cover and up into the throat of the eye. Now pull on the bight of the twine (arrow) until the eye of the needle or splicing tool is drawn through the cover to protrude at the throat.*

Figure 3-14. *The Shuttle Splice in action. Thread one load of core yarns through the cover to emerge at the fourth mark, then go back for more.*

the splice. But it's generally worth it because it makes rendering easier, makes a fairer splice, and eliminates internal chafe that a bulky, square-ended cover can cause.

But with light-duty, soft-laid lines you can skip this step, or just cut the end at an angle with a knife or scissors. The splice will go much faster, and without all those little ragged taper ends to deal with it doesn't even matter if you tuck the core or cover first.

C. Thimbles When splicing around thimbles, the length between the second and third marks is what fits around the thimble. So make your first two marks on the cover, then bend the line around the thimble to measure for the third mark. Try for a semi-snug fit; too tight, and the fibers will chafe against the thimble, and there'll be little room to run a shuttle. If the eye ends up a little loose, you can always stitch it at the throat or hold it for a few minutes in hot water to shrink it.

D. Splicing Through Clew and Shackle Rings Splicing through clew rings is a good way to attach sheets. The ring provides a wide bearing surface, so no thimble is necessary. Splicing through the closed ring of a snapshackle can be done with or without a thimble. If without, you can make the eye extremely small. This looks tidy and can give you a couple of extra inches of hoist on a tall sail. But the line is prone to chafe, so for long-lived lines thread the shackle onto a thimble first. Open the thimble a bit by driving a fid into it, then close it in a vise if it is on the shackle.

With or without a thimble, you must thread the shackle onto the cover before you begin splicing. Have I ever gotten a splice completely finished and then had to cut it off and start all over again because I'd forgotten to thread the shackle on? Nahhh.

LUGSAIL

A lugsail is a wonderful rig—weatherly, simple, and easily handled. But sometimes in a chop or off the wind, the yard swings around and bangs into the mast. To prevent this, splice the standing end of the halyard around the mast, then lead it by way of strap eyes to the usual attachment point.

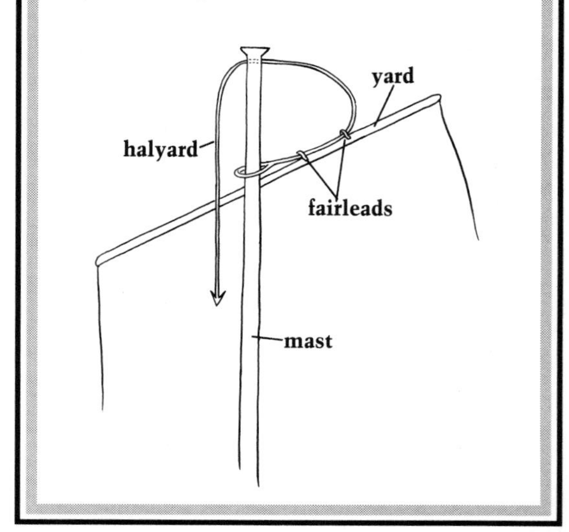

E. A Very Good Idea When you know how to splice this stuff quickly and well, people are going to ask you to do their lines for them. It's great work, as a favor or for pay. But sooner or later someone is going to ask you to splice a line that is not new. When they do, you can save yourself a lot of trouble by answering with a firm "no." Old double-braid, broken down by sunlight, salt water, and use, is vicious, unrenderable stuff.

If you must try: Wash the line first and use fabric softener; after excising the core, bunch the cover back fiercely against the loop knot to loosen the fibers; and pound on the cover with a rubber or wooden mallet before you go to finish.

With the unpleasantness of aged rope in mind, I'll close with an updated Rigger's Blessing: Large blocks and new line to you!

CHAIN SPLICES

Like a wire-and-rope halyard, a chain-and-rope anchor rode combines two materials in order to gain the virtues of both.

An all-rope rode of appropriate size would be plenty strong to make a boat stay put. And if it's made of nylon—as it should be—it will be elastic, so that staying put would not involve hull- and teeth-jarring shockloads as the hull fetched up in a swell. But an all-rope rode would be subject to chafe; it would chafe away on rocky bottoms, and it would chafe away at the boat's roller or hawse, particularly from storm-induced side loads.

PLYWOOD SLING

Ever had to carry a sheet of plywood? It forces you into a wrist-twisting, view-obscuring, back-straining, top-heavy carrying position. If that doesn't appeal to you, make up a sling about 8½ to 9 feet (2.6 m to 2.8 m) long, with an Eyesplice or Bowline at either end. Slip an eye over each bottom corner. Grasp the middle of the sling with one hand and rest the other hand on the top edge of the plywood. *That's* how to do it.

An all-chain rode is ultimately chafe-resistant, but it is also ultimately inelastic. It compensates somewhat for this inelasticity because in use its very weight causes it to describe that elegant sag we call a catenary. But in moderate wind and swell, the chain goes not quite straight and BAM!, instead of riding over waves, you're colliding with them. (As a friend of mine puts it, "There's no catenary at 50 knots.") This is why, with an all-chain rode, a nylon snubber is a good idea: After the anchor is set, a heavy-duty length of nylon rope, 15 to 30 feet long, is hooked or tied to the rode. The other end of the nylon is made fast to the Samson post. The rode is then slacked until the nylon takes the strain, acting as a snubber. And of course the chain is belayed as insurance should the rope fail. For more details on this system, see Roger Taylor's *Knowing the Ropes.*

With a snubber in place, you gain the virtues of chain and nylon, so who needs a rope rode at all? But there are two other factors to consider. They don't have anything directly to do with anchoring, but they can nonetheless be crucial to rode selection for your boat. The first is cost: Three-strand nylon rope costs only a fourth as much as comparable-strength chain. For the yacht-poor sailor, this alone would make rope attractive—if there were a way to prevent chafe.

The second factor is weight. Particularly in light-displacement boats, anchor chain when stowed will trim the hull down at the bow. This cuts speed, exaggerates weather helm, and makes nose dives more likely. To compound matters, the chain when deployed will leave the bow too buoyant, so that the boat will be inclined to yaw and sail around on its anchor.

If your boat is heavy enough and full enough forward that chain doesn't affect your trim, then an all-chain rode with snubber can be a good (if expensive) way to go.

But for the majority of boats afloat, there's a strong argument for combining rope and chain. Put enough chain at the lower end to provide catenary and abrasion resistance down there, and let the rest of the rode be strong, resilient, light, inexpensive nylon. To avoid chafe at the upper end, see to it that your hawse or roller has smooth, wide-radiused sides, and that the rope cannot jump free under side loads. You can also cushion the rope with a length of heavy-duty hose positioned to take the chafe. Another good practice is to seize a one-foot length of hose or leather onto the rode every 20 feet. You get built-in chafe protection and length markers all in the same package. Finally, you can always hitch a separate rope snubber onto the rode for insurance, just as with chain.

This brings us at last to the Chain Splice, for we must have a way to join these two materials. The most often-seen way to do this is

BILL PAGE'S ANCHOR TAMER

Boatbuilder Bill Page's cutter carries two Fisherman-style anchors, in rollers on either side of his bowsprit. A line seized to the crown of each anchor is led to a cleat. These lines, hauled taut, keep the anchors from jumping around under sail.

A small buoy is spliced to the end of each line, and when an anchor is deployed, line, buoy, and all go down with it. This gives you a convenient, comfortable "handle" if you have to dive on the anchor to shift it. But Bill uses it when bringing the anchor up: When the buoy clears the surface, he snags it with the boathook. Then he can control the anchor's ascent to the roller, preventing it from dinging the topsides.

to eye-splice the rope around a thimble, then shackle the thimble to the chain. It's an easy method, but it has several drawbacks:

1. The thimble is inclined to chafe the rope, or to pop out, or both. You can seize the thimble to the rope, but seizings can chafe away, too.
2. The shackle can also chafe the rope, despite the presence of the thimble. In any event it's one more piece that can fail; a primary rule of rigging is to eliminate all possible links.
3. When hoisting the anchor, you run into trouble when the thimble hits the bow roller or hawse, and again when it reaches the winch—you have to wrestle it past both points. This is why some authorities recommend a maximum chain length of approximately the draft of the boat plus the freeboard at the bow. With that length of chain your anchor will have broken out by the time the thimble reaches the hangup points, and it'll be easier to wrestle the thimble past them.

But for most boats, this means only 10 feet or so of chain, too little for adequate catenary or chafe protection. I recommend a chain section of about half your average scope. So if the anchorages in your area are typically about 30 feet, you might veer a rode of 150 feet (5:1 ratio), 75 feet of which would be chain. If you're in deeper water, or if the weather picks up when you're in shallower water, the extra rode you'll deploy will be shock-resistant, low-weight nylon. And now back to the Chain Splice. This is, after all, a section about the Chain Splice.

The trick is to join the rope directly to the chain, so that there are no hangups when you weigh anchor. Figure 3-15 shows one method: separating the rope into four equal bundles and weaving the bundles into the chain. Two bundles go back and forth through the links in one plane, and two go

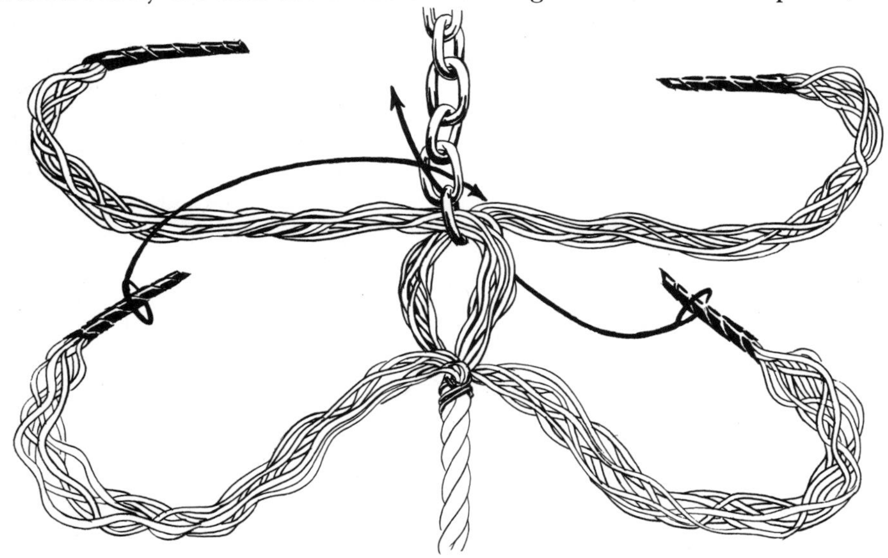

Figure 3-15. *The "Shovel Splice"—Part One. It'll work when splicing three-strand rope to chain, and is the only alternative when splicing double-braid to chain. Apply a Double Constrictor or other seizing about 2½ feet from the end of the rope, then separate the three strands (or double-braid core and cover) into four equal bundles of yarns. Weave two bundles back and forth through every other link, and the other two bundles through the intervening links, endeavoring to pull all yarns evenly tight as you go. Don't pull so tight that you put slack in the chain.*

Figure 3-16. *The Shovel Splice—Part Two. Tuck each bundle of yarns six or seven times, double the ends back on themselves, and seize thoroughly. It ain't elegant.*

up and down through the links in the other plane. For maximum strength, adjust all the rope yarns as you go, so they all bear an even strain. Each pair is tucked six or seven times, then the ends of all four bundles are very securely seized to the chain (Figure 3-16). Although this splice is easy to do, you need to be careful with it, since everything hangs on that seizing. Also, if the rope ever touches the bottom, this splice brings up such prodigious amounts of mud that a friend of mine calls it "the Shovel Splice." It's the only way to splice double-braid to chain—you put a seizing on the cover, separate core and cover into four bundles, and splice away—but since three-strand is stretchier and cheaper, it's better for rodes.

The traditional alternative to the Shovel Splice is the very tidy, very secure Chain Splice for three-strand shown in Figure 3-23. It is trickier to make, but well worth the effort.

Prepare the rope by soaking six feet of one end in a solution of 16 parts water and 1 part laundry starch (or use bottled liquid starch full strength). When the rope dries, the individual strands of nylon will hold their spiral shape, or "lay better," while you splice.

To start the splice, unlay the strands about 2½ feet. Tuck two of the strands through the last link of the chain (Figure 3-17). You only put two through because chain that is matched to rope strength is too small for all three strands to fit through.

Pull the two strands through until the link reaches the odd strand, then hold the link at this point. Lay the odd strand out a little farther, leaving a groove. Bend one of the two other strands down, give it a twist, and lay it firmly into this groove. With the chain on your left, this will probably be the nearer of the two strands, as it usually leads more directly and fairly to the groove.

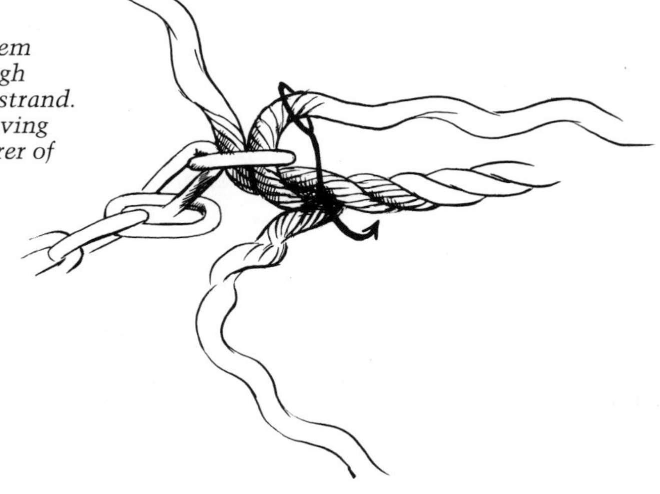

Figure 3-17. *Beginning the more comely Chain Splice for three-strand. Unlay the strands about 2½ feet and pass two of them through the end link, pulling them through until you reach the junction of the third strand. Now begin unlaying the third strand, leaving a vacant groove to be filled with the nearer of the two other strands (arrow).*

Figure 3-18. *Continue "laying out and in" until 6 to 8 inches of the "in" strand remain. As you go, give the "in" strand a firm twist and pull at each turn to make it lie fair. It should be indistinguishable from the other two strands.*

Continue laying out the odd strand and laying in the other strand until 6 to 8 inches of the latter remain (Figure 3-18).

Before proceeding farther, look back the way you came. Can you tell by appearance which strand you just laid in? If you can, that strand is imperfectly tensioned, which means it will bear more or less strain than the other two strands. Either way, the splice is weakened. Put everything in reverse and head back to the link. Go back and forth a few times until you can do it right. Remember, the secret is first to twist the strand clockwise while holding a little tension on it, and then to pull it firmly into the groove. This same technique is also used for making rope grommets, long splices, and single-strand repairs, so it's a skill well worth having.

When you're a competent layer-in, return to the position in Figure 3-18 and cut off all but 6 to 8 inches of the laid-out strand.

Here you have options. For the smoothest splice, divide both the laid-in and laid-out strands into two equal bundles of yarns, right down to the rope, then tie an overhand knot with two opposite bundles, left over right. There should be just enough space between the strands for the Overhand Knot to fill (Figure 3-19). A simpler, slightly bulkier option is to knot the whole strands together (not shown).

With either method, the next step is to tuck the knotted ends against the lay, over one and under one, four or five times. If you split the strands, just leave the unknotted ends hanging out. Figure 3-20 shows the left-hand end already tucked and the right-hand end being tucked. (The splicing tool shown is the remarkable Fid-O.)

To finish this part of the splice, roll it underfoot to fair it, then cut the ends off, leaving at least a rope-diameter of end showing. This will pull back and wear off in use.

Figure 3-19. *Cut off all but 6 to 8 inches of the laid-out strand, then divide it into two equal bundles. Divide the final 6 to 8 inches of the "in" strand similarly. Overhand Knot two opposing bundles, left over right. The knot should just fill the space between the strands.*

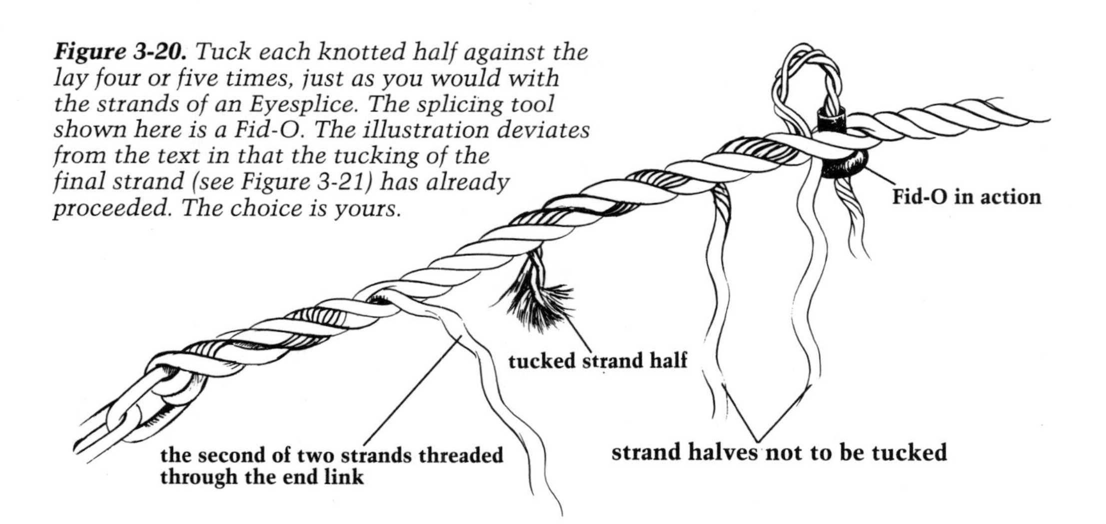

Figure 3-20. *Tuck each knotted half against the lay four or five times, just as you would with the strands of an Eyesplice. The splicing tool shown here is a Fid-O. The illustration deviates from the text in that the tucking of the final strand (see Figure 3-21) has already proceeded. The choice is yours.*

Almost done now. Go back to the link and that lone, uncommitted strand. Tuck it under itself, against the lay (Figure 3-21). Pull this first tuck snug, but not so much as to distort (i.e., weaken) the rope. Tuck another four or five times, roll to fair, and cut. (Figure 3-23).

It isn't crucial to splice strength, but for insurance you can add seizings, whippings, or Double Constrictor Knots just behind the end link and where the strands exit the rope.

Many people wonder how this splice can be strong enough when only two strands pass through the chain. But it's really two going in and two coming out, so the load is split four ways, like a line going through a two-sheave block. The link radius is small, but both strands bear fully on it.

You will have noticed that only about half

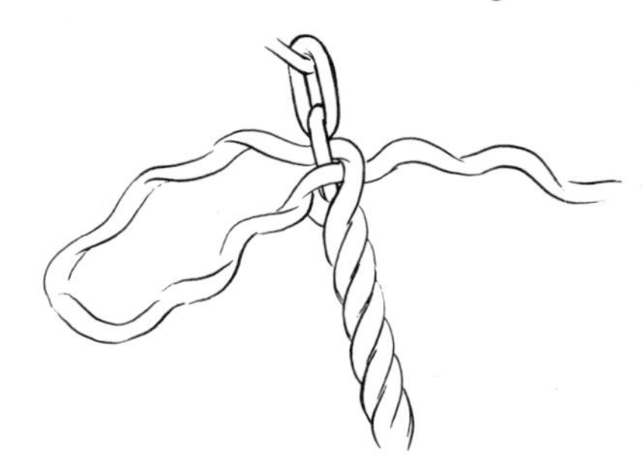

Figure 3-21. *Return now to the lone untucked strand, which with any luck still hangs forlornly from the end link of the chain. Pass it under its own part to form a Half Hitch . . .*

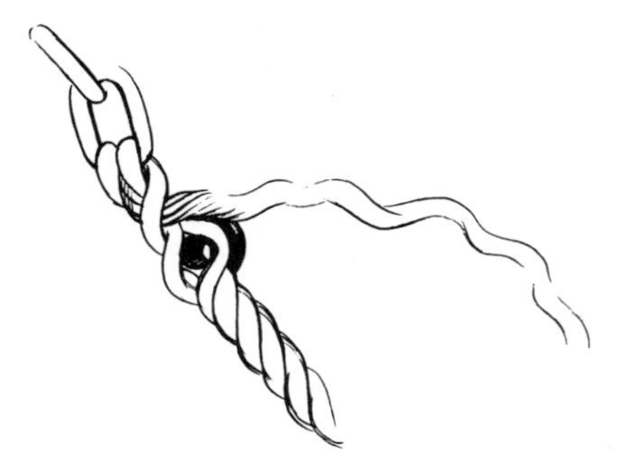

Figure 3-22. *. . . then tuck it over and under, against the lay, five or six times.*

Figure 3-23. *Roll the splice underfoot to fair, then trim the ends so they stick out at least one rope diameter. Presto! A finished Chain Splice.*

of this section was devoted to splicing per se. This is because any technique is only a reflection of the real business of rigging: understanding the relationships among boat type, strength requirements, sailing efficiency, cost, and convenience. Whether or not your boat can use a Chain Splice, you can use it as a way to understand the rest of your rig.

THE MENDING SPLICE

Every so often a boat will slash away at a piece of running rigging with a burr on a masthead sheave mortise, for example, or an unsuspected sharp porthole corner. You can smooth out the edges that did the damage, but what about the line? You can darn a lightly chafed yacht braid cover with needle and thread, but a deep cut is impractical to mend. So about all you can do is replace it and hope there's a long enough piece of whole rope left to use where you need a shorter line.

But if you damage three-strand rope with a sudden slash, it's likely that most or all of the damage will be done to one strand. And if that's the case, the Mending Splice will repair it.

You'll need about 3 feet of the same diameter, construction, and material as the damaged piece. If the damaged piece is plenty long, cut 3 feet off its end. Soak this section in a mixture of 16 parts water and 1 part laundry starch, as for the Chain Splice. This is to keep the line from losing its spiral shape, or "lay," as you work. Let the line dry, then gently unlay one strand out of it. This is your mending strand.

Now go to the damaged piece and cut the wounded strand the rest of the way through. Undo, or "lay out," the two resulting ends a full turn each, and set the middle of the mending strand into the space they leave. Lay out one of the wounded ends another turn, and lay in behind it with one end of the mending strand. Twist the mending strand clockwise, with a little tension on it, and pull it firmly into the groove. It should be indistinguishable from the two strands next to it. If it is tighter or looser than they, or has more or less twist, it will take more or less strain than they do, and the rope will be weakened. So practice until you can lay in smoothly (Figure 3-24).

Lay out and in until there's 6 to 8 inches of the wounded strand you've been following, and split both strands in half down to the standing part, leaving a small space between the two pairs of strands (as in Figure 3-19 or 3-26). Tie an Overhand Knot, left over right, with half of each pair. The knot should just exactly fill the space. If it fits poorly, undo it and lay up or unlay the halves until you get two that fit well when knotted. Leave the other two halves hanging out, and tuck each of the knotted ends against the lay four or five times. As a sim-

Figure 3-24. *Mending Splice. Cross and finish the ends as for a Chain Splice or Long Splice.*

pler but lumpier option, you may tie the whole strands together and tuck them.

Repeat the procedure with the other two ends, wash the starch out, and you're done.

With a little practice at delicate handling, you will find you can skip starching all but the softest-laid line. This will qualify the splice as not only clever and economical, but also valuable as an emergency procedure.

It's worth noting here that boats are always doing non-emergency damage to rope. They chew on it with fairleads, sheaves, winch drums, and especially with chocks, stoppers, and self-tailing gear. They chew on it gently, but steadily; about the only thing you can do is to try to blunt their teeth. Use bigger fairleads and sheaves. Carefully angle leads to winches so you get no wraps. File gentler curves into the edges of chocks and bow roller side-keepers. Minimize the use of stoppers and self-tailers. And end-for-end lines when wear becomes noticeable, to get the most life out of them.

All of these little details are a lot less dramatic, but in the long run a lot more valuable than the wonderful Mending Splice.

THE LONG SPLICE

In all of rigging, there is no knot more often asked about nor less often useful than the Long Splice. There's something undeniably fascinating about it, about the way it leaves a line's appearance and diameter almost unchanged. Perhaps it reminds people of those ever-popular Cut-and-Restored tricks—"And now I will take this severed line and make it whole again!"

Handy as that ability might seem, most of the times I've been called upon to make a Long Splice have been because of mistakes (for instance, a halyard that was made too short), or in emergencies (for instance, a halyard that used to be long enough that suddenly became too short). In either instance, a Short Splice might serve as well, so long as the splice doesn't have to go through a sheave or stopper. And in that case, just having some spare rope as a replacement is the best solution.

Making a Long Splice requires quite a bit of rope and skill. And unlike the more practical Mending Splice, which only disturbes one-third of the rope, the Long Splice requires that you deal with all three strands of two pieces and that you get them all evenly tensioned to preserve rope strength.

Still want to learn it? Well, okay. It is worth knowing, and is essentially just an elaboration of the Mending Splice. Here goes:

First, the ropes must be of the same diameter, material, and construction.

Next it's a good idea to starch the ends, to help the strands to hold their spiral shape, or "lay." This is particularly important with nylon rope. Starch at least 4 feet (1.2 m) of line for every 1 inch (25 mm) of diameter. So a ½-inch (13 mm) diameter piece would be starched at least 2 feet (0.6 m) from the end.

When the starch is dry, gently unlay the two lines almost to the untreated section. "Marry" the ends just the way you'd lace your fingers together (Figure 3-25). Take any convenient strand from the line on the left and "lay it out" of the marriage without disturbing any of the other strands. Hold it off to the left, leaving a space where it had been. Now adjust the position of the two lines, pushing them slightly closer together or pulling them slightly farther apart so that the corresponding strand from the right side falls neatly into that just-vacated space. This is the most difficult part of the splice; it's tricky getting a good close marriage and keeping everything together while you make these initial setting-in moves. But just as with a real marriage, you only need to be careful, attentive, and to take your time.

With that right-hand strand neatly in the

REEFING PENDANTS

Wire-and-rope halyards are desirable because they are cheaper, lighter, longer-lived, and present less windage than all-rope alternatives. They are also extremely inelastic, which means they maintain consistent luff tension for optimal sail shape. Ordinarily, the splice is located just above the winch. But when the sail is reefed more rope is exposed, and rope is relatively elastic. With a deep reef, halyard tension can be compromised, resulting in a light-air sail shape (baggy, center of effort moved aft) when you least need it. So for deep reefs and storm trysails, consider making up a wire rope pendant that will compensate for the loss of luff length when reefed. Always shackle the pendant to sail and halyard *before* releasing the regular halyard shackle, so you don't risk losing the halyard.

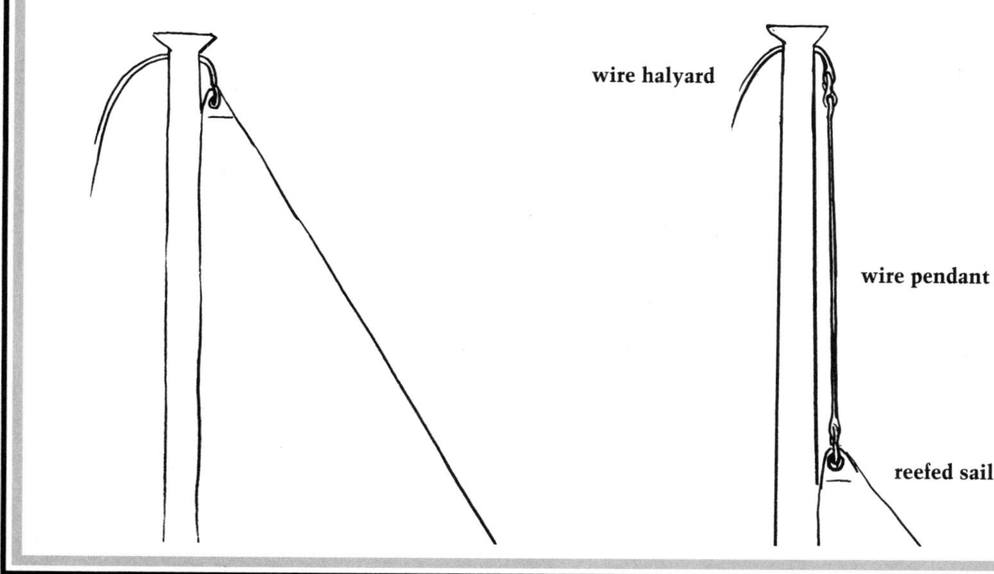

groove, seize or Constrictor together the four strands that are still married. This part is easier with an assistant.

Leaving the married strands for the moment, go back to the two working strands and lay them out and in, just as with the Mending Splice. Either knot and tuck whole strands, or, for a more compact splice, split them first (Figures 3-26 and 3-27).

Come back to the married pairs and cast off their seizing. Overhand Knot two whole opposing strands together—with just enough tension to hold them in position for the moment—and lay the other two strands out and in, off to the right (Figure 3-28). Splice 'em.

Return once more to the original marriage site and undo the Overhand Knot. Make any adjustments necessary to a fair lead (twisting, untwisting, tightening, slacking). Reknot and splice, splitting the strands first if you prefer. Wash the starch out and you're done (Figure 3-29). In an emergency, I'd be inclined to do an eye-to-eye splice if that part of the line didn't have to get through a sheave or stopper.

Figure 3-25. The Long Splice. Marry the two ends and Constrictor four of the six strands together. The other two, which lie in the same groove, are laid "in and out," with the one on the left being removed from the groove and the one on the right taking its place.

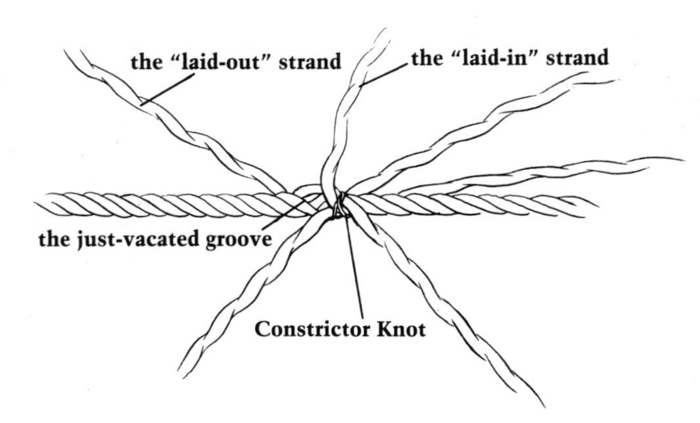

Figure 3-26. Lay out and in as for the Mending Splice and Chain Splice until the "laid-in" strand is just long enough to make four splice tucks, then halve the two strands (optional), Overhand Knot, and . . .

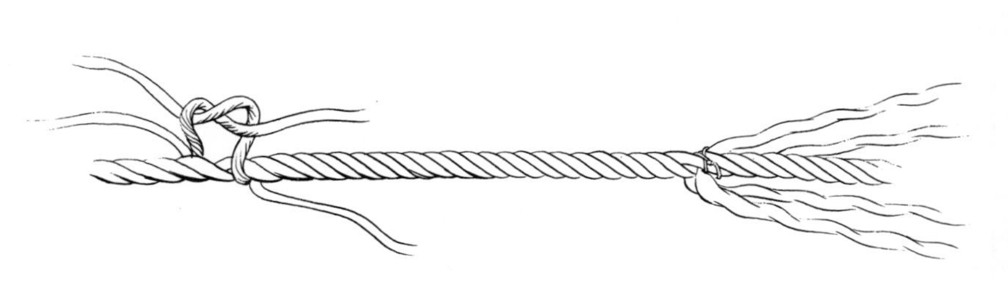

Figure 3-27. . . . tuck the ends to finish, again just as for the Chain Splice.

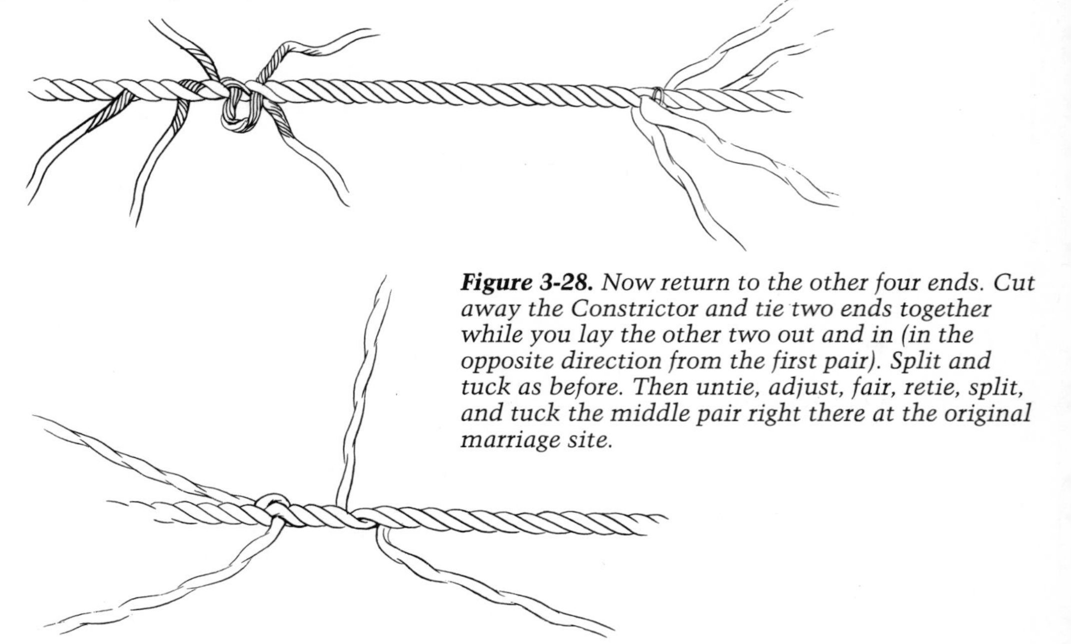

Figure 3-28. Now return to the other four ends. Cut away the Constrictor and tie two ends together while you lay the other two out and in (in the opposite direction from the first pair). Split and tuck as before. Then untie, adjust, fair, retie, split, and tuck the middle pair right there at the original marriage site.

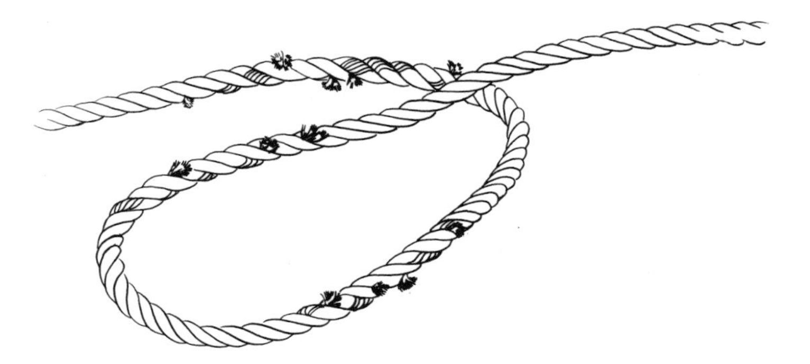

Figure 3-29. *A finished Long Splice. Leave all ends a half-inch or so "proud," or whip over all junctures and cut the ends flush.*

THE STITCH SPLICE

For emergency repair of double-braided rope, where eye-to-eye bulk is objectionable, a compact, quick, and easy solution would be to cut about a foot off the end of each line's heart, lay the cover ends alongside one another, and stitch through-and-through like crazy with stout sail twine. If you cut the hearts on a long angle to taper them you can maximize this Stitch Splice's strength without making it too bulky (Figure 3-30).

THE 1 X 19 WIRE SPLICE MADE POSSIBLE

In the days of hemp rope, if riggers wanted an eye in the end of a line, they spliced or seized it there. As iron wire rope came to prominence in the latter part of the 19th century, riggers treated it pretty much like hemp, splicing and seizing as they were used to doing. It was more difficult to hold and work the stiffer, tougher material, but they managed—once they had devised some specialized tools such as rigging vises.

Figure 3-30. *To begin the Stitch Splice, pull out a foot of core from each rope end and make a rough taper by cutting the marked strands.*

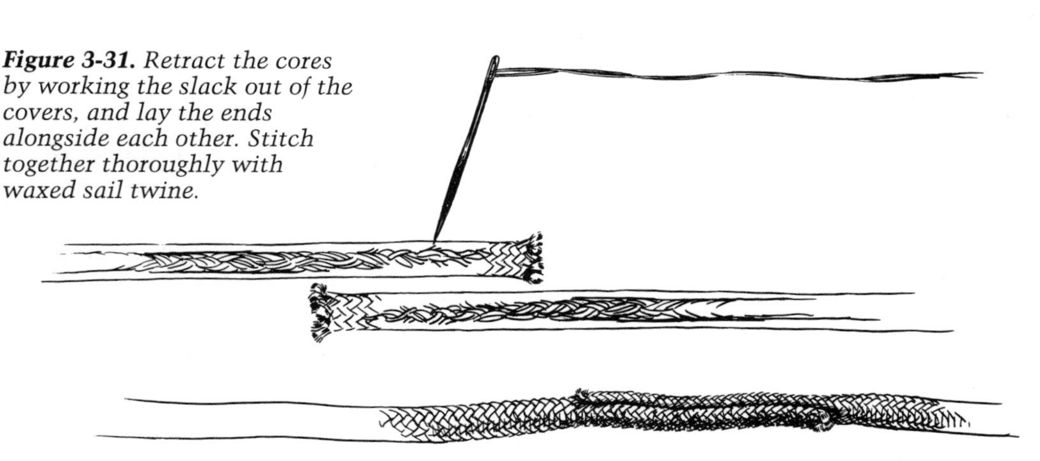

Figure 3-31. *Retract the cores by working the slack out of the covers, and lay the ends alongside each other. Stitch together thoroughly with waxed sail twine.*

Figure 3-32. *The finished Stitch Splice.*

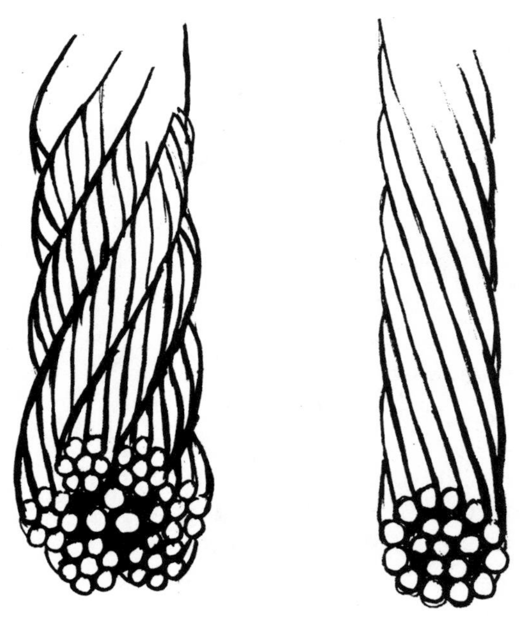

Figure 3-33. *The old 6 x 7 wire rope (left) has given way to 1 x 19 (right).*

With progress, iron gave way to plow steel, and plow steel (at least for yachts) to stainless steel. Wire construction was also changing, with the old 6 x 7 being replaced by 1 x 19 wire (Figure 3-33). These changes were calculated to reduce (even if only slightly) weight, windage, maintenance, and stretch, to aid the performance of 20th-century sailing craft. Unfortunately for riggers, wire evolution also increased stiffness, intricacy, and the need for precision in measurement and fabrication. With the development of mechanical terminals, particularly swages, a lot of riggers breathed sighs of relief. Today, splices in 1 x 19 wire are rarely seen, and so most sailors, and even many riggers, assume that the stuff is impossible to splice.

Why Splice?

Nevertheless, if you want to use 1 x 19 wire because of its structural advantages of low windage and elasticity, there are some strong, practical arguments for splicing it. First of all, there's cost: Even if you pay a professional to do it, the price of a 1 x 19 splice is competitive with that of a good mechanical terminal. If you do the work yourself, you're only out time and the cost of a few simple tools.

Another advantage is fatigue resistance: Because a splice is flexible along its entire length, there are no fatigue-inducing hard spots, like the point where a flexible wire enters a rigid fitting. North of the tropics, fatigue resulting from cyclic loading is even more of a problem for stainless steel than atmospheric corrosion. On modern vessels, particularly fin-keelers and multihulls, high initial stability and inelastic sails translate into hard "shock-load" conditions every time a puff of wind hits the sails, so, fatigue-resistance is an even more valuable virtue.

Finally—and this is most important to cruisers—there are the linked virtues of ease of inspection and replacement. By periodically lifting the service from a sample or two, you can see right into the splice to check for evidence of corrosion or fatigue. If something has gone wrong, you don't need a multi-ton press or an expensive screw-on fitting to fix things—just a marlingspike, a rigging vise, and the skill to use them.

It is only fair to warn you that the level of skill required to produce a proper 1 x 19 splice is quite high. Structurally, it is actually simpler than a 6 x 7 wire splice, so figuring out where to tuck the strands will be easier. But 1 x 19 wire is much stiffer, and easier to distort or kink while splicing. So it's up to you to get to know the wire as much with your hands as with your head. Practice until you can turn out consistently smooth, strong splices.

Materials and Tools

Thirty feet (9 m) of ¼-inch (6.5 mm) diameter 1 x 19 stainless-steel wire rope will be enough for a few practice splices. You'll also need about 15 feet (4.5 m) of 1/32-inch (0.8 mm) annealed stainless seizing wire, either single- or multi-strand construction, and a 1/4-inch (6.5 mm) solid-bronze thimble. This last item (Figure 3-34) is designed to accommodate 1 x 19 wire. The thimble's wide radius suits 1 x 19's bend-resistant nature, and its solid mass will stand up to extremes of loading, yet it is still thin enough to fit into the jaws of the appropriately sized turnbuckle.

A rigging vise clamps the wire around the thimble while the splice is made. In a pinch, [the author assures that this gripping phrase is an intentional pun—ed.] you can cobble one out of Vise-Grips and blocks of wood, but a model like the one shown in Figure 3-37 will be faster, surer, and easier, and will enable you to handle a variety of wire sizes.

Scratch awls (Figure 3-35) make perfect marlingspikes for wire up to 5/16-inch (8 mm) in diameter. Snap-On's scratch awl has a notably superior taper. File the tip flat for easier entry into the lay of the wire.

A rope tail and an unlaying stick (shown in Figure 3-38), optional for yacht-sized 6 x 7 wire, are essential for working with ornery 1 x 19 wire.

You'll also need nippers, pliers, a hardwood or soft-metal mallet, and—please—a pair of safety glasses.

The Splice

The first step is to put on a wire seizing a short distance from one end to serve as a stop for unlaying strands. The distances for yacht-sized wires are as follows:

Wire	Distance
1/8" and 3/16" (3.5 mm and 5 mm)	1' (0.30 m)
1/4" (6.5 mm)	1'3" (0.38 m)
5/16" (8 mm)	1'4" (0.41 m)
3/8" (9.53 mm)	1'6" (0.46 m)
7/16" (11 mm)	1'8" (0.51 m)
1/2" (12 mm)	1'11" (0.58 m)

These distances will give you a comfortable length of wire to work with while splicing, although you might find, as you gain proficiency, that you prefer to work with more or less wire than suggested here.

To apply the seizing, wrap a 1-foot (0.30 m) length of seizing wire onto the 1 x 19 wire, in our case 1 foot 3 inches from the end. Wrap tightly, against the lay, covering the short end with the working part, until

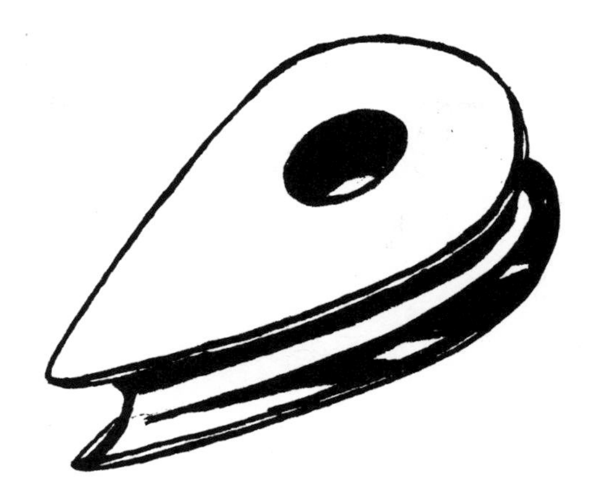

Figure 3-34. *A 1/4-inch solid bronze thimble works well for the eye.*

Figure 3-35. *A scratch awl makes a good marlingspike for 1 x 19 wire.*

Figure 3-36. Applying a seizing.

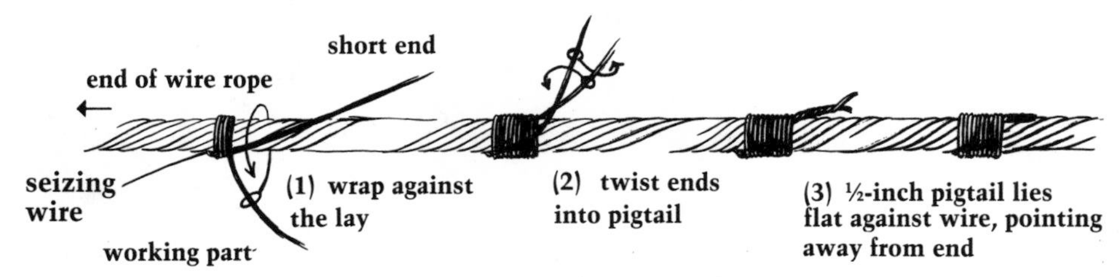

the seizing is approximately square. Then twist the two ends around one another to make a "pigtail." Tighten this pigtail with a pair of pliers, nip off all but ½-inch (12 mm) or so, and then press the tail down against the wire as shown in Figure 3-36. Note that it points away from the end of the wire rope.

Secure the vise at elbow height, and open its jaws as wide as they will go. Measure the circumference of your thimble—not in the score, but around its outer edge—and make a mark on the wire half this distance above the seizing. Bend the wire by hand at this point, so that the tail of the seizing is on the inside of the curve. Bend it just enough so that you can horse it into the vise (end on the left as you're facing the vise). Put the thimble in place, and crank the jaws in until there's no slack around the thimble, with the seizing just outboard of the lefthand jaw, and the hollow at the pointed end of the thimble just past the seizing. (Figure 3-37 shows this setup in detail.)

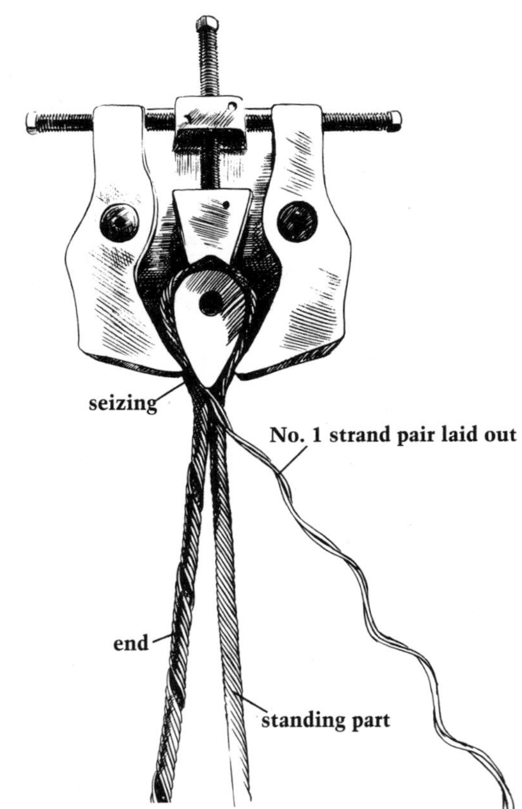

Figure 3-37. Wire positioned in rigging vise; No. 1 strand laid out.

Tie the standing part of the wire out horizontally under light tension, then wrap on the unlaying stick against the lay (as shown in Figure 3-38), about 3-feet (0.9 m) from the vise. Take out a single turn. If the stick is not heavy enough to hold the turn with its own weight, brace it securely.

When the stick is set, the lay of the wire should be opened up just a little. If it looks like a loosely woven basket, you've taken out too much lay. The idea is to have it just loose enough for a reasonably easy spike entry.

Now for a little spike practice. Stand on the right side of the wire, brace your spike arm against your body, lay the tip of the spike on the wire parallel with the lay, and with your free hand grip the wire and the spike as shown in Figure 3-39. Lean on the

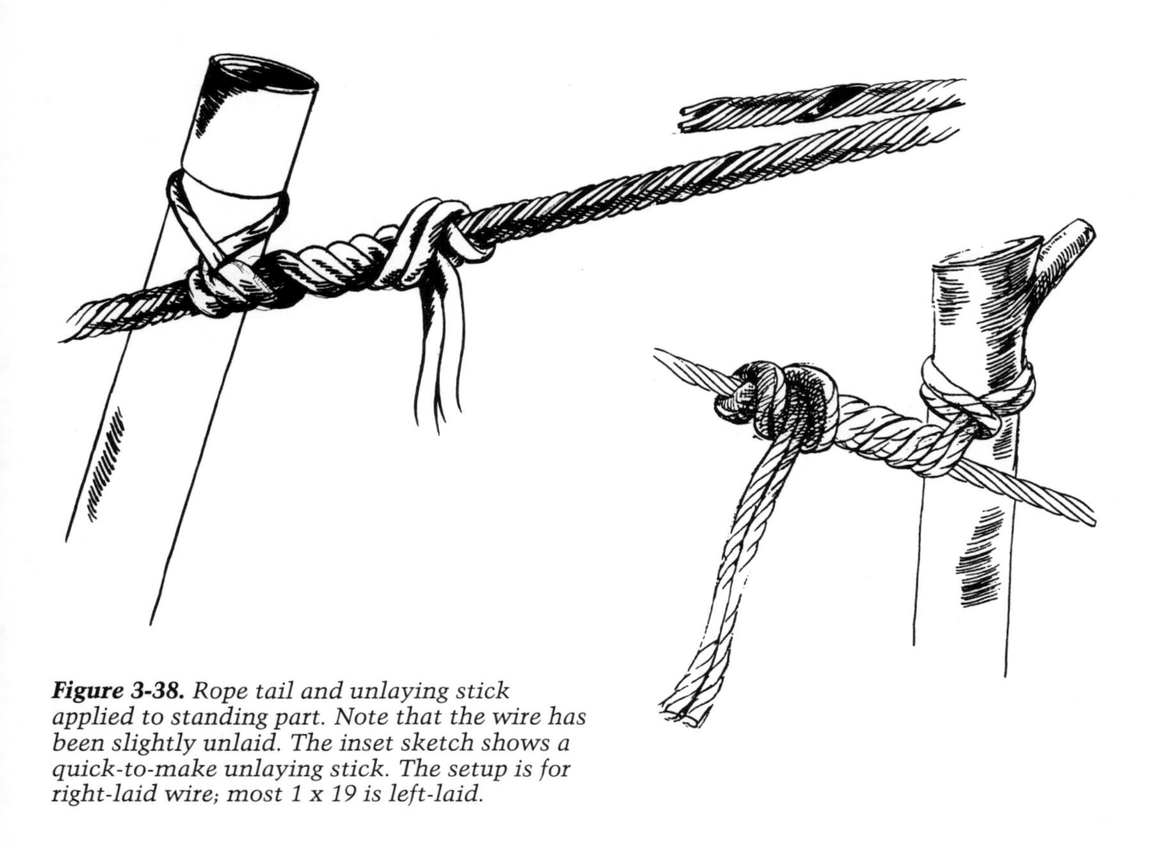

Figure 3-38. *Rope tail and unlaying stick applied to standing part. Note that the wire has been slightly unlaid. The inset sketch shows a quick-to-make unlaying stick. The setup is for right-laid wire; most 1 x 19 is left-laid.*

Figure 3-39. *Entering the spike into the wire.*

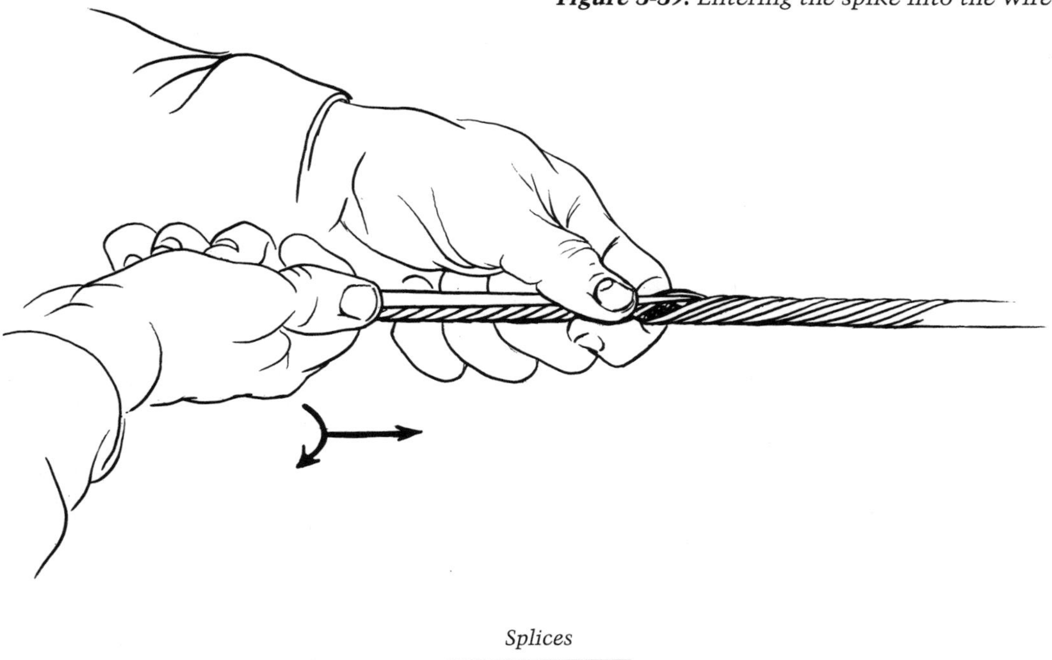

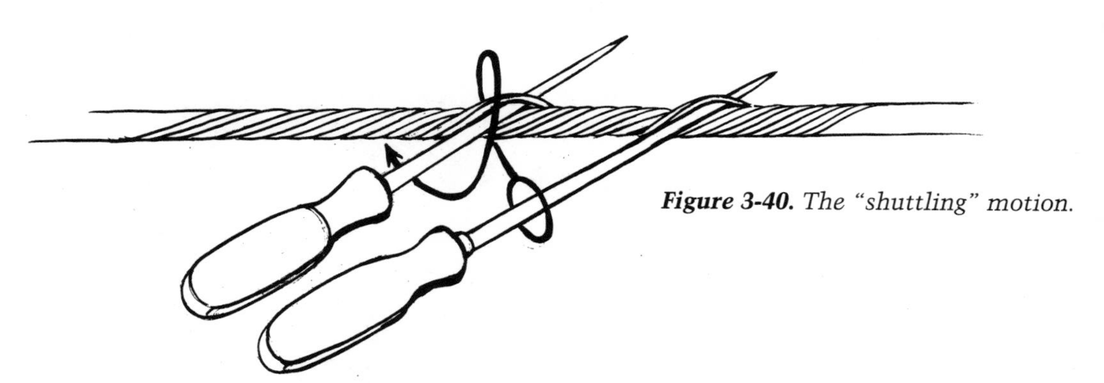

Figure 3-40. *The "shuttling" motion.*

wire, using your free hand's thumb as a fulcrum to ease the tip down between two surface yarns.

Keeping your spike parallel with the lay, roll it up and down the wire a few times. This is the shuttling motion that will send the strands home as you tuck them. Practice keeping the same length of spike in as you travel—about 1½ inches (38 mm) of the spike's tip should be showing (see Figure 3-40). If, as you travel, strands jump over one another or kink, then the spike is at the wrong angle, in too far, or both.

Once you're at home with entry and shuttling, it's time to prepare the strands. Notice that 1 x 19 wire is composed of a left-laid outer layer of 12 yarns, a right-laid inner layer of 6 yarns, and a single, straight heart yarn (Figure 3-33). Nearly all 1 x 19 wire is made this way, though you might stumble across some with opposite lay; if that's what you have, mirror-image these directions.

The trick is to convert this unspliceable mass of yarns into six spliceable strands. This is done by dividing the 12 surface yarns into pairs, and matching each pair with an interior yarn. In the interest of a fair entry, we want particular yarns joined with one another, and tucked into particular points of the standing part.

To find the No. 1 pair, look at the base of the thimble and find the two end yarns that are just to the right of center where they exit the seizing. That is your No. 1 pair for the end. Mark or make note of where your No. 1 end pair should be, then lay out a randomly chosen pair from the end. You just might get the right pair (as shown in Figure 3-37), but if you don't, count how many yarns it is away from the No. 1 pair. If it's an even number, proceed to lay out all the other surface yarns in pairs. If it's an odd number, lay your random pair back up, shift over a half-step, then lay out all the pairs. Bend each pair slightly outward from the seizing, just enough so they won't tangle with one another (see Figure 3-41). Watch out for sharp ends.

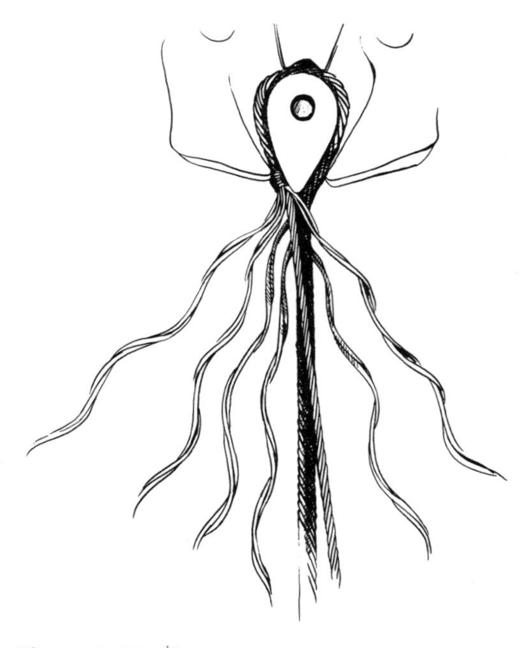

Figure 3-41. *The surface yarns unlaid in pairs.*

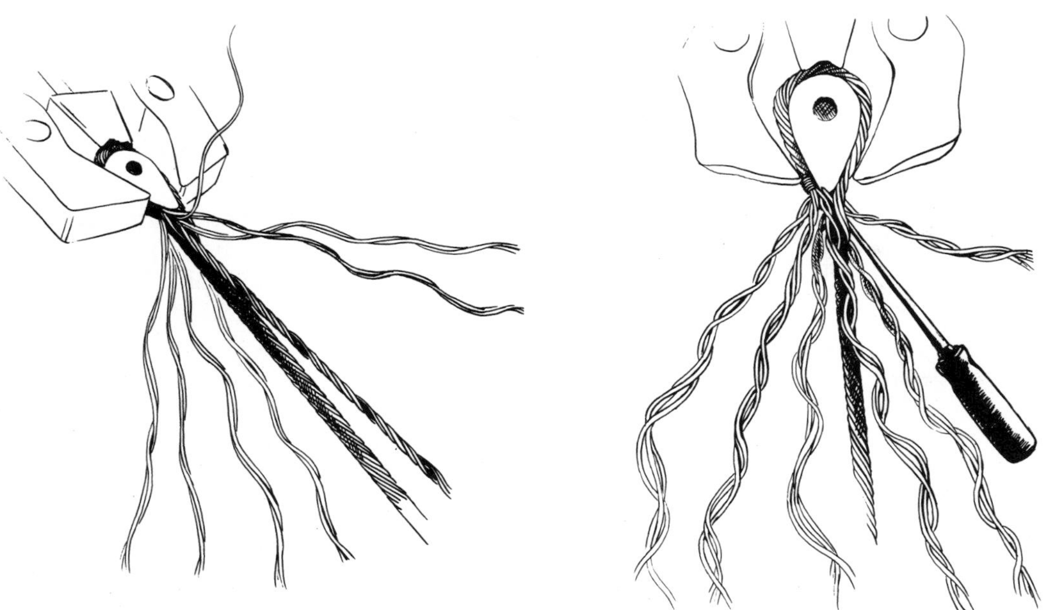

Figure 3-42. *Laying in the first core yarn with its appropriate surface pair.*

Figure 3-43. *Entering for first tucks of the No. 1 strand. Once the spike enters as shown, roll it back a full turn to the position shown in Figure 3-44.*

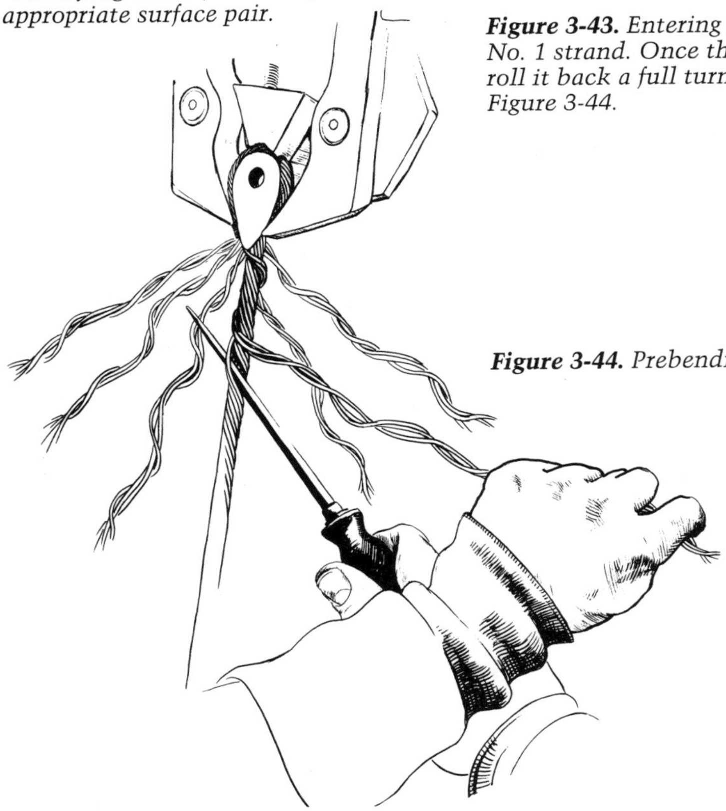

Figure 3-44. *Prebending the No. 1 strand.*

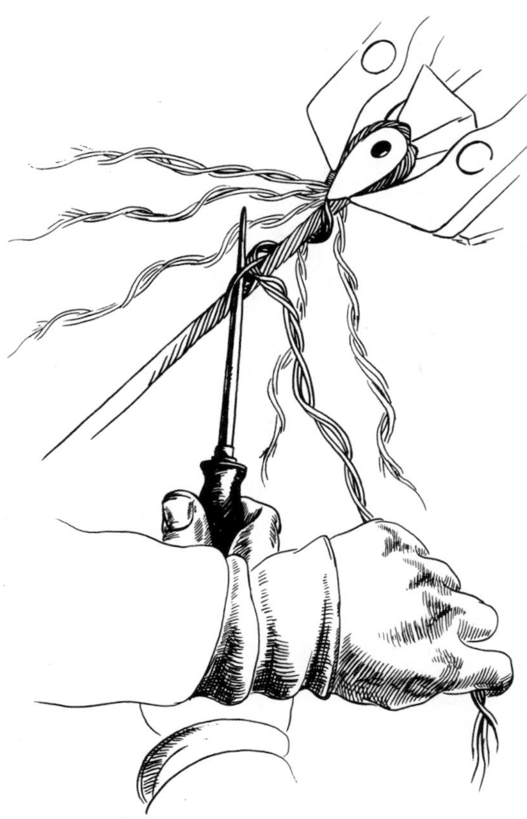

Figure 3-45. Tuck the No. 1 strand through the opening made by the spike as shown, then roll the spike toward the vise one turn to push the tucked strand home.

Now lay out a yarn from the interior group. You'll want to match it with the pair of surface yarns to which it will lead most cleanly (see Figure 3-42). Ideally, the pair of surface yarns will be offset slightly clockwise from the single yarn. Lay the single yarn in with this pair, twisting clockwise. Lay out successive yarns from the bundle and lay them in with successive surface pairs. If you choose the first match well, and avoid crossing leads, then all will lead fair.

This leaves the heart yarn to deal with: Break it off at the seizing by bending it sharply and rapidly back and forth several times.

Now it's time for the first tuck. As shown in Figure 3-43, enter the spike under the standing part's No. 1 pair (just to the left of center, corresponding to the end's No. 1 strand) and roll the spike back a full turn. Next, take your No. 1 strand—remember, it's just to the right of center at the seizing—and bend it to the right and completely around the standing part, following the path of the spike. Pull on the strand as you pass it (Figure 3-44) so that it lies flat against the standing part, prebending it so it will be fair when rolled home.

To tuck, pass the wire through the space the spike makes, passing from tip to handle (Figure 3-45).

Roll the spike toward the vise, keeping the strand alongside it and under tension. Let the spike push it home, and it will settle into place at the thimble. Be gentle. Holding the strand up toward the thimble, roll the spike back to fair out any distortions you might have caused.

As shown in Figure 3-46, enter under the next pair of standing-part yarns to the right,

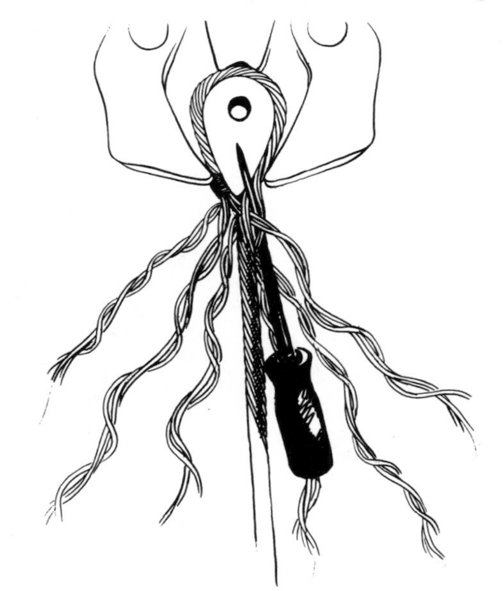

Figure 3-46. Enter the spike under the next two standing-part yarns to the right, then roll back one turn as before.

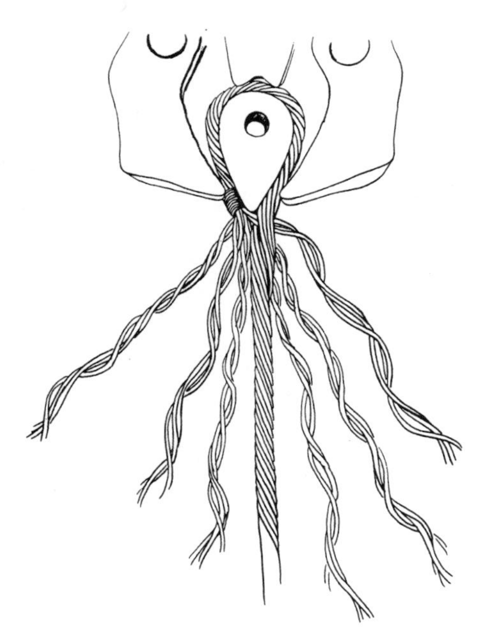

Figure 3-47. *The No. 1 strand is now tucked under four standing-part yarns.*

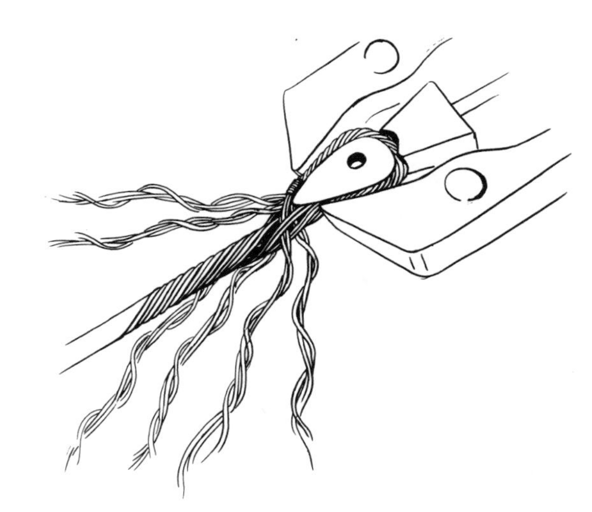

Figure 3-48. *The No. 2 strand tucked once.*

and roll back. Pass the No. 1 strand under four adjacent standing-part yarns (Figure 3-47). You could have tucked under all four at once, but probably not without distorting the lay.

Roll back to fair, then return the spike to the original pair of standing-part yarns. Enter there again, and roll back. Select the No. 2 strand—it's immediately to the left of No. 1. Give it a prebending spiral, tuck it, and roll it home. You'll now have the No. 1 strand going under four yarns, and No. 2 entering at the same point but emerging after going under only two yarns (see Figure 3-48).

Roll back to fair, then enter the spike under the next pair of standing-part yarns to the left (Figure 3-49). Roll back, prebend, and tuck the No. 3 strand—it's just to the left of No. 2—and roll home.

Tuck strands 4, 5, and 6 under successive pairs of standing-part yarns. Concentrate on minimizing distortion and prebending the strands so that they lie fair. To aid fairing, give each strand a counterclockwise twist before prebending, to open the lay a little more (see Figure 3-50).

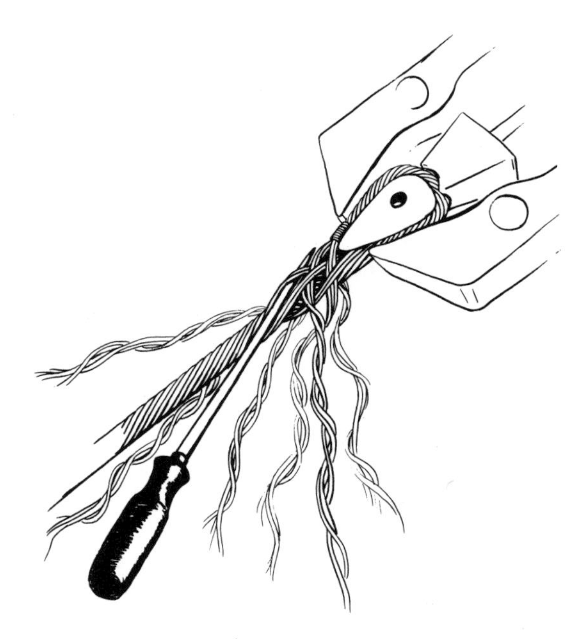

Figure 3-49. *Here the No. 3 strand has been tucked, and the spike is entered for No. 4.*

A more time-consuming but more easily

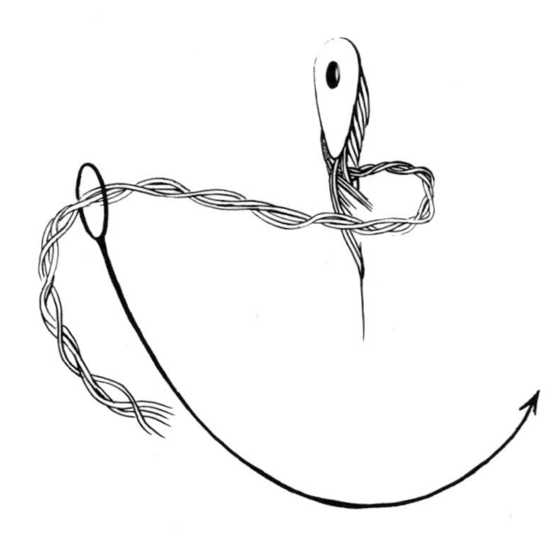
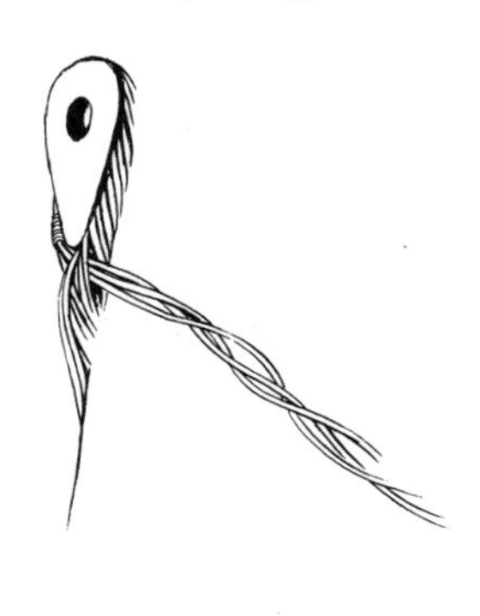

Figure 3-50. *Breaking the lay of a strand to aid fairing.*

faired technique involves tucking each yarn separately. Separate the three yarns in each bundle, then prebend and tuck the yarn in each bundle that is nearest the vise. Roll it home, then tuck the middle strand, and finally the strand farthest from the vise.

The standing-part yarns may tighten up as you work, so pause after each set of six tucks, if necessary, to take out an extra half-turn or so, using the unlaying stick.

As you splice, keep the strands clear of one another and the vise. It will be less confusing visually, and will prevent strands from springing or whipping at you.

To begin the second row of tucks, find the No. 1 strand. (It will be closest to the vise and coming out from under four yarns.) Enter the spike where the strand emerges, but only under two yarns. Roll back and tuck the No. 1 strand under these yarns. If you prebend well, the strand will lie down smoothly when you send it home, wrapped around those two yarns as if it wanted to be there. If it doesn't lie smoothly, work at untwisting and prebending subsequent strands; the knack will come to you.

Now the No. 2 strand is closest to the vise, emerging— as all strands now will be—from under a pair of yarns. Enter the spike under those yarns, roll back, and tuck.

Proceed with the rest of the strands, moving counterclockwise. It is very easy to mis-tuck, particularly if the wire is distorted by careless spike-work or excessive unlaying. If a strand doesn't look right, trace it back visually. Working from the end toward the vise, it should go under two yarns away from you, back over the same two yarns toward you, then back under them away from you, for as many tucks as you make.

If you have mis-tucked, put everything in reverse: roll the spike back toward you with the strand alongside it, but push rather than pull on the strand. After a half-turn or so, the strand should spring loose so that you can push and pull it out to untuck. Hunt for the proper location and then re- tuck.

After the second row of tucks is done, taper by laying the third yarn out of each strand, leaving the two surface yarns. Get the third yarn completely clear of the other two at their base, so that it will not be trapped between them and the standing part during the next tuck (Figure 3-51). If necessary, tug the third yarn toward the vise to get it out from under its pair, then bend it out of the way.

Tuck the remainder of each strand twice more, and you'll be ready to taper again. This time, separate each pair and lay one of

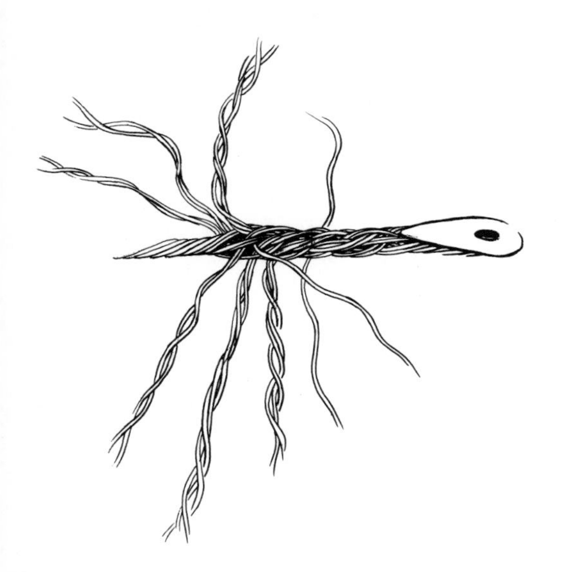

Figure 3-51. *Second row of tucks done, taper started.*

them out. If they twist around one another at the standing part, pull them gently apart until they spring clear of each other. Figure 3-52 shows a splice that is about as rough as it can be and still be acceptable once faired.

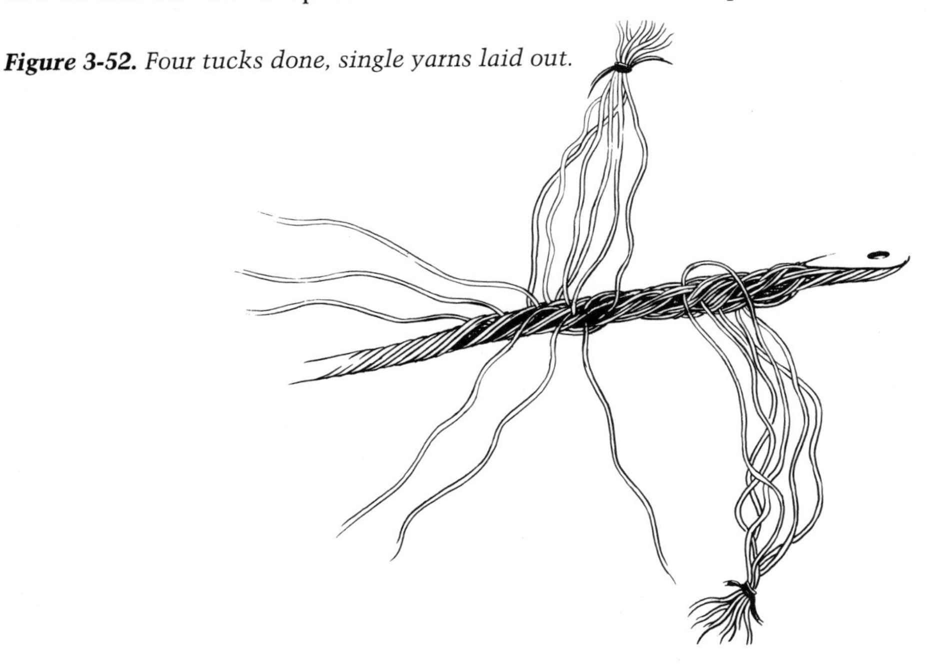

Figure 3-52. *Four tucks done, single yarns laid out.*

Your splices must look at least this good at this point before you can consider yourself a splicer. Practice.

Tuck the remaining single wires twice more, and you've finished tucking.

Before fairing, tie all the ends down to the standing part, then take the splice out of the vise and over to a stump. The stump will serve as a soft, organic anvil. To fair, beat the splice with your mallet, working away from the thimble, smoothing out any irregularities so that when a load comes on all the yarns will take an even strain. Strike with an L-shaped stroke (Figure 3-53), coming almost straight down, then ricocheting away from the thimble. Use enough force so that the mallet isn't just bouncing and skidding off the wire; on the other hand, if you're working up a sweat, you're striking way too hard. Be moderate, and go over the work twice. (If that doesn't yield something that resembles the finished splice shown here, there's more wrong with your splice than fairing can cure.)

Put the splice back in the vise, stretch the

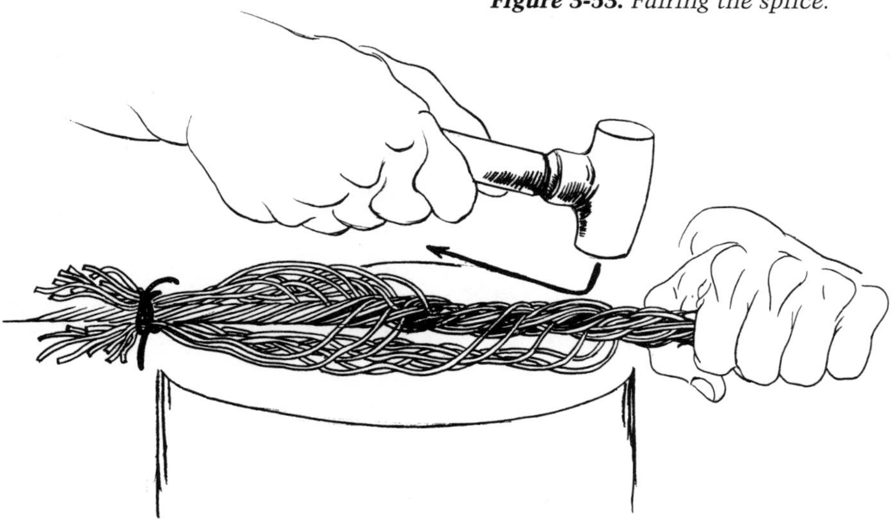

Figure 3-53. *Fairing the splice.*

standing part out again, and pull from the bundle the yarn closest to the vise. Bend it sharply toward the vise, parallel with the lay, then swing it in a low, counterclockwise arc (Figure 3- 54). It should break off after less than a full turn. The part that's left will have a little hook in its end, just at the surface. This keeps the yarn from coming untucked, but it won't protrude far enough to be a "meathook."

Work your way down the splice, breaking off all the yarns. Then, if necessary, you can fair again, very gently, to settle any recalcitrant spots. It's easy here to dislodge ends, so watch how you strike.

The splice (Figure 3-55) is now complete and ready for tarring, parceling, and service. "Tarring" in this case can be done with a coating of anhydrous lanolin. The lanolin, available at your local pharmacy, will pre-

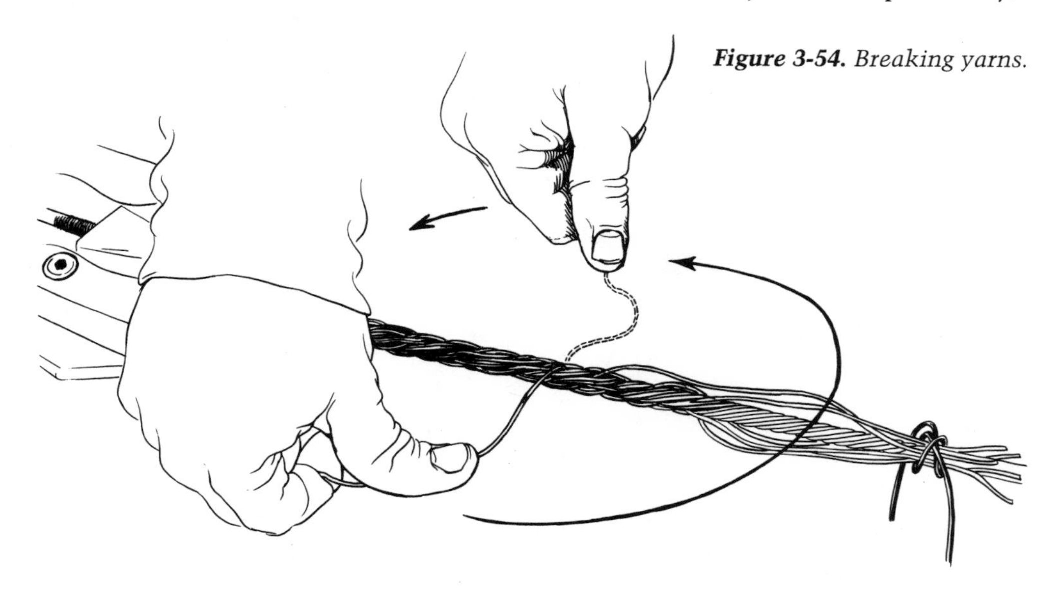

Figure 3-54. *Breaking yarns.*

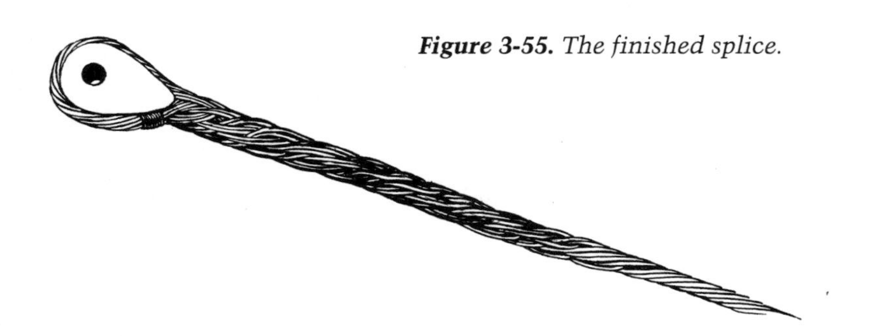

Figure 3-55. *The finished splice.*

1 X 19 OVERSIZED EYE ENTRY

1 x 19 wire is usually spliced up "hard" to a thimble, locking the thimble in place. But occasionally you'll see 1 x 19 splices where the eye is served with annealed stainless steel 1 x 7 seizing wire before being bent around the thimble, and in this case the eye is slightly oversized so that the splice service can meet it smoothly. This oversized eye results in a slightly different-looking splice start: Rather than jumping across from one side of the thimble to the other, the ends are right next to the standing part. To get a fair start choose the pair of standing part yarns and end yarns that are almost touching each other, as far inboard as you can easily reach.

Served, oversize eyes look real slick—they were often put into the rigs for Concordia yawls—and the wire service helps support the wire yarns as they lay around the thimble, perhaps preventing distortion under extreme loads.

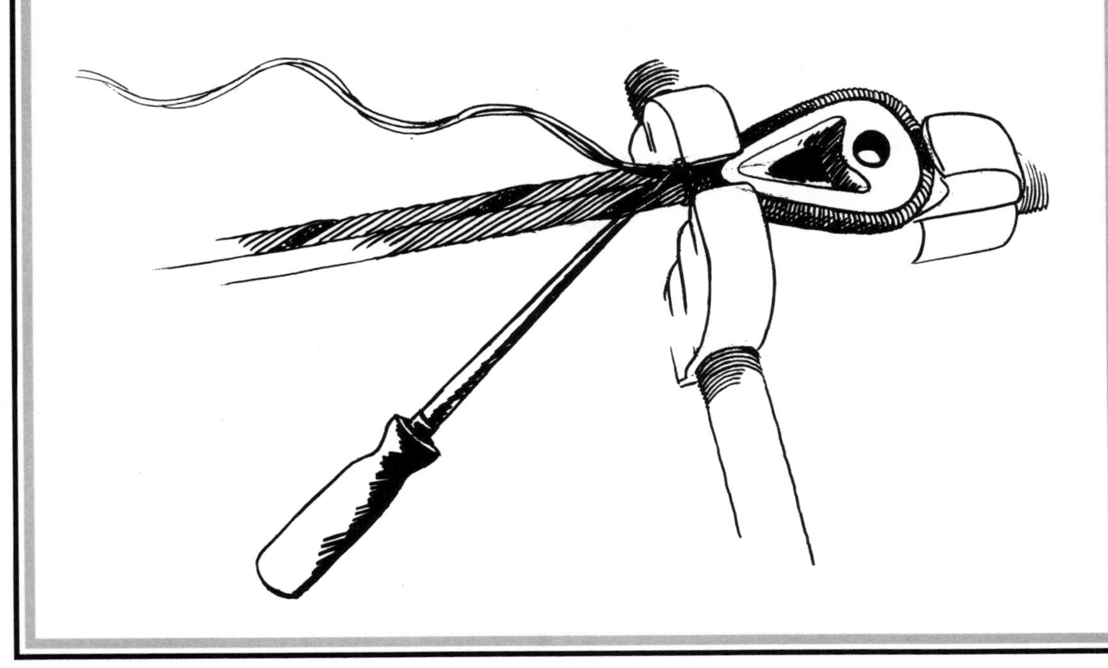

vent both rust and oxygen starvation under the parceling and service.

The parceling can be of friction or adhesive tape, with more lanolin applied to waterproof the splice. If you're going to serve with wire, use parceling made from polyester bed sheets or soft, light sailcloth instead of, or in addition to, the tape. The cloth parceling won't get chewed up by wire service.

Use wire service of 1⁄16-inch (1.5 mm) annealed stainless 1 x 7 seizing wire for stays with sail hanks on them, or any other splices that receive a lot of chafe, as from sheets. Chapter 4 shows how to go about it. Otherwise, nylon seine twine of size 30-36 thread makes an excellent, inexpensive, durable service. Paint a finished nylon-served splice with a mixture of one-third black paint, one-third varnish, and one-third net dip (available at fishery supply stores) for a tough, resilient, handsome finish.

Once you get the idea of the splice firmly fixed in your mind, turn out 15 or 20 of them for practice, to get the splice firmly fixed in your hands. When your work consistently looks good, it's probably consistently high in strength. But before you consider using one, send a few of them off to be destruction-tested. Wire rope manufacturers and distributors in your area will know the location of the nearest testing facility.

To test the greatest number of splices with the least work, splice one end of each piece and have the tester swage the other end. They'll pull each piece until it breaks and send you a certificate that says how much strain was on each splice when it broke. By comparing these figures with the wire's ultimate strength, you can determine the strength of your splices.

The actual breaking strength of good domestic wire rope can be 5 to 15% more than its rated strength; the specific percentage will vary from batch to batch. The best manufacturers, notably MacWhyte Wire Rope and Carolina Steel and Wire, test every run of their product. When you buy wire, get the production number of the spool it came from, then call the manufacturer's quality-control department for the breaking-strength figures for that production run. That way, you can get precise numbers for your own destruction tests.

Swages and other mechanical terminals are almost universally rated as being stronger than splices, but bear in mind that: (a) a rig's design and safety factors are based on the wire's rated strength; (b) the anti-fatigue splice will degrade more slowly than mechanical terminals will; and (c) mechanical terminals aren't always as strong as their manufacturers claim. You might break a swage or two in your tests.

TIGHTENING THE LAY OF SEIZING WIRE

When applied as a seizing, seizing wire sometimes will unlay and flatten under load, preventing it from lying fair. To prevent this, tighten the lay of the wire before using: clamp one end in a vise, the other end in a pair of Vise-Grips, and twist the lay tighter.

Possibilities

The 1 x 19 wire splice enjoyed an all-too-brief vogue in the first half of this century, before an expanding yacht population, high labor costs, and a shortage of skilled labor all conspired to make swages and other terminals the preferred alternatives. But this splice just refuses to die. People seek it out for new boats, for born-again classics, and for cruising vessels of all descriptions. It isn't likely to dominate the market ever again. On the other hand, if a splice in 1 x 19 wire is possible, anything is.

THE WIRE ROPE LONG SPLICE

A long splice in wire is analogous to one in rope, but instead of three flexible strands, you're dealing with six intractable ones plus a heart that may be wire or fiber. It is much more difficult to get a uniform distribution of load over all six strands, but just as vital as with the rope's three strands; proceed with infinite fussiness.

To start, unlay the strands in pairs to a length indicated in the accompanying table for your diameter wire rope. Cut the hearts off at the unlay-to point.

"Marry" the two lines. Lay out and lay in two opposing pairs until the laid-in pair is about half used up. For example, if you're splicing with ½-inch wire, you'll have 10-foot tails, and you'll lay the pair in until about 5 feet is still hanging out. The top illustration shows this first step in progress off to the right, with the other four pairs still married in the middle.

Now open up the two pairs of strands you've been working with, leave one member of each pair where they are, and lay in and out with the other two a little farther, until the lay-in strand has about a foot showing. This will give you the bottom illustration, with two pairs of strands meeting at separate points on the wire.

Go back to the marriage and lay out two more opposing sets, this time off to the left, again stopping halfway out and sub-splitting.

This leaves a lone set of strands in the middle. Split them where they lay, and lay each pair away from the middle, in opposite directions, until they're about one-quarter of the way to the farthest pairs.

You should now have six pairs of strands more or less evenly distributed along the wire. To finish, trim all ends about 1 foot long, and bury them as for the wire rope grommet shown in *The Rigger's Apprentice*.

Extra length, *each end*, for long splice

diam.	length	diam.	length
¼	5'	¾	15'0"
5/16	6'3"	⅞	17'6"
⅜	7'5"	1	20'0"
7/16	8'9"	1⅛	22'6"
½	10'0"	1¼	25'0"
9/16	11'3"	1⅜	27'6"
⅝	12'6"	1½	30'0"

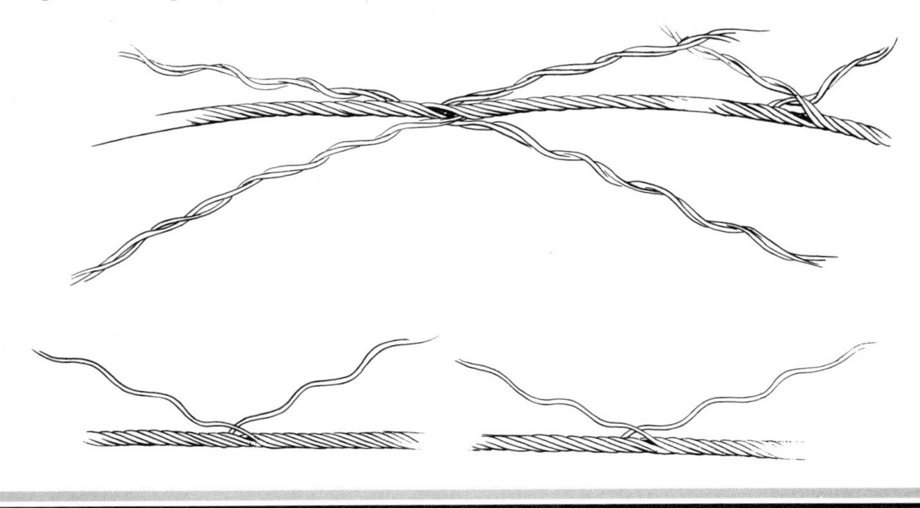

Chapter 4

Procedures and Hardware

"...the blessing, amnesty, and encouragement that good climbers requisition from the thin air."

– Mark Helprin

Rigging is based on simple things: knots, and the principles of friction and tension they imply. But the practice of rigging is infinitely complex, involving procedures that can be mentally and physically challenging, fussy, and even dangerous. The practice of rigging is, in short, an art. This chapter contains descriptions of some of those procedures, as well as some of the hardware that riggers and sailors depend on.

SERVICE

Service, a protective coat of twine around rope, is an ancient procedure, dating back to the days of hemp rigging. But it didn't go out with galleons, because it also prevents galvanized wire rope from rusting. Now, since most boats today are rigged with stainless wire, you might think service has no place today. But cost-conscious sailors are discovering that a little maintenance time is a fair trade-off for wire that costs one-third what stainless does. Service can make sense even on stainless wire, to cover splice ends, as a firm "bed" to lash to, or to lessen chafe on sails. Here are some details to inform your serving jobs. You will find full instructions for applying service in *The Rigger's Apprentice*, but I'll summarize those instructions here.

Service is properly applied, as shown in Figure 4-2, over a bed of twine "worming" and tarred "parceling," usually with a specialized tensioning device called a serving board or mallet (Figure 4-3), but a marlingspike will do in a pinch. One might say it is used with absence of mallet.

As turns of service are taken, the hauling part shortens. When it becomes too short, the Marlingspike Hitch (Figure 4-1) is capsized back into a straight length that in a few more turns becomes part of the service itself. This capsizing calls to notice a hidden characteristic of the Marlingspike Hitch. Notice that the direction from which strain comes on the knot minimizes any tendency for it to jam. To prove this for yourself, make the knot and anchor both ends. When it is pulled on from the wrong direction, it tends to slip around to one side of the spike and jam there. But if pulled the other way, it is more likely to remain stable and to disappear without any fuss once the spike is removed. Be careful, then, to make the hitch as shown.

There are two ways to stop service: If you're using a spike, make the last three or

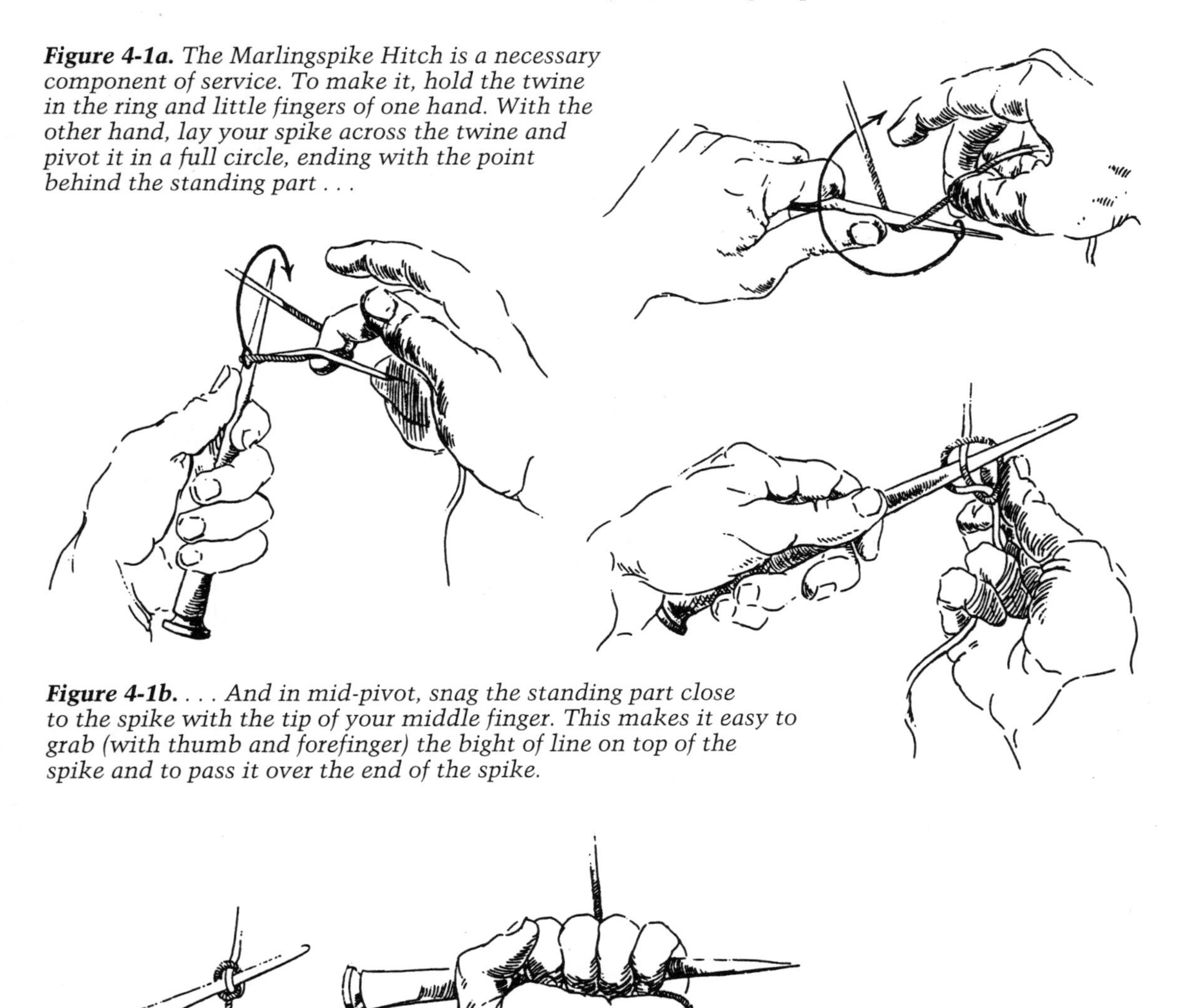

Figure 4-1a. *The Marlingspike Hitch is a necessary component of service. To make it, hold the twine in the ring and little fingers of one hand. With the other hand, lay your spike across the twine and pivot it in a full circle, ending with the point behind the standing part . . .*

Figure 4-1b. *. . . And in mid-pivot, snag the standing part close to the spike with the tip of your middle finger. This makes it easy to grab (with thumb and forefinger) the bight of line on top of the spike and to pass it over the end of the spike.*

Figure 4-1c. *The completed Marlingspike Hitch.*

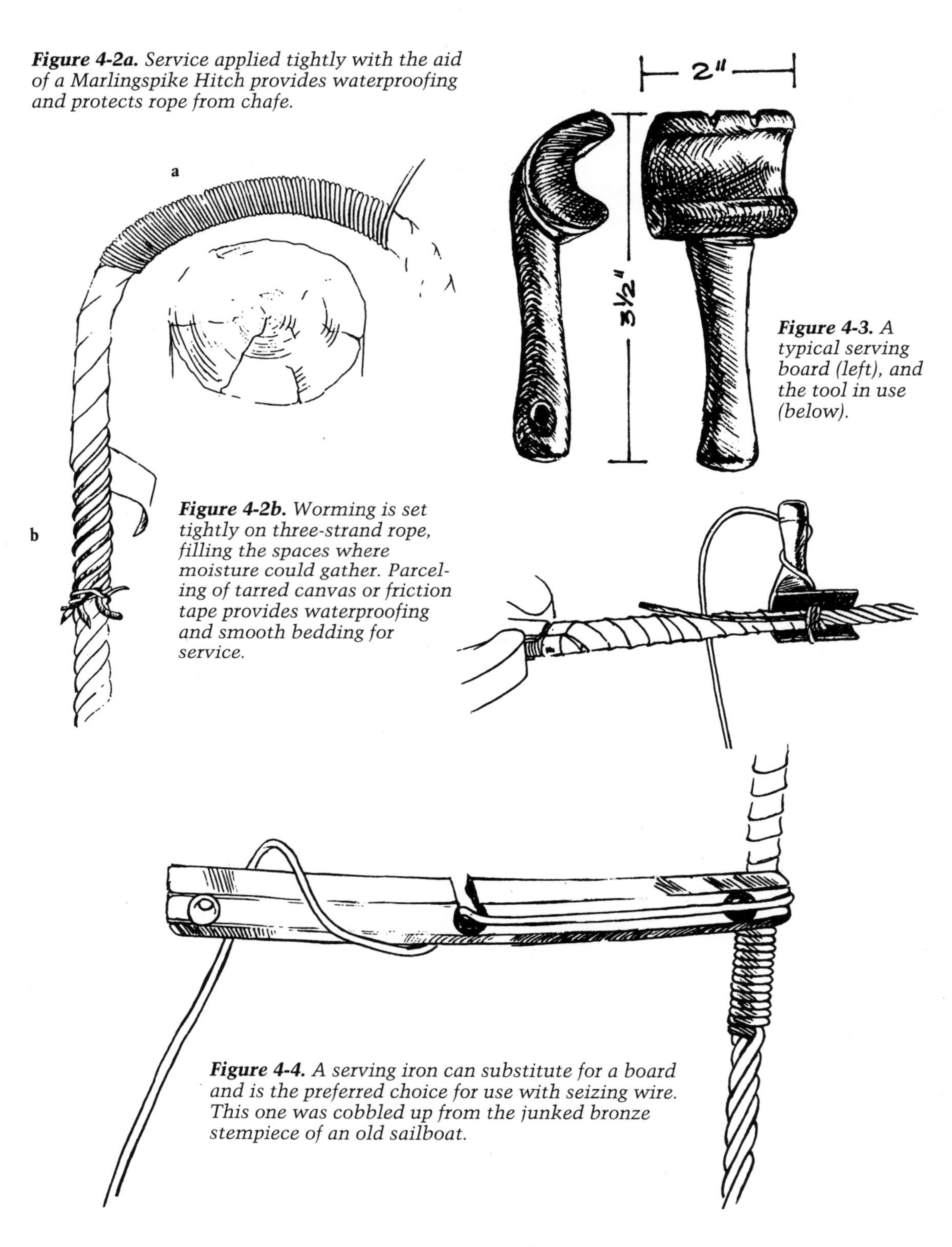

Figure 4-2a. *Service applied tightly with the aid of a Marlingspike Hitch provides waterproofing and protects rope from chafe.*

Figure 4-2b. *Worming is set tightly on three-strand rope, filling the spaces where moisture could gather. Parceling of tarred canvas or friction tape provides waterproofing and smooth bedding for service.*

Figure 4-3. *A typical serving board (left), and the tool in use (below).*

Figure 4-4. *A serving iron can substitute for a board and is the preferred choice for use with seizing wire. This one was cobbled up from the junked bronze stempiece of an old sailboat.*

four turns loosely around both it and the wire. Pass the end under the turns, then tighten the turns by working them around with the spike. Hitch onto the end and pull all slack out, then give a few sharp jerks to snug things completely down before trimming.

The serving board's tidier finish involves stopping at the same point, but making the three or four turns on the other side of the board from the end of the service, quite loosely. Then tuck the end of the twine under the last turn of service and begin serving over it, thus undoing the turns you made and ending up with a large bight at the end of a smooth, tight service. Remove the serving board and hold onto the service with one hand, hitch onto the end with a spike and pull with the other hand, and guide the slack out, keeping the twine from twisting, with your third hand. This procedure is the same when made with seizing wire, as in Figure 4-4.

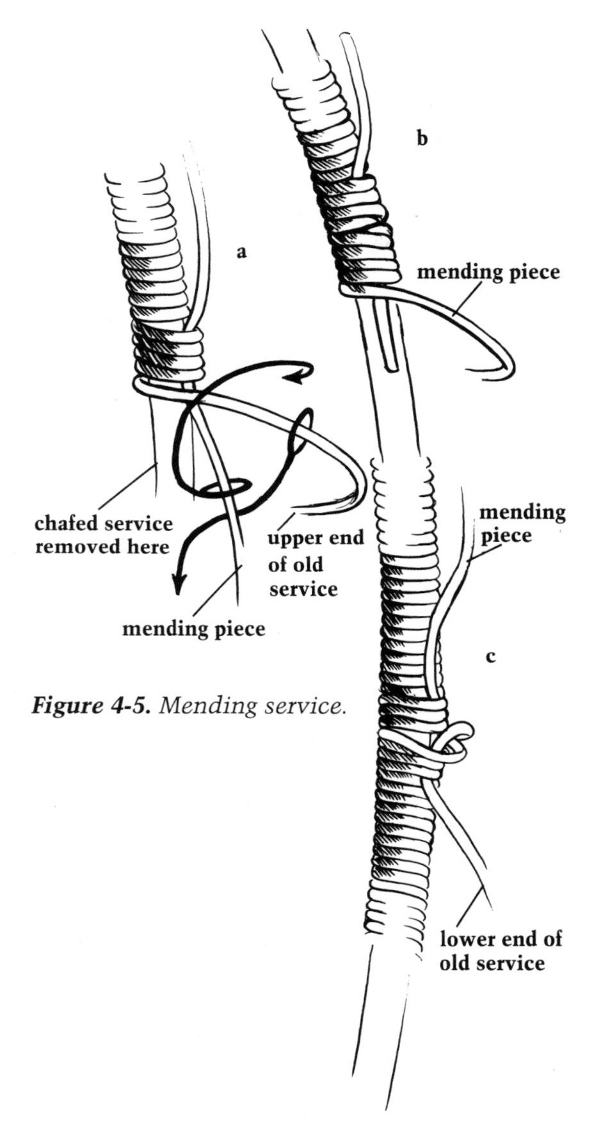

Figure 4-5. *Mending service.*

Mending

The oldest rigging afloat is not stainless steel, which has a lifespan limited by fatigue, but parceled and served galvanized wire, which is susceptible neither to fatigue nor rust. Here are instructions for the mending that chafe occasionally necessitates.

Unwind the service from the affected area and trim the ends to about 2 feet (0.6 m) in length. Inspect the parceling and replace as necessary. Plain old friction tape makes excellent parceling, or you can rip up polyester bed sheets (they're no good for sleeping on anyway) into inch-wide (25 mm) strips and tar them. This is a good time to check the condition of the wire under the parceling. Retar as necessary, particularly if the service is over stainless steel, as the tar (or anhydrous lanolin) excludes both air and water, preventing crevice corrosion.

When the parceling is set, make up a hank of twine. If the area to be repaired is a long stretch, calculate how much twine you'll need as follows: Lightly serve a 1-inch (25.4 mm) length, then remove this service and measure how much twine it took. Multiply this by the total number of inches (or mm) to be served, and add a couple of feet (or 0.6 m) for the tails at the finish (see below).

If you are working aloft, thread the end of the twine through the lanyard hole in the end of your serving iron or board. Once you get started, this will trap the iron, so you can't accidentally drop it.

Using a spike, serve over the end of the new twine with the upper tail of the old twine (Figure 4-5a). This anchors the new twine's end. Then twist the two pieces together, lay the old twine's end down on the wire, and begin serving over it with the new twine, using the iron or board. Start carefully, so there's a minimum gap where the two pieces are twisted together (Figure 4-5b). Serve over the old piece for four or five turns, then trim its end flush. Continue to serve the rest of the bare area.

SHIELDED MALLET

After a few thousand feet your serving mallet will have worn itself into a deeply grooved, organically distorted artifact. Wear is particularly pronounced at the leading edge. So to prolong head life, tack on a tin "shield."

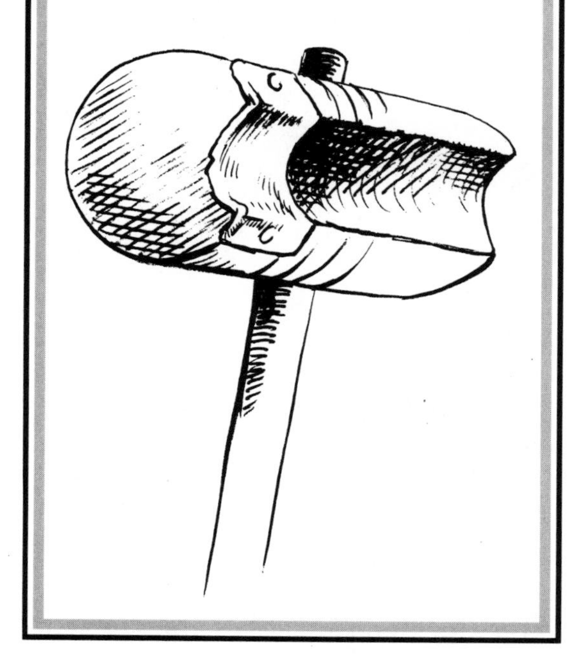

Simple so far. The tricky part comes when you reach the other end of the service; how do you get a smooth, tight join down there? The best way is to serve right up against the old service, then unthread the iron and stow it carefully away. Now undo exactly three turns from the old and new service. Lay the tip of a spike down and lightly wrap three turns back on with one of the ends, around the spike and the wire both. Remove the spike and thread the end under those three turns. Lay the spike back on the wire opposite these three turns, and restore the other end's three turns. Remove the spike, and thread the end under the last turn of the other end, then under its own final three turns (Figure 4-5c).

Begin tightening the turns on one side, slowly and carefully, one turn at a time, with the tip of the spike. When the turns are good and tight, pull the end to draw the last turn down. Tighten the turns on the other side, and pull that end down. Finish by jerking on both ends to tighten the crossover. The two last turns should mesh into one another, leaving a barely discernible join.

Use this same procedure for finishing service on a grommet.

If chafe in an area is a recurring problem, double-serve and/or leather. And check your running rigging leads to see if there's a way to lessen the chafe.

Two-Way Service for Eyes

The best time to serve an eye is before you splice it, so you can work on straight wire. But if you're re-serving an eye, you have no choice but to work within the confines of the eye. And it's very difficult to keep the turns of the twine from separating as you go around the curve of the eye. So start in the middle of the eye, serve down one side, then come back and serve down the other. You'll get fairer service and work with shorter lengths of twine.

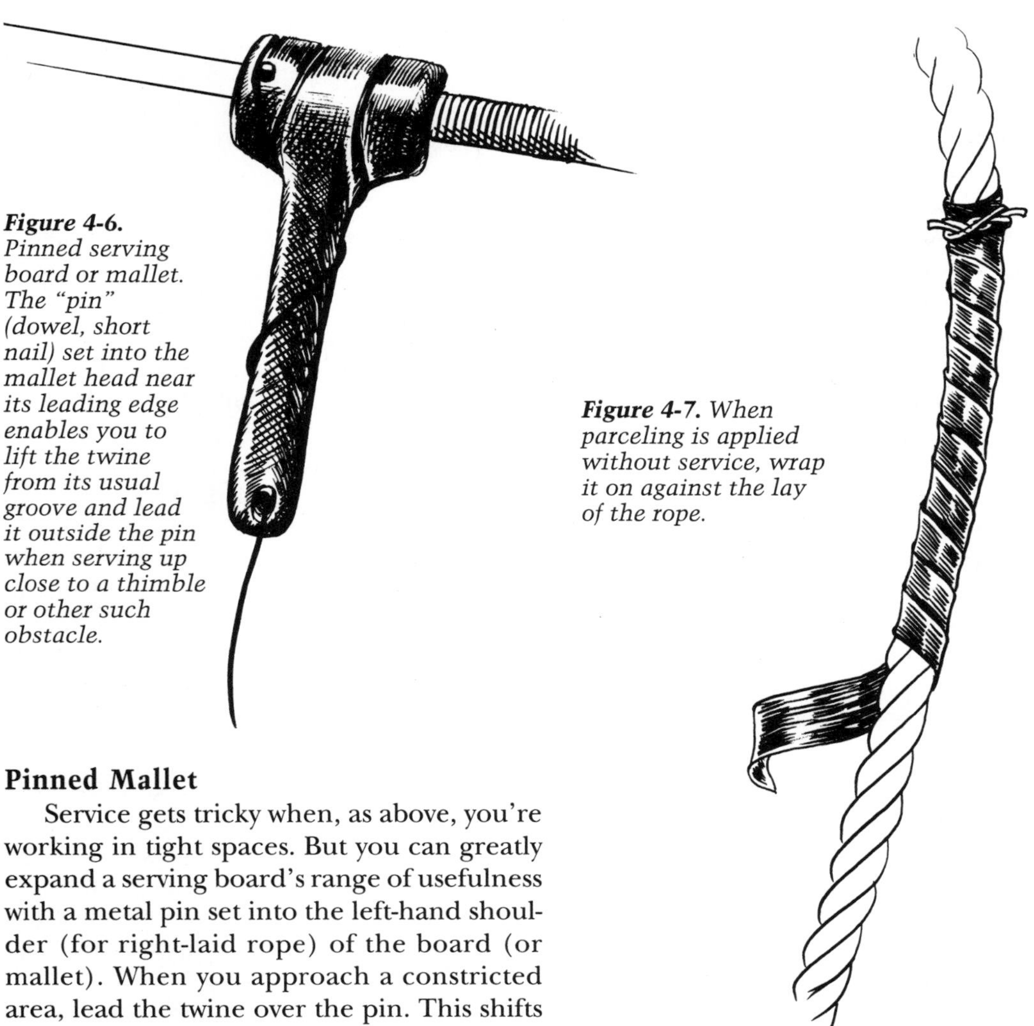

Figure 4-6. *Pinned serving board or mallet. The "pin" (dowel, short nail) set into the mallet head near its leading edge enables you to lift the twine from its usual groove and lead it outside the pin when serving up close to a thimble or other such obstacle.*

Figure 4-7. *When parceling is applied without service, wrap it on against the lay of the rope.*

Pinned Mallet

Service gets tricky when, as above, you're working in tight spaces. But you can greatly expand a serving board's range of usefulness with a metal pin set into the left-hand shoulder (for right-laid rope) of the board (or mallet). When you approach a constricted area, lead the twine over the pin. This shifts the body of the mallet away from the obstacle, allowing you to take more turns in clear air.

Parceling Sans Service

Service is a good chafe preventer, but heavy-duty parceling is often as good and much quicker for chafe protection on mooring lines.

The old saying is to "parcel with the lay," that is, to mimic the rope strands' spiral with the parceling. This lets the parceling lay down smoother under the service, which is put on against the lay. But parceling in the absence of service should go on against the lay. That way, when the rope untwists a bit under load, the parceling will tighten and stay put, instead of loosening and slipping out of place.

Birdcage Warning

A serving tool is a form of lever. When it is used with excessive force, this lever can distort the lay of the wire, causing the strands to separate into a strength-destroying "birdcage."

Back when marline was the material of choice, it used to be hard to make a birdcage, simply because the marline would break before sufficient force could be exerted. But nylon, today's service material of choice, is far stronger than hemp. It's also much more decay-resistant and far superior to hemp in every way, except that it allows an inattentive server to destroy the wire.

The question, then, is how tight is tight enough? The answer is that service should feel hard, but not so hard that friction tape parceling squeezes out between the turns. So take it easy on your parceling, your service, and yourself.

Nylon Slush

Slush is the thick, paint-like substance that goes on service to keep it from drying out. Ideally it is hard enough that it won't rub off on sails or crew, but soft enough that it won't crack and let water in. With marline, slush was traditionally made with a mix of Stockholm tar, linseed oil, and Japan drier, with maybe a little varnish thrown in. This works well with nylon service, too, but there is an even better alternative: equal parts black paint and net dip. Net dip, available from fishery supply stores, is an asphalt-based paint in which twinemakers dip nylon twine to tar it. It's a little too thin to hold up on standing rigging by itself, hence the addition of a good anticorrosive black paint. Test paint a short stretch of service, since some paints dry harder than others. If the slush dries too hard, thin with Cuprinol and varnish to taste.

Wire Service

Seizing wire is the serving material of choice for splices where hank or sheet chafe is severe—staysail stays, and the bottom ends of shrouds. It seems logical to serve galvanized wire with galvanized seizing wire, and stainless with stainless. But in practice, annealed stainless seizing wire of 1 x 7 construction is nearly always the better material.

For one thing, galvanized seizing wire, which is made from iron, not steel, is less than half as strong as stainless. And corrosion is most severe at deck level, where most wire service is needed; stainless seizing wire lasts longer here. Fatigue is not a problem, because the wire is annealed, and is also not structural. Finally, stainless wire is much more widely available, since it has wide commercial use.

Corrosion arising from mixing dissimilar metals is generally not a problem with rigging materials—given the insulation of service, or even just paint—since (if all goes well) rigging is not kept immersed in the electrolytic medium of water. As insurance, it is customary to double-parcel under wire service, mixed metals or no: one layer of friction tape to waterproof, and one layer of lanolined polyester bedsheet to insulate the wire and keep it from chewing up the friction tape.

Wire and twine service are the same in principle, but differ in tools and technique. At the start of wire service, you may need to clamp the end down with Vise-Grips to get started. Always use a serving iron.

To finish wire service, come within five or six turns of the thimble (wire is slicker than twine, thus the greater number of turns), make and undo the turns as with twine, pull the slack out with a pair of pliers, and get out your heaving mallet. Figure 4-8 shows "Mallet de Mer" with the seizing wire belayed to its head and handle, ready to pull the last turn taut—too much of a job for a marlingspike. Apply just enough tension to bring the turn down snug, then put the mallet over on the other side so that the strand will be pulled to that side, wedging itself permanently between the rope and the turns of service. If the service is very tight the strand

__Figure 4-8.__ To finish a wire seizing, use a heaving mallet to pull first left (facing vise), then right. The wire should snap at the exit point.

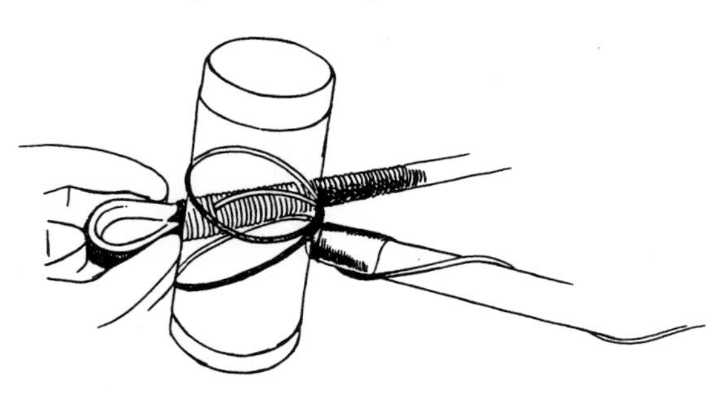

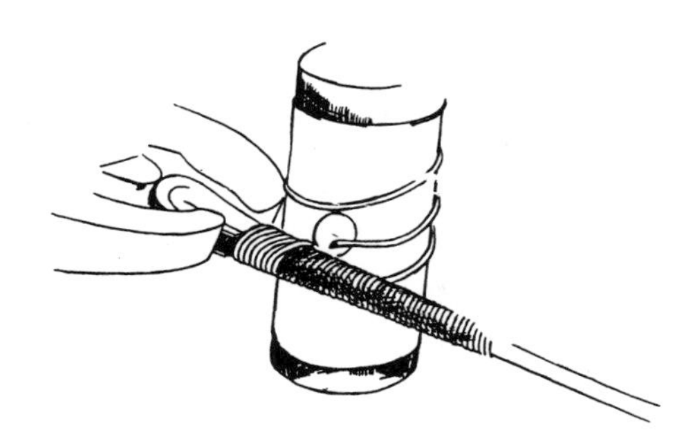

will shear off cleanly at its exit point with a good haul on the mallet.

If it won't shear, but you're sure the fit is snug, just bend the seizing wire sharply back and forth repeatedly until it breaks off at the exit point.

The one place you shouldn't mix rigging metals is on splices that do spend a lot of time under water, like the lower ends of bobstays and boomkin shrouds. Also check to see that the wire rope itself is compatible with hull fittings. If they're bronze, either use stainless wire or bronze rod, or very carefully double-served galvanized wire, and check regularly for chafe.

THE BASEBALL DIAMOND STITCH

For cylindrical objects such as rope, wire rope, oars, and stanchions, the best leather-affixing stitch is the Baseball Stitch. Its double-needle crisscross pattern simultaneously provides a smooth, tight seam and a built-in redundant strand in case of twine chafe. It's also a pretty stitch, pretty enough to turn mere chafing gear into an art form.

It's such a good stitch that I hesitate to point out three small flaws, let alone presume to suggest solutions. But I will. It's the little details that matter with handwork; they add up and make a difference in the quality of the finished product. And I've come up with a nifty, small- detail stitching variation that I just have to tell you about. But first, the flaws:

(1) The corners of the leather at the starting point tend to get pulled diagonally, distorting the start. This is because you're pulling two corners together instead of two sides, and it's a problem you'll encounter no matter what stitch you use. This initial distortion is hard to prevent, and can result in an unsightly, chafe-prone lump (Figure 4-9).

(2) The leather tends to "shrink" a little in length from the tension of the stitches. If, for example, you put a 1-foot-long (305 mm) leather on an oar, you might find that it's only 10 or 11 inches (250-280 mm) long when you're done. This shrinkage is particularly galling when you're leathering eyes for standing rigging. You lay the work out oh-so- carefully, stitch with consummate skill, and then find when you bend the eye around that it's lopsided.

(3) You have to pull all that thread through. I don't mean to whine, but for every foot of leathering you do there's three feet of doubled twine

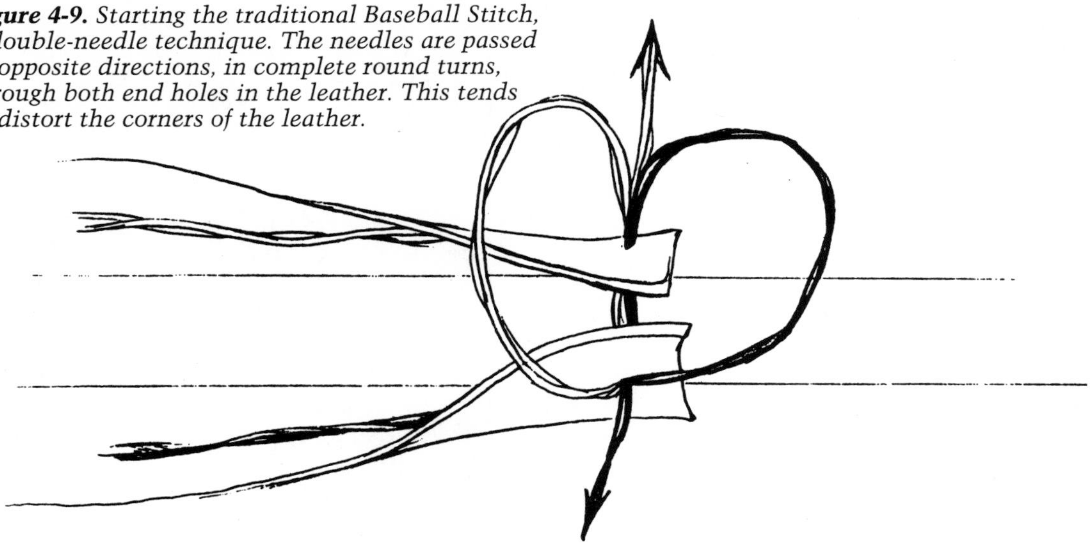

Figure 4-9. *Starting the traditional Baseball Stitch, a double-needle technique. The needles are passed in opposite directions, in complete round turns, through both end holes in the leather. This tends to distort the corners of the leather.*

on each needle, and they're always getting tangled, and pulling all that stuff through just takes so much time. Oh, it shortens up eventually, but by then a lot of the wax has rubbed off in all those holes. Wouldn't it be nice if there were a way to work with shorter ends?

Well, sailors, there is! As well as a way to deal with asymmetrical shrinkage and foul starts! There are some extra setup steps, but they result in shorter overall job time! Introducing, the evolutionary new Baseball Diamond Stitch!

And here's how it's done:

(Have you measured your leather? If not, Figure 4-10 shows how.)

Cut a piece of waxed #7 twine 12 times the length of the leather. So for a 1-foot (.30

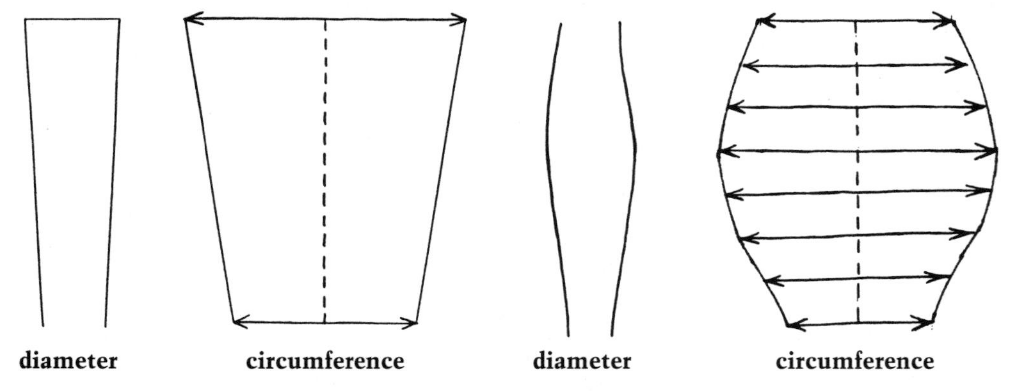

Figure 4-10. *To leather a straight taper, such as an oar (left), measure the circumferences of the oar bole at the positions where the leather top and bottom will lie. Draw a midline of the required length on the leather, and lay out the half-circumference distances on either side of the midline, top and bottom. Connect the top and bottom corners to make the taper. To shape a piece of leather for an irregular taper, such as an Eyesplice (right), measure circumferences at regular intervals along the splice and plot the respective half-circumferences on either side of the leather's midline. In either case, deduct 1/16 to 1/8 inch for elasticity, depending on the leather.*

Figure 4-11. *Preparing for the Baseball Diamond Stitch.*

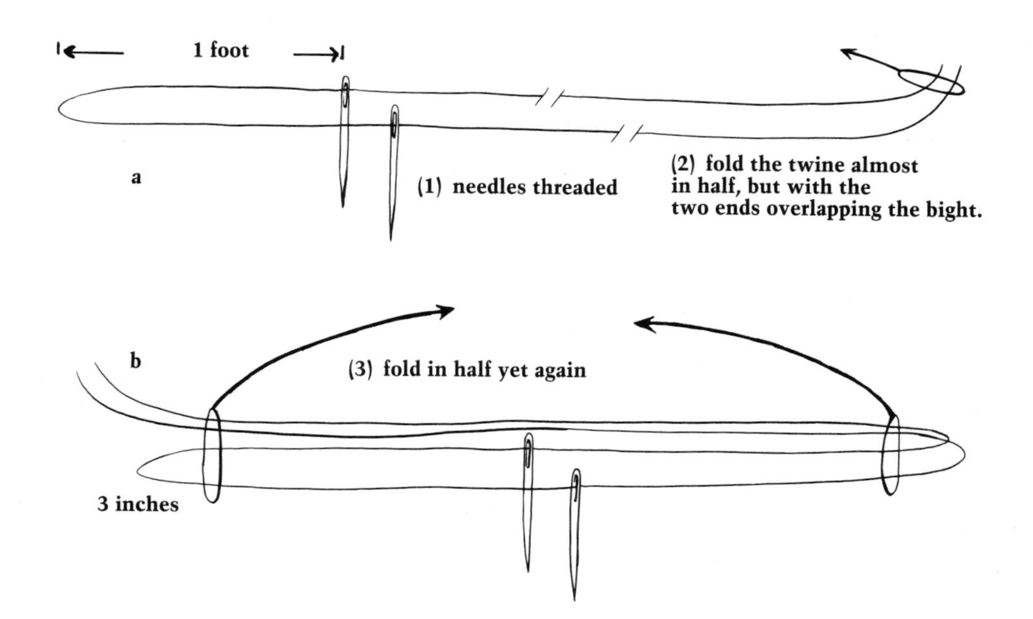

m) leather you'd need 12 feet (3.66 m). This assumes stitching holes ⅜ inch (9.5 mm) from each edge and ⅜ inch (9.5 mm) apart.

Fold the twine in half and thread a sail needle—#14 is a handy size—onto each end (Figure 4-11). Leave the needles hanging a foot or so on either side of the middle for now. Fold the twine in half again, but not quite exactly in half; leave the ends sticking past the bight about 3 inches (76 mm).

Now let go of the ends and fold in half yet again the length that remains in your hands. Move the needles to the middle of this length (Figure 4-12). The needles are now in position and ready to sew. All that folding rigmarole was just a way of locating them, without measuring, one-quarter of the total length on either side of the middle. With the needles in this position, you can sew half the length of the job, then come back and use the long ends to sew the other half.

And so to work. Pre-punch the holes into the leather with a .00 punch ("double-aught" is what they call it at the leather store), and rub some neatsfoot oil onto the inside of the piece. Fold the leather in half (crosswise) to locate the middle holes, mark

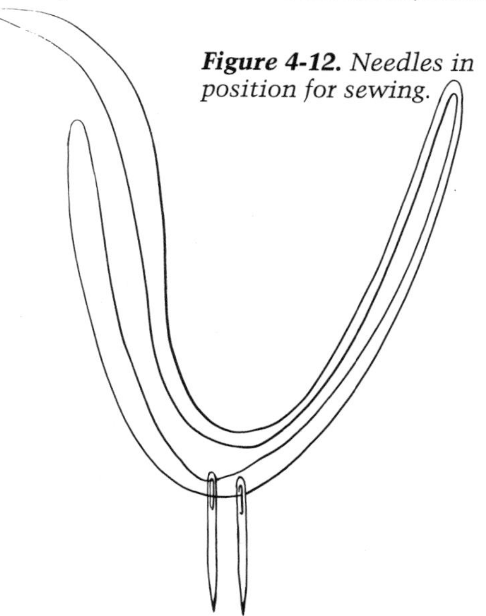

Figure 4-12. *Needles in position for sewing.*

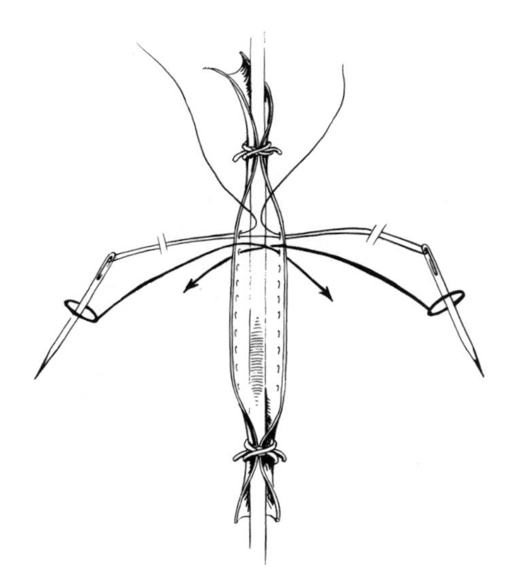

Figure 4-13. *Stitching down the first half of the leather.*

them with a pen, then straighten the leather out and seize or Constrictor it lightly to the leatheree. Thread the needles from the inside out through the center holes (Figure 4-13). Take up the slack so that the needles are evenly extended. Be careful to avoid letting thread slip through the needle's eyes at this point, as that would leave you short for stitching the second half.

When everything is even, thread one needle diagonally off to one side, then the other needle through the corresponding hole on the other side (Figure 4-13). Firmly draw all the slack out after each set of stitches, first pulling the threads straight out to the sides to tighten previous stitches, then crossing them over the top of the seam and hauling to the sides again to tighten the current stitch. Wear roping palms or heavy gloves to avoid hand damage.

Keep the seam straight as you go. Always enter the same side first—i.e., if the first stitch you took was with the needle on the right, always begin a set of stitches with the needle on the right. It looks better.

For an idea of what a good seam should look like, get a baseball. You'll see that the leather edges are firmly butted together, with a little ridge of leather bunched up on either side.

Avoid, at all costs, pulling so hard that you rip the leather. If the piece is the right size, you shouldn't need to pull that hard.

When you get to the end, make a com-

Figure 4-14. *Finishing off the first end of the stitch.*

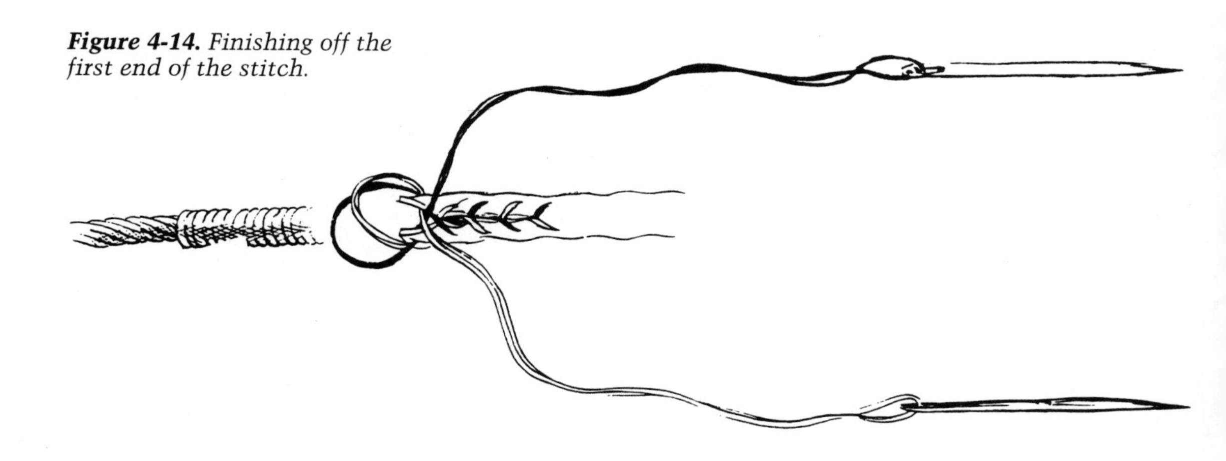

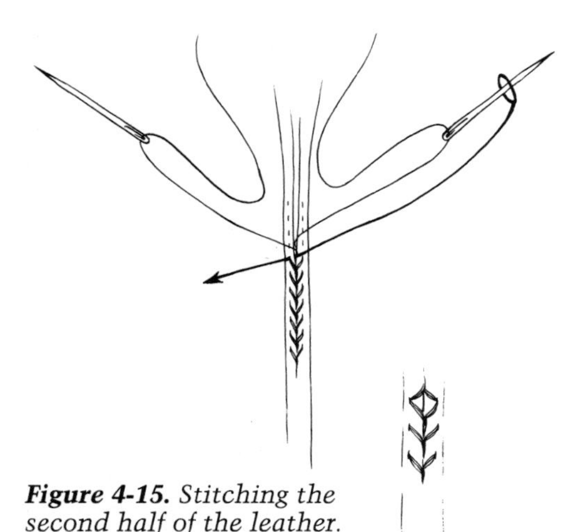

Figure 4-15. *Stitching the second half of the leather.*

plete turn through the last set of holes with each needle. They'll come out from under the leather at the end (Figure 4-14). Haul taut. Reef Knot the ends together, snapping each half-knot smartly to set it back under the leather. Trim the ends short.

Now remove the leftover thread from the needles and thread them onto the long ends hanging out of the middle of the work. Turn and face toward the unstitched side and begin sewing through the same two holes you started in before (Figure 4-15). Leave 3 to 4 inches (76 to 102 mm) of thread ends behind, and sew over them to anchor the stitches. For the most pleasing appearance, begin with the opposite needle that you did for the other half. Sew to the end and finish as before.

Some notes on the application of this stitch:

- Latigo is the best material to use where chafe is severe (anchor rodes, spar chafe, oars, etc.). It's the thickest and longest-lived material you can get, and a little boot dressing now and then will keep it healthy.
- "Synthetic leather"—rubber, plastic hose, and the like—can be quite tough to work with and becomes brittle with age.
- For light duty, as where sails chafe on shrouds, the supple, pale-gray leather that sailmakers use is excellent stuff. It's tough, elastic, and durable, but thin enough that it won't look clunky in place. If you apply it over stainless, rub a generous amount of anhydrous lanolin (available from druggists) into the wire first to minimize oxygen starvation. Tar and serve galvanized wire before leathering.
- Always turn the stitches away from chafe. When leathering nonfeathering oars, run the seam in line with the blade edge. With feathering oars, run it a little aft of the edge so that it won't chafe against the forward side of the oarlock as you feather. Row for a bit before leathering; the chafe marks will show where the middle of the leather should be. If the boat has more than one rowing station or a sculling notch, you'll want an extra-long leather to accommodate the different bearing points. Especially for feathering and sculling oars, it's good to glue the wood and leather together with contact cement before stitching.
- If your twine does chafe through sometime, anchor the repair strand(s) and the original strand(s) under the first several stitches you take. This is far better than tying a knot in the end of the twine, as that knot is liable to pull through under strain.
- And finally, for a few more leathering details, see *The Rigger's Apprentice*, pages 117-119.

RATLINES

"Arr, the crew swarmed nimbly up the ratlines as the vessel approached her mooring." The crew, it seems, was forever swarming nimbly up the rungs of this indispensable ladder, to set or furl sails, to assist in navigation, or to perform maintenance and repairs. Sadly, ratlines now suffer from the dreaded "Anything-that-salty-must-be-useless-these-days" syndrome.

And that's a pity. Although most contemporary yacht sails are set and furled from deck, ratlines still offer a clear view from aloft for spotting coral heads, windshifts, or land. And since maintenance by no means went out with sprits'l tops'ls, you can still use ratlines to get aloft with varnish and paint. And they're the fastest way up in an emergency. And it just feels so good up there on a warm, breezy day, with the world spread out beneath you, and the mast tracing gentle arcs in the sky....

Where was I? Oh yes, ratlines. The modern, more expensive equivalent is a system of metal steps affixed to either side of the mast. These are relatively easy to climb when you're in a calm harbor, but have you ever tried going up them while the boat is pitching and heeling? A more versatile, safer arrangement for cruising yachts employs ratlines on the lower shrouds to get you as far as the spreaders, and a good bosun's chair for anything above that (see "Living Aloft" in this chapter). With the ratlines, you can ascend unassisted to deal with the forestays'l halyard, jammed sail track, slipping spreaders, and the like. If the problem is at the masthead, you can save your crew half the time and effort they'd ordinarily expend getting you all the way up.

Wood or Rope?

All-wood ratlines make a comfortable, stable ladder. But they're a lot of work to make, a lot of weight and windage in place, and they're often visually clunky. All-rope ratlines are inexpensive, quick to make, and physically and visually unobtrusive. But they're awkward to climb and uncomfortable to stand on. I find that a sequence of two rope ratlines and one wood works best. The wood ones are close enough together that there's always a solid place to stand nearby. And they act as struts to hold the shrouds apart, so the rope rungs don't sag as much when you step on them. So you get an optimal combination of quick production, low cost, and low bulk without sacrificing too much comfort.

Fabrication

Rattling down can be done alone, but it will go much faster if a hand aloft splices and installs rope while a hand alow measures and cuts wood.

Unless your shrouds have sheerpoles on them, the lowermost ratline will be of wood. Commonly, this would be an overlong piece lashed outboard of the shrouds. This is easy, but "internal" ratlines—those which fit between the shrouds—are appreciably lighter, more handsome, and have no line-snagging projecting ends.

The first step in layout is to hold a bevel gauge against one of the shrouds, adjusting it so that the blade is level with the horizon (Figure 4-16). Mark this angle on a miter box or other cutting jig. Now get the angle off the other shroud you'll be rattling, and mark this angle on the miter box. Cut these angles on the ends of a wooden ratline of the right length, and it will fit level between the shrouds. Since these angles are constant all the way up, you need only measure them once, and at any convenient height. Shrouds that are served will hold ratline lashings more readily. But sufficiently tight lashings will hold on bare wire. If you do serve, it is easier to do so the full length, as this saves

Figure 4-16. *To set the angle of wooden rung ends, trim the boat level, then sight parallel to the horizon with a bevel gauge. Rungs cut to the resulting angle will seat level when installed. Note: This technique only works for shrouds in the same plane, as two lower shrouds. If you are rattling between a lower shroud and another at a different angle—say an intermediate—a spiraling ladder will result. To level these rungs, you'll need to use a builder's level.*

the considerable labor of measuring, starting, and stopping patches of service for each ratline. Full-length service also helps galvanized wire last forever.

The ratline stock should measure at least 1½ inch (38 mm) from top to bottom edge, and at least 1¼ inch (32 mm) thick. This will be adequate for up to ½-inch (12.7 mm) wire; above that, increase the dimensions proportionately. The grain of the wood must always run parallel with the longer dimension, as this makes a stronger, stiffer rung. Rout the corners for a comfortable-to-stand-on, lighter, smaller-looking rung.

With a drill bit that matches the wire diameter, drill a hole through one end of a piece of stock, at the angle of the forward shrouds. Next drill a 5/16-inch (7.9 mm) hole parallel to the first one and about 1¼ inches (32 mm) in from it. This is for the lashing line. Finally, drill for and install a rivet right between these two holes and at right angles to them (Figure 4-17). This will keep the end of the wood from splitting.

Cut the first hole you made in half, leaving a deep, angled groove. Round the corners with a file and sandpaper. Fair the lead from the lashing hole with a small gouge or penknife so that the lashing won't have to go over any sharp corners.

With some typist's white-out, mark the height at which you want your first ratline. This will probably be just above the splice or fitting on the forward shroud. (Because of sheer, a start on the after shroud would result in the forward end of the ratline landing on the turnbuckle. The opposite may be true on mizzen shrouds.) Hold the half-

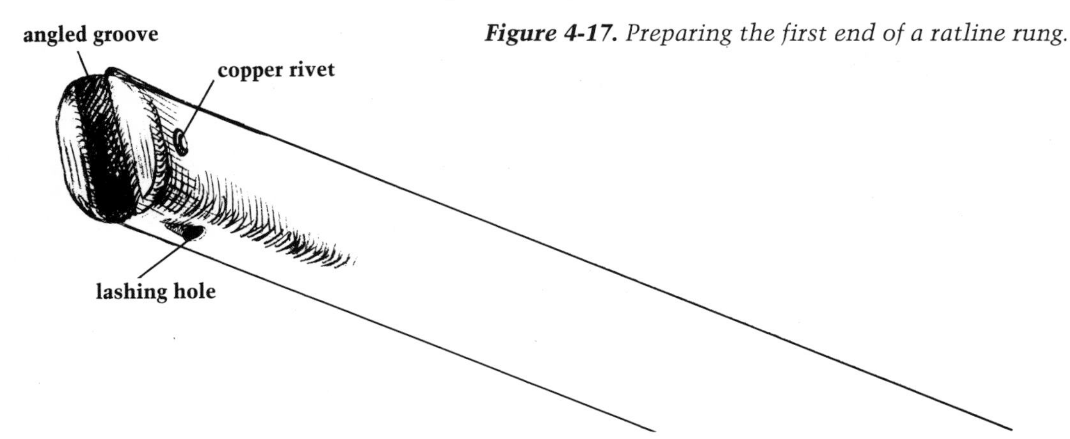

Figure 4-17. Preparing the first end of a ratline rung.

done ratline against the shroud, setting the wire firmly into the groove. Level it against the horizon, then mark the top of the after end, exactly where the middle of the shroud touches it. Take the piece back down, drill three more holes at the appropriate angles, install the rivet, and cut the first hole in half. Smear a little oil, varnish, or other grain-sealer on the ends, go back to the shrouds, and set it in place.

Lashing

You should now have a perfectly fitted ratline, lacking only secure attachment to its new home. Cut 8 feet (2.44 m) of 1/16-inch (1.5 mm) or so tarred nylon twine and hitch it to one of the wires, right at the top of the ratline. Pass the working end down through the lashing hole, around the wire below, and back up through the hole (Figure 4-18). Continue this "square lashing" until there are four or five turns on either side. Finish with Figure-Eight turns made over the end, making a Half-Hitch with each turn for security (Figure 4-19). Pull each of the square and Figure-Eight turns very tight as you go, using a Marlingspike Hitch. Secure the end with a couple of Half-Hitches made around the wire below the rung. Tying a Figure-Eight knot in the end and working it up very

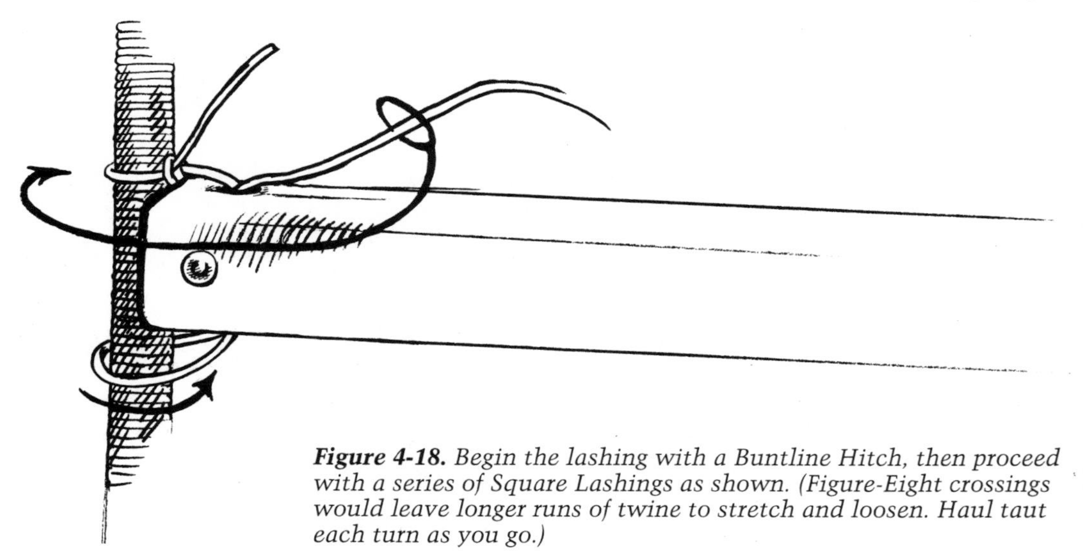

Figure 4-18. *Begin the lashing with a Buntline Hitch, then proceed with a series of Square Lashings as shown. (Figure-Eight crossings would leave longer runs of twine to stretch and loosen. Haul taut each turn as you go.)*

close to the hitches will assure they never come undone (Figure 4-20).

Repeat the lashing procedure at the other end. Don a climbing harness, tether it to a shroud, and step onto the ratline. Jump up and down. Ain't lashings grand?

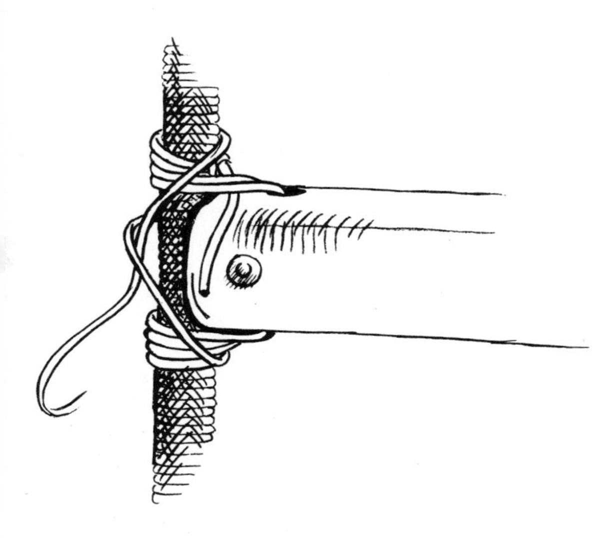

Figure 4-19. *When you can no longer fit turns of Square Lashing through the lashing hole, begin Figure-Eight turns, finishing each with a Half-Hitch. This "fraps" the underlying turns and gives chafe-protection and backup strength.*

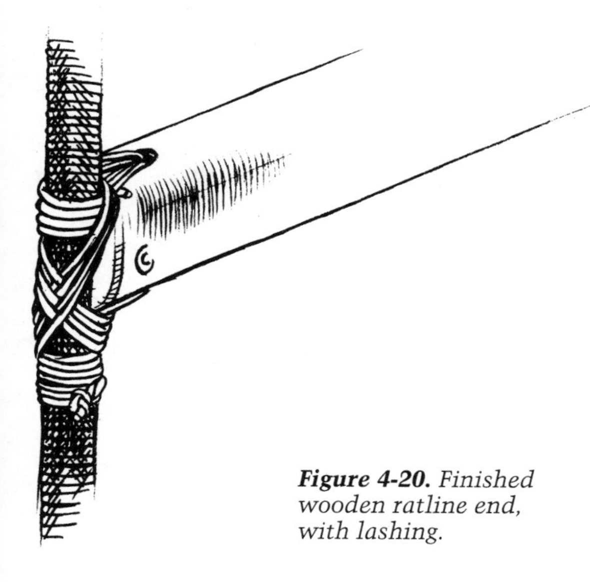

Figure 4-20. *Finished wooden ratline end, with lashing.*

From where you're standing it is easy to measure up, mark for, and install the second wooden ratline. Space between rungs is a matter of taste and leg length, but 16 inches (406 mm) is standard. With two intervening rope ratlines, the second wooden one will be 4 feet (1.22 m) above the first. Set the tape measure on top of the first rung and run it up alongside the most nearly vertical of your two shrouds. This will probably be the forward one. Make a mark at 16, 32, and 48 inches (406 mm, 813 mm, and 1,219 mm).

Drill, cut, rivet, and install the second wooden rung. Before moving on to rope, carefully measure the difference in length between these two ratlines. Then simply deduct this length from successive wooden ratlines as you go.

Rope Rungs

To avoid the dreaded "sagging step" syndrome, prestretch the rope you'll use. (One-half-inch (13 mm) three-strand Dacron or Roblon is ideal.) Tie one end to a masthead halyard, belay the other end on deck, and tighten all with a winch. Or stretch it with a come-along ashore. This will remove the "initial elasticity" caused by all the rope yarns settling into place.

Splice a small eye in one end of the rope and lash it in place. Use a Square Lashing as before, but this time finish off with frapping turns between wire and rope (Figure 4-21). Stretch the standing part out level and make a bight to form an eye that just touches the opposite shroud. Mark the middle of the eye, then cut the line, leaving enough to splice with.

When you go to lash the second eye in place, you'll find that the splice has "shrunken" the line—it won't quite reach the shroud. Lashing it tightly now will, ideally, result in a snug-fitting ratline, but not so snug that it pulls the shrouds together. Adjust your lengths to suit the shrinking effect

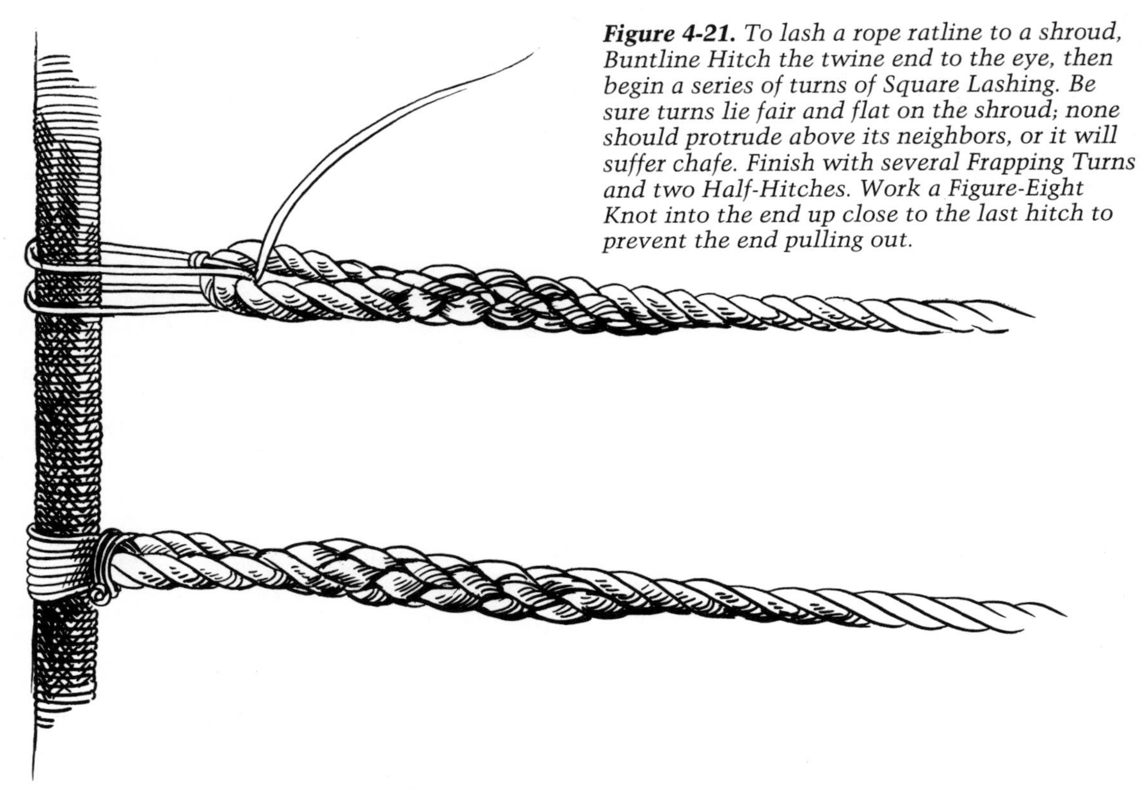

Figure 4-21. *To lash a rope ratline to a shroud, Buntline Hitch the twine end to the eye, then begin a series of turns of Square Lashing. Be sure turns lie fair and flat on the shroud; none should protrude above its neighbors, or it will suffer chafe. Finish with several Frapping Turns and two Half-Hitches. Work a Figure-Eight Knot into the end up close to the last hitch to prevent the end pulling out.*

of your splices and the rope's own characteristics.

Rope ratlines fit best if you finesse each one as you go, rather than deducting a specific length each time. Be slightly more generous with your initial measurements as you ascend; shorter ratlines aloft stretch less, and splice shrinkage consumes proportionately more of their length. As you proceed, you'll find it easy to fall into a rhythm of splicing and lashing while your partner is drilling and cutting.

A semi-utilitarian historical aside: "Rattling down" originally meant just that—installing ratlines from a bosun's chair, starting at the top and working down. Most or all of the ratlines were rope, and standing on a rope ratline will pull even very tightly set-up shrouds toward one another. "Rattling up" will produce the dreaded "Siamese sine curve" syndrome (Figure 4-22). That problem is prevented here by setting the wooden rungs in place first, then filling between them with rope. But while making your fittings, be very careful to avoid leaning either on the shrouds or the rope ratlines. This requires a good sense of balance, a calm day, and, to repeat, a safety line.

So, you've installed the first two wood and first two rope ratlines. The third wooden one comes next, and you're getting up around 10 feet (3.05 m) off the deck. At this point, have succeeding pieces come up to you on a light gantline. I shackle a small block to my harness for this purpose. Always leave the gantline hitched to the wooden rungs until the first lashing is finished.

Climb up, splicing and lashing as you go, first one wood, then two rope, on and on. Keep everything level; skewed rungs are glaringly obvious to the most casual observer.

As you climb, get firmly in the habit of

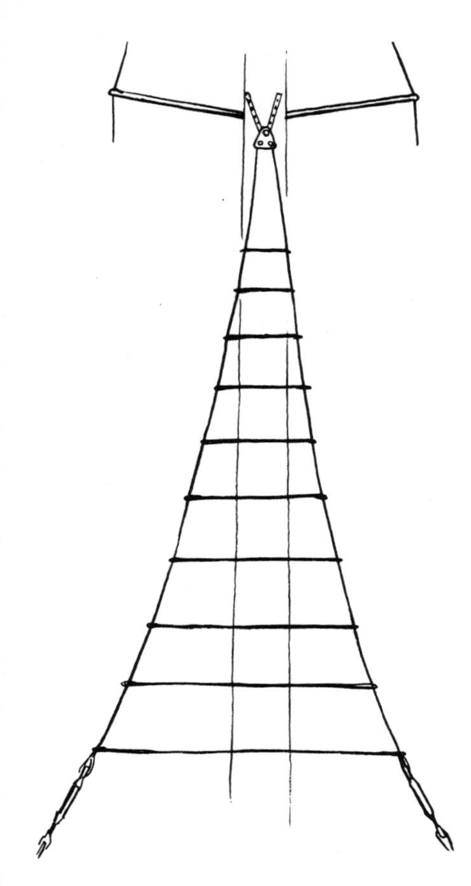

Figure 4-22. The dreaded Siamese Curve syndrome.

holding onto the shrouds, not the ratlines; the former are less likely to give way than the latter. Ratlines are for feet.

As you near the top and lengths decrease, you'll find that the rope splices will begin to back into one another. When this happens, the sweetest procedure is to switch to rope grommets, making them out of line one-half the size you've been using. Proceed until the space between the wires is narrower than your foot.

Options and Variations

Especially on rigs that have shroud eyes seized in place around the mast, the space

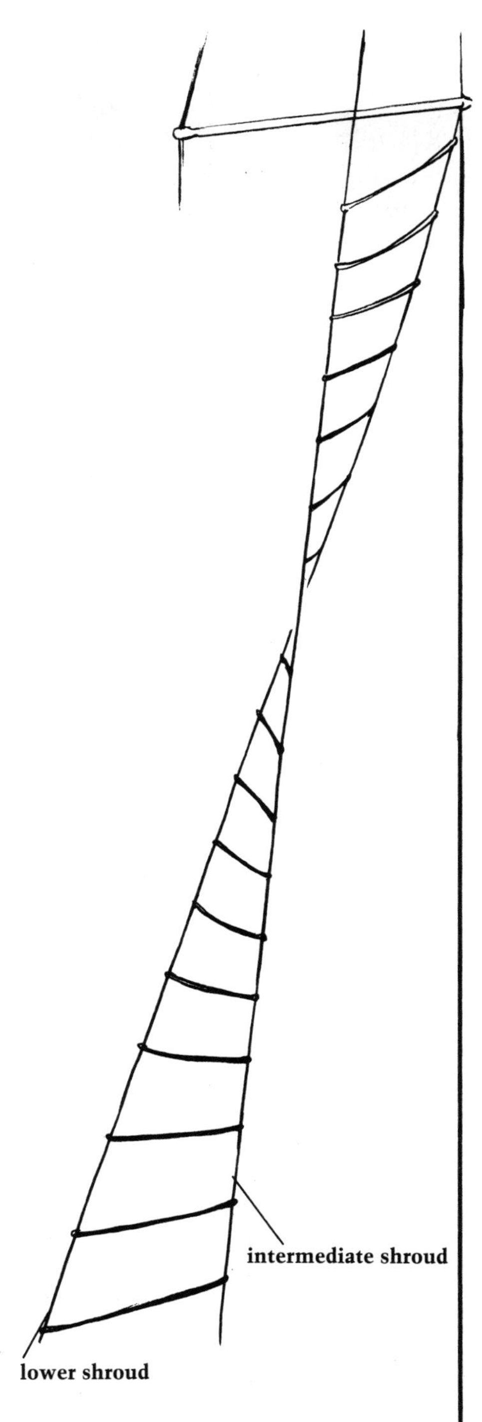

Figure 4-23. Forward lower shroud rattled to intermediate.

between the wires will get baby- foot wide well down from the spreaders. One option to gain a little footroom is to slack the shrouds and lash a grooved triangular block in place just below the seizing. The angle at the apex should be quite shallow to avoid imposing excessive lateral stress on the seizing. Even so, this option should only be used on low-stress rigs, and only with seizings, never with tangs or mechanical terminals.

Another technique is to rattle the forward lower and intermediate shrouds (Figure 4-23), thus escaping the "Narrowing Gap" syndrome altogether. This results in a helical ladder that is quite pleasing in appearance. Unlike mast steps, this ladder is never athwartships, even at the top. The only complication lies in leveling the rungs; your axis shifts as you go up, so you cannot use the horizon as a level reference. This is one place aboard where you can use a builder's level. Get the boat plumb, even if it means shifting gear on deck to balance your weight in the shrouds. And put a lanyard on the level.

A second variation, for vessels with topmasts, is to rattle the lowers until the gap narrows, then to jump over to the topmast backstay for the rest of the distance.

Sometimes the space between the lower shrouds is quite large at the bottom. If you're worried about a wood rung's ability to bear weight over a long span, install a sheerpole at the bottom to double as the first step (Figure 4-24). Or install a "jack shroud," lashed to the middle of the bottom ratline at one end and to the second or third wooden rung at the other (Figure 4-25). The jack shroud can be of ⅛-inch (3.5 mm) plastic-coated wire with a thimble in either end. Make it a little shorter than the space between rungs. Lash to the lower rung first, then pull all snug with the lashing to the upper rung. By Clove Hitching any intervening rope rungs to the jack shroud, you support

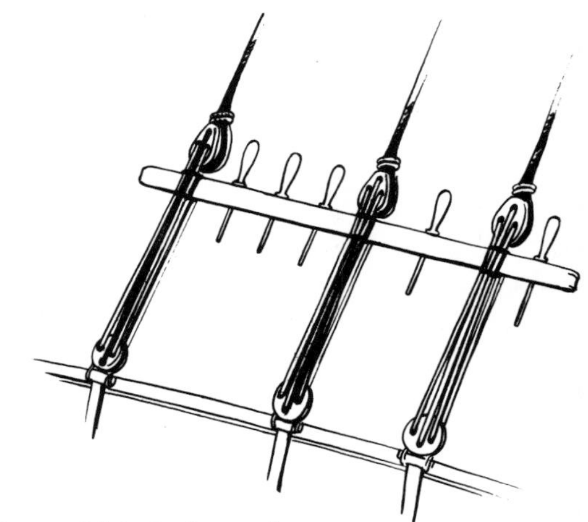

Figure 4-24. *A sheerpole on a traditional rig, lashed under the deadeyes and between the lanyard legs. Lashings must be very secure to prevent the sheerpole from riding up and bearing on the wire eyes. If this happens, the service and even the wire strands can be chafed away.*

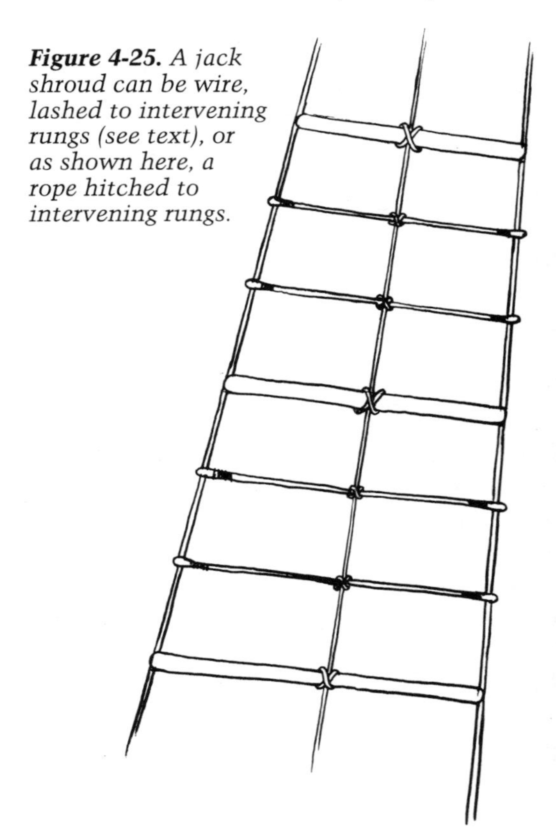

Figure 4-25. *A jack shroud can be wire, lashed to intervening rungs (see text), or as shown here, a rope hitched to intervening rungs.*

their middles, too. Traditionally, this knot is made with the crossing turn outboard, lower end leading aft. If the Clove Hitch slips on the plastic, make a Rolling Hitch instead.

On big, old-style vessels with multiple lower shrouds, rope ratlines are Clove-Hitched to the intervening shrouds, and any wooden ratlines must be lashed outboard. The forwardmost or "swifter" shroud is often left bare, as yards and lines chafe against that shroud. Likewise the aftermost shroud is often left unrattled if the middle shrouds provide a wide enough ladder for proper swarming. A nice touch is to run every fifth ratline to the aftermost or the swifter shroud, to get out of the way of traffic while conning or sightseeing.

With a large crew in a large ship, it is prudent to make the first several ratlines heavier than the rest, as there's usually a packed crowd at the bottom.

For vessels of any size, it's safest to climb ratlines when they're on the weather side, so they form a sloping ladder with the wind holding you on instead of a vertical ladder with the wind trying to knock you off.

A final aesthetic/practical note: One frequently sees wooden ratlines secured to shrouds with cable clamps. Lashed ratlines are graceful and tidy. Cable clamps are just a heavy, ugly, wire-damaging, shin-scraping way to avoid doing the job right. Avoid them and you'll avoid the dreaded "dancing-Swan-Lake-while-wearing-snowshoes" syndrome.

BLOCKS

Blocks, those ancient indispensable sail-control tools, have been high-tech'd. For so many years they were simple, stolid assemblies of wood, bronze, and steel. But nowadays they're likely to evidence Spielbergian gee-whiz design and obsessively close-tolerance engineering of multisyllabic plastics.

And more has changed besides materials and appearance; old-time blocks were mostly variations on a simple theme, but modern ones are specialized as to function, load, and (often novel) rig detail. On a typical production boat today you'll find spreacher blocks, two-line turning blocks, ratchet-sheave blocks, backstay adjusters, etc. And old-style blocks, adapted and updated, are still very much in the picture. Far from being outdated, they add to the rich variety of hardware that today's sailor has to choose from. What follows is information about the details of blocks, old and new, to help you pick the gear that best suits you and your boat.

The Body of the Block

There are two components to every block: the body, or shell, and the sheave. It is helpful to consider these separately.

The body is a combination of sheave housing, load bearer, and rope protector. In a traditional block, all of the load is borne by a metal strap that runs down the inside of the block, on either side of the sheave (Figure 4-26). Wooden "cheeks" are set outside of this strap, and are fastened together with spacers, called "swallows," in between.

Modern blocks are aluminum- or steel-cheeked to save weight and bulk. On some models the cheeks are a weight-bearing structure, but most often that job is done, as with traditional blocks, by a metal strap, though in this case the strap is exterior (Figure 4-27). The strap is preferably of type 316 stainless steel, a particularly corrosion-resistant alloy. Bronze straps are also excellent, since they are both corrosion- and fatigue-resistant, but many sailors just like the look of shiny steel.

When choosing between wood and metal bodies, the first consideration for many sailors is weight—will the cumulative mass of a lot of blocks make the boat topheavy, compromising sailing efficiency? On pure race

Figure 4-26. A Golden Dove block.

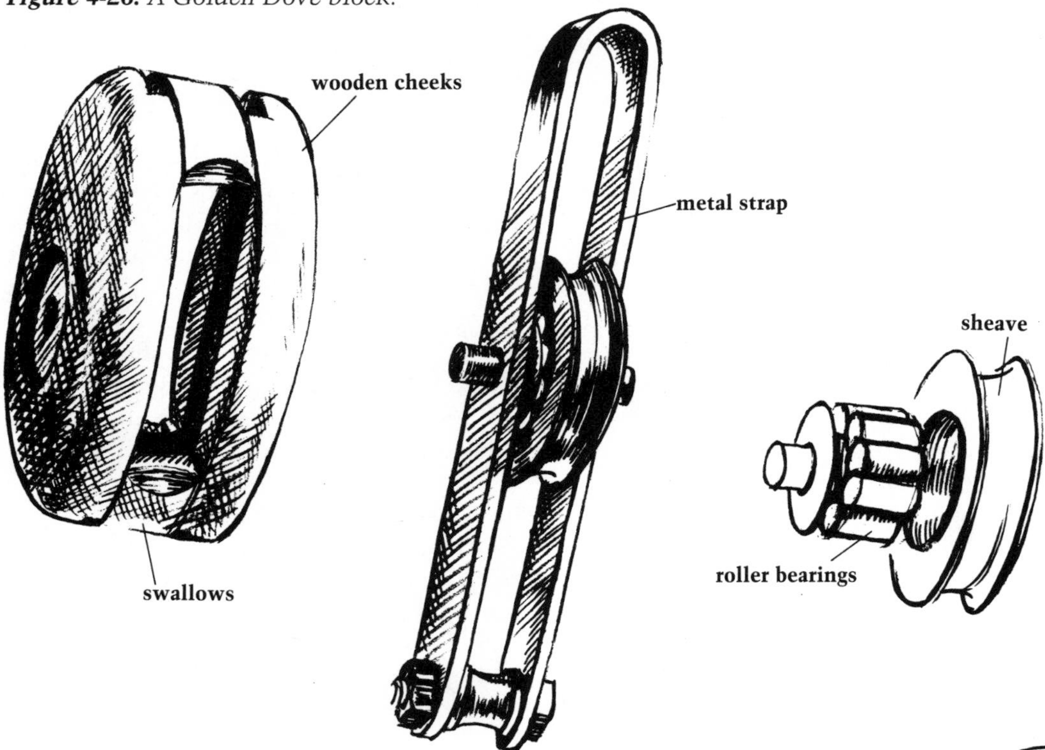

boats, at least inshore, the answer is a qualified "yes"; you want the nearest thing you can get to a weightless block, as long as it has enough meat to hold itself together.

On other than pure racers, lightness becomes less of an issue. Cruisers, for instance, prize toughness, and toughness means weight. And it turns out that too little weight can have an adverse effect on sailing efficiency; you can reduce a vessel's roll moment of inertia (essentially, the amount of force it takes to start a vessel rolling) to the point where every little puff of wind kicks at the boat and shakes the sails. The jerky motion of an underweight rig, exacerbated by stiff modern hulls, is also hard on crew and shortens rig life. As C.A. Marchaj points out, (*Seaworthiness: The Forgotten Factor*, Adlard Coles Nautical, 1986), typical rig weights have been reduced by almost 30 percent

Figure 4-27. A Schaefer block.

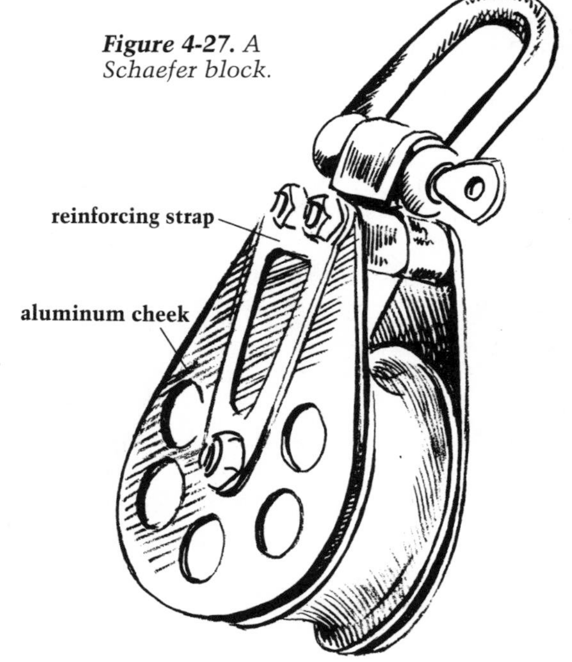

in the last two decades with nothing like a corresponding increase in performance.

Owners of classically styled vessels in particular are faced with a significant nonstructural consideration: appearance. It is natural and pleasing to complement a classical hull with classical, wood-shelled blocks. But owners of any kind of sailboat will find that these blocks are more easily repairable and are easier on your mast and decks than blocks with metal bodies.

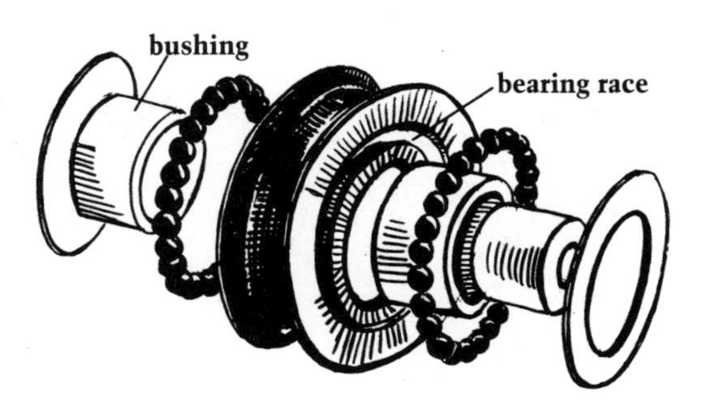

Figure 4-28. A Schaefer block with N.T.E. bushing and thrust- load ball bearings.

The Heart of the Block

The sheave, that little grooved wheel, that axle-transfixed puck, is the heart of the block. It exists to direct and share the load on a line. But every time a line runs over it, friction siphons off some of the force that the line is trying to deliver. The amount of friction at design load level can vary from 2 to 10 percent per sheave, depending on bearing efficiency. For a 100-pound load, this translates into 8 to 40 pounds of extra force you must exert to overcome friction in a typical four-sheave mainsheet. Technology has proven most useful in the effort to minimize friction.

Sheave friction is generated at four bearing points: where the rope passes over the grooved edge; the two sides of the sheave where they rub against the side of the block, and where the hub of the sheave bears against the block's axle. Given a fair lead and an adequate-diameter sheave (see below), it is the axle bearing that generates the most friction.

Bearings The simplest axle bearing, called a "bushing" or "plain bearing," is essentially a reinforced hole. It is closely matched to axle size to assure the broadest distribution of load on the pin, and thus less friction and wear, as in Figure 4-28. Today's slickest (3 to 5 percent), most expensive bushings are made from an epoxy-coated carbon fiber–reinforced blend of Nomex and Teflon (N.T.E.). This exotic mix is durable, stable under load, and actually produces less friction with use, as bits of it are smeared onto the axle. Most manufacturers, including Harken, Schaeffer, and Mariner, offer N.T.E. bushings. Bear Blocks, of Port Townsend, Washington, specializes in a similarly slick, stable bushing made of graphite-impregnated nylon. A nylon bushing without graphite is too soft and sticky for all but the lightest loads.

Delrin, an acetal resin, makes a good, inexpensive, medium-duty bushing. And bushings made from oil-impregnated bronze are another good, low-tech option. They're not as slick as the plastics (5 percent), but hold their shape well under static loads.

More Elaborate Bearings Ball, pin, or roller bearings reduce friction by reducing the surface area of contact with the axle and/or block sides (Figures 4-26 and 4-28). Bronze roller bearings, a specialty of Golden Dove Marine, are simple and bulletproof enough for prolonged deep-water use, yet slick enough (3 to 5 percent) for the performance-minded.

But racing gear, and even the light-air gear on a cruiser, demands the absolute minimum of weight as well as friction for maximum ease, speed, and smoothness of adjustment. Shearologists have come up with some exquisite variations on bearing themes for these applications. Lewmar makes stainless steel roller bearings for free-turning toughness. Harken makes its bearings from a pricey plastic called Torlon—axle roller bearings, and two sidewall ball bearing races for thrust loads—and gets resistance under moderate loads down around 2 percent. Schaeffer's Circuit Sheave uses an N.T.E. bushing for the axle and keeps the thrust-load ball bearings, making for a simpler, less expensive sheave with comparable friction characteristics.

Plastic ball, needle, and roller bearings are extremely light and easy to maintain (just keep them away from solvents and grit and rinse them regularly with fresh water), but they will distort under heavy, static loads. This is most often a problem with halyard sheaves for mainsails and jibs, a good spot for metal roller bearings or perhaps an N.T.E. bushing if weight is an anorexic issue.

Sheave Material Loads on the sheave itself are not so concentrated as those on the bearing, but sheave material can be important relative to cost, weight, and stability under load. Nylon, for instance, is very cheap and light but can distort under heavy load. Worse, it swells when wet, increasing sidewall friction. Delrin sheaves are more stable and nearly as cheap, making a good medium-duty sheave. Bronze, stainless, and aluminum sheaves are heavy and expensive but will stand up to extremes.

Sheave Shape A consideration peculiar to modern blocks is that of sheave score profile (Figure 4-29). It used to be that all sheaves had a semicircular profile, and that shape is still appropriate for three-strand and standard braided lines. But the introduction of Kevlar-cored lines has necessitated the "flat-bottom sheave." The reason for this is that when a line passes over a sheave, the fibers at the sheave surface travel less distance than the ones farther away. The elasticity of most rope fibers, natural or synthetic, compensates for this difference. But Kevlar is both extremely inelastic and extremely fatigue-prone. With a flat-bottomed sheave, the Kevlar fibers can spread out and flatten, so they all travel more nearly the same distance around the sheave. Even so, the sheave must also have a relatively large diameter to reduce internal friction and fatigue still fur-

Figure 4-29. *Sheave score profiles.*

ther. All major manufacturers produce Kevlar-scored sheaves.

With wire-to-rope halyards, the long-accepted standard was a semicircular profile with a notch cut out of the bottom. The semicircle was for the rope, the notch for the wire. Because serious loads only come onto these halyards after the rope part is clear of the sheave, the little notch doesn't tear up the fibers. All-wire halyards simply used a very skinny sheave with a semicircular profile. But the modern V-groove sheave supports the wire much more completely, reducing fatigue and the onset of "meathooks." And it's even easier on fiber rope.

Most sheave-related problems occur when sailors change halyard materials or diameters without changing sheaves. Running Kevlar over a normal sheave is asking for trouble, and so is running an all-rope halyard over an old-style wire- and-rope sheave (when the sail is fully hoisted, the rope will chafe away in the notch). A different problem can occur if you do replace a sheave to match halyard type, but the new sheave doesn't fit snugly in the block or mast mortise. The halyard can chafe on entrance and exit points or the top of the mortise, or can jump out of the sheave groove and jam between the sheave and the mortise wall, an eventuality that can be prevented with "keeper bars" (Figure 4-30).

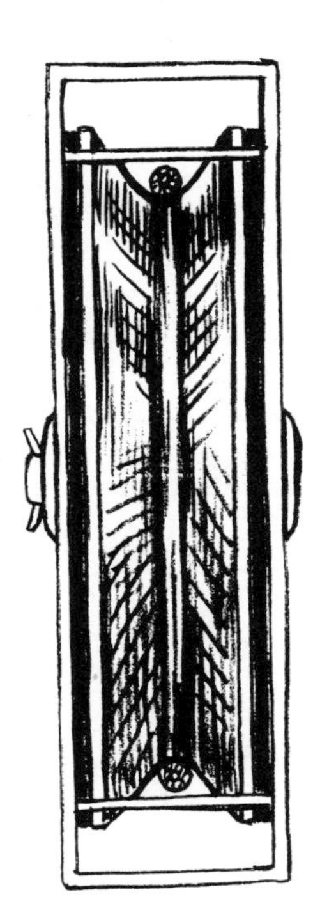

Figure 4-30. *Keeper bars can be welded on across a halyard sheave opening to prevent the wire halyard from jumping out and jamming between the sheave and mortise walls.*

Sheave Size Related to the question of sheave score profile is sheave size relative to rope diameter; different materials are more or less susceptible to fatigue, and increased sheave diameter reduces fatigue by reducing the sharpness of the bend rope takes in passing over a sheave. Standard three- strand and braided rope is happy on a sheave four to six times the rope's diameter—i.e., a ½-inch (13 mm) line needs a 2- to 3-inch (52 mm to 78 mm) diameter sheave. As mentioned above, Kevlar needs a larger sheave; rope manufacturers recommend that sheave diameter be at least 12 times the rope diameter. And wire rope should have a sheave at the very least 20 times the wire diameter.

To get an idea of how much difference this ratio can make, take a look at Figure 4-31, which shows the effect of sheave diameter on wire-rope life (graph courtesy of the MacWhyte Wire Rope Co.). Wire life is almost doubled in going from a ratio of 15:1 to a ratio of 20:1, and more than tripled between 15:1 and 25:1. A larger sheave, because of its gentler curve, also produces less surface friction as the rope passes over it. Furthermore, axle and thrust bearing friction is relative to the ultimate strength of the sheave; by using an oversize sheave you reduce the relative load, and thus further reduce friction. (This is the source of that old rigger's blessing, "Big blocks and small lines to you.")

Considering all the above, it obviously makes sense to use the biggest sheave you

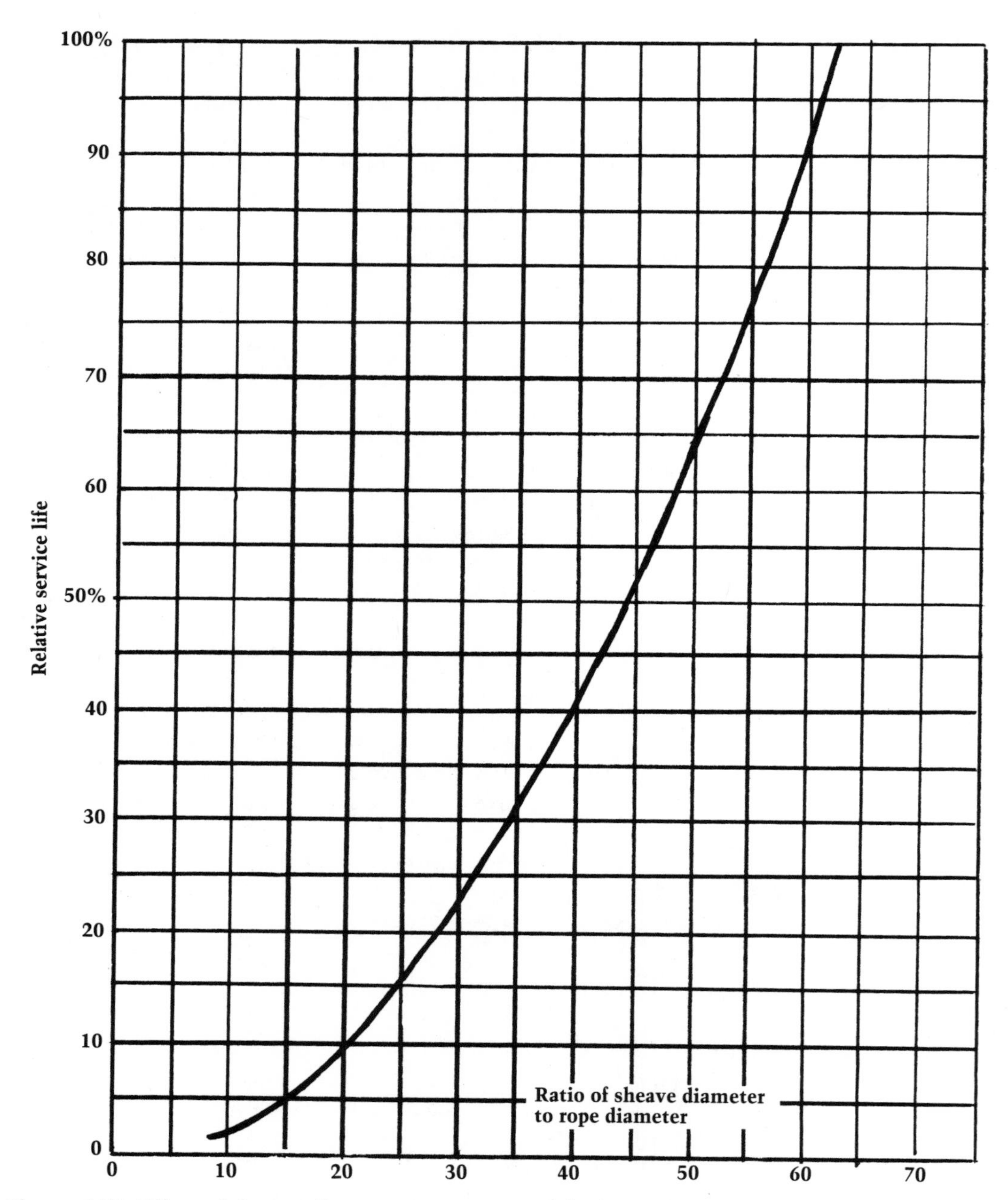

Figure 4-31. *Effect of sheave diameter on wire-rope life. (Source: MacWhyte Company handbook)*

can, no matter what your halyard material. But there are practical limitations. For one thing, bigger blocks cost a lot more than smaller ones. For another, an oversize block might just plain not fit in a tight space, especially at the masthead, trapped between mast and stays. And of course big means heavy. Altogether, we're left with the com-

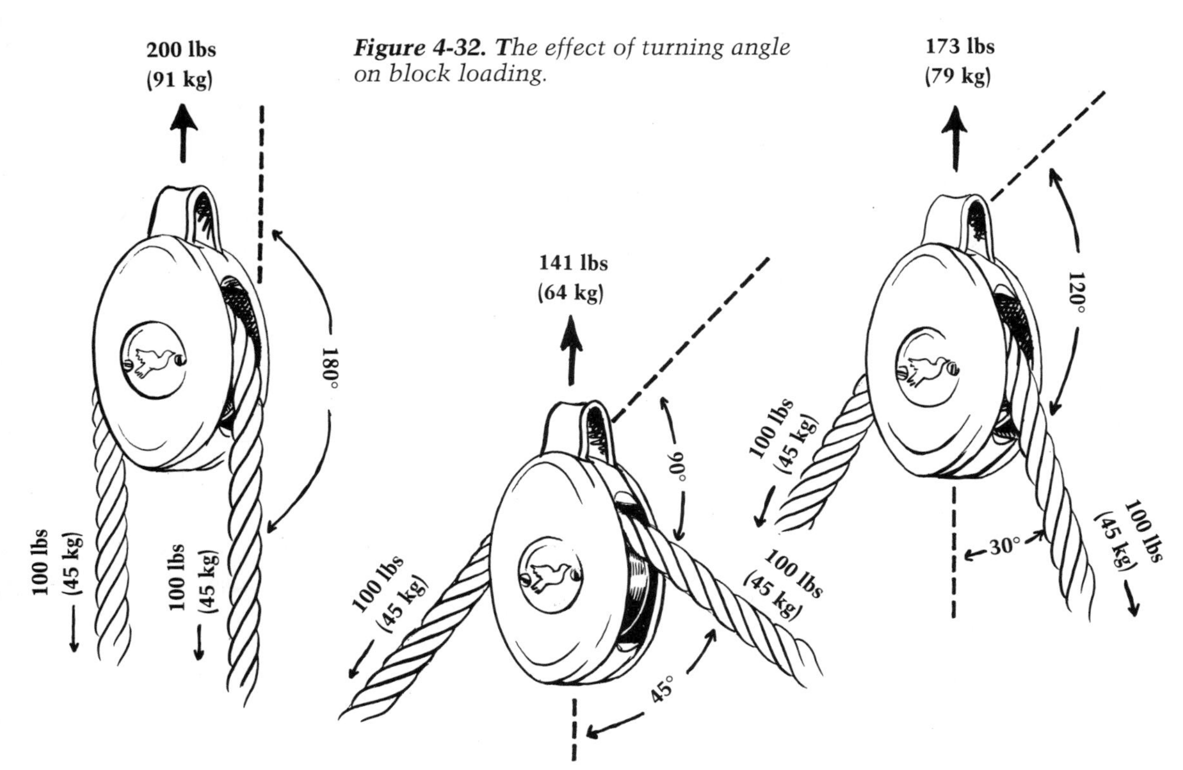

Figure 4-32. *The effect of turning angle on block loading.*

promise of using the biggest sheave that will fit, is light enough, and won't bankrupt us.

Block and Rope Strength One more consideration: block strength relative to rope strength. In the days of Manila and hemp, blocks were built to match the strength of the rope that fit them. But since synthetics are at least twice as strong as natural fibers and can be used safely at a lower safety factor since they are less prone to rot, this is no longer practicable. For example, take a modern block with a 2-inch (50 mm) diameter sheave and a breaking strength of 2,500 pounds (1,136 kgs), designed for ½-inch (13 mm) line. That line has a breaking strength of 6,300 pounds (2,864 kgs). Apply just one-third of that line's breaking strength and you'll have a load of 4,200 pounds (1,909 kgs) on the block, 2,100 pounds (955 kgs) on each side. You'd have to go down to an uncomfortable-to-grip 5/16-inch (8 mm) rope before block strength would be in scale with rope strength. But nowadays rope size is scaled for lowest stretch range, ease of handling, and fitting cam stoppers and winches. You'll probably never put even a 1,000-pound (455 kgs) load on that ½-inch (13 mm) line. So instead of trying to match block and rope strength, the procedure is to match block sizes to design loads—the actual loads they'll bear. This is most easily done from rope and block manufacturers' tables, available in catalogs and chandleries.

Also note in Figure 4-32 that as the angle of incidence of a line to its block widens, the load on the block lessens. So fairlead blocks can be considerably smaller than blocks through which the rope takes a 180-degree turn.

Miscellaneous Practicalities

Cruisers, for whom a broken block can be more than a simple inconvenience,

Figure 4-33. Snatch blocks.

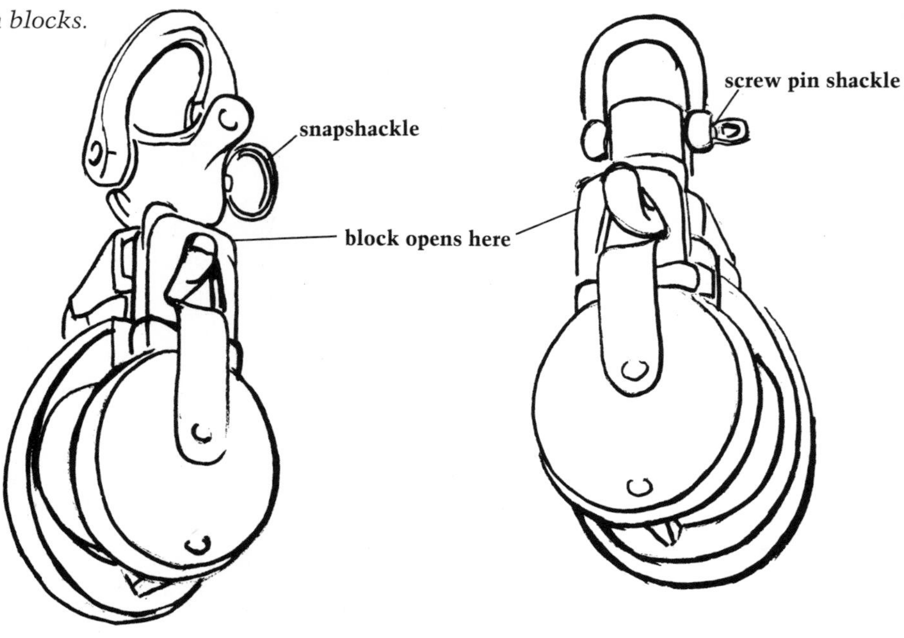

should always use significantly oversize blocks for a high safety factor. Reduced friction and prolonged rope life are other consequences of prudent oversizing.

All sailors should bear in mind that swivel blocks fail most frequently; use front-bail, side-bail, or upset-bail blocks whenever possible, and use extra-stout swivel blocks when you have to use them at all.

Poor lead is the primary cause of block failure; be sure that all lines lead fairly to and from all blocks for even strain and low friction.

Snatch blocks (Figure 4-33) can be stupendously handy for impromptu fairleads and hoisting, but they do have their limitations. Avoid using them on a line that is subject to violent flogging—like genoa sheets—as this can cause them to spring open. Holding a snatch block up with a short bungy cord will help hold it still while you open and close it, will keep it from clattering on the deck when the load is off, and will help to dampen motion from a flogging line.

Nearly all snatch blocks available today come with a snapshackle attachment. So flogging or inattentive closing can result in not only the block but also its snapshackle coming open. One form of snatch block, developed for mountain rescue, is positive-locking (Figure 4-34). It's a little slower to open and close, but you can trust it with your life. It's perfect for use as a turning block when hauling someone aloft (see "Living Aloft" below).

With that advice, I'm edging away from changeable technology and toward a timeless, personal responsibility: that each sailor assess and make skillful use of the available gear. One can, for example, combine light, free-running plastic sheaves with tough, lovely, wooden blocks if that's what's needed or preferred. No matter how high-, low-, or middle-tech the gear is, it must be mated with given jobs on a given boat, and it is the sailor who plays the matchmaker.

Figure 4-34. *Mountain climber's block, for use as a turning block on a halyard that is taking someone aloft or in other mustn't fail situations. A screw pin shackle or locking carabiner both holds the block closed and connects it to its attachment point. Available at climbing and camping supply stores, in aluminum with a plastic sheave or stainless with stainless sheave.*

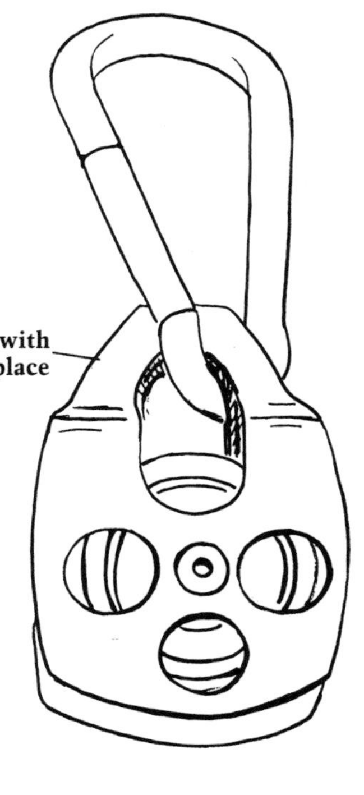

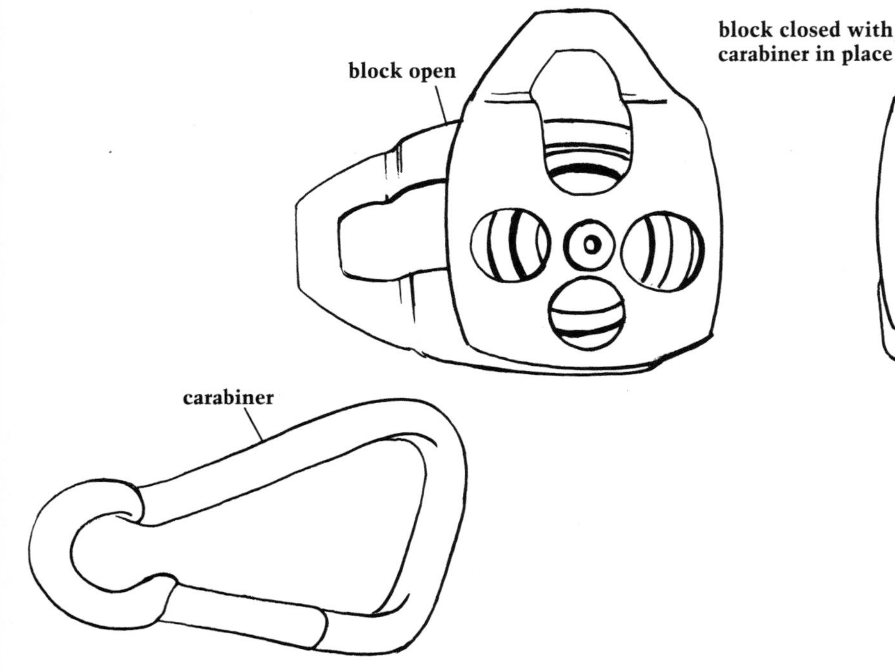

LIVING ALOFT

Once I was 100 feet or so aloft, up by the main truck of the bark Elissa, installing some gear. Discovering that I hadn't brought enough shackles, I called down to the deck for extras. A new hand ran to get some, ran back with them, then smartly cast off what he thought was an unused gantline (messenger line). The chair dropped out from under me, and only a primal grab at the backstays kept me from following it down.

And once a friend was doing a routine masthead light replacement on a big sloop. His chair was hanging from the hook of a dockside crane. When he was finished, my friend signaled to the crane operator to bring him down. The boom swung away from the boat and stopped over the dock. Just then someone from the shipyard office called to the driver, who turned to respond, inadvertently releasing the halyard clutch, causing the chair and its occupant to begin an unnoticed freefall. A chance peripheral glance by the driver and a quickly applied brake saved the day, the chair stopping so close to the dock that my friend just stepped out of the chair and walked, unsteadily, away.

When working aloft, gravity is your enemy. But near-death dramas are almost invariably the result of poor planning, poor

communication, poor attentiveness. My near fall was a lesson to the deckhand: Look aloft before you let go a line, to be sure you have the one you want. Then take the turns off slowly, so you can feel and control any surprise load before it's too late.

My near fall was a series of lessons for me: Ensure that things on deck are in good, organized hands before you go up; go up on two halyards in case one fails; tie safety tethers to something solid once aloft, for additional backup support; and bring lots of spare shackles.

My friend on the crane had the advantage of being on gear far stronger and more easily and precisely adjusted than a sailboat halyard. On the other hand, the operator was the only thing between safety and a Wile E. Coyote–style landing. Redundancy, however primitive, might be preferable to helplessness.

So much for the scary stories and general lessons; now let's get down to the details that can make life aloft a pleasant, relatively safe experience.

First of all, preface any job with an on-deck conference in which you go over the job in detail, including likely material needs, alternative scenarios in case things go wrong or differently than you expect. (For example, what will you do if a tool falls in the water? Do you have a magnet? A spare tool?) You can also take this time to be sure you agree on nomenclature and hand signals, to avoid confusion later. And you definitely want to be sure that you've allowed a generous amount of time to do the job and that your deck crew can hang around that long. Ever been stuck up a mast at suppertime?

Next, lay out the primary halyard, the one your deck crew will be hauling on, in a way that will keep them out from under you, both to protect them from falling gear and so that they can see you clearly at all times. If the hauling part of the primary halyard is forward of the mast, lead the halyard to the foredeck. The crew can pull by hand or use the anchor windlass. If the hauling part is aft of the mast, lead it through deck blocks toward the cockpit (Figure 4-35) for either

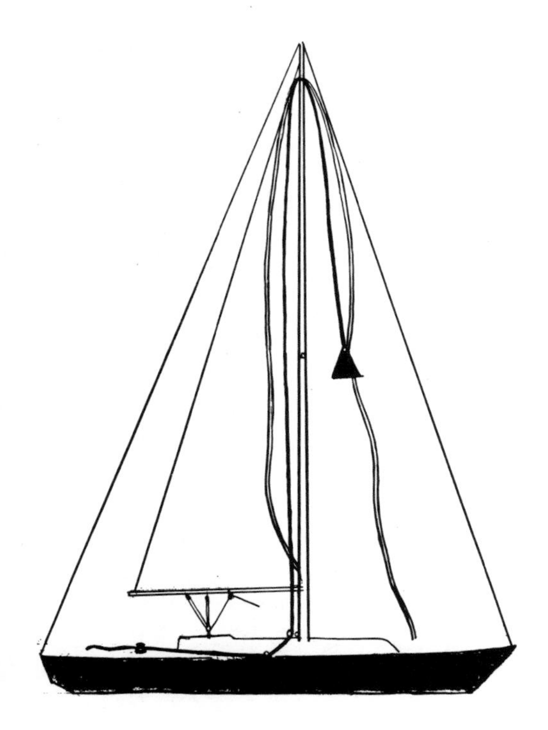

Figure 4-35. *The primary halyard leads from the chair, through the masthead sheave, down to the deck, and aft to a sheet winch. A second halyard belays at the base of the mast and leads through the masthead sheave and down past the chair. The occupant of the chair goes hand over hand up the secondary as the chair is hoisted, in case the primary should fail. Once aloft, the secondary can be tied to the chair as a safety line, or used for self-lowering.*

hand power or one of the sheet winches.

Next, lay out a second, safety halyard. You will be handling this one in case something goes wrong with the primary. One way to arrange a safety halyard is to belay one end of a spare halyard and lead the hauling part to your chair. You just hand-over-hand up this halyard while the deck crew does all the work. Alternatively, you can attach one end of the safety to your chair or harness (see below) and haul on the other end. This is a little more complex to set up, but gives you a 2:1 mechanical advantage, easing the deck crew's job. If you are strong enough to haul yourself aloft with this configuration, then this can be the primary halyard, and the deck crew handles the safety, only needing to keep the slack out and a turn around a belay in case you should tire or slip.

With either safety halyard configuration, lead the hauling part through a carabiner that is clipped to your chair or harness. Then, even if you temporarily lose your grip, the hauling part can't fly out of reach. And with either configuration, belay the safety halyard to the chair, or, better yet, to your safety belt once you are in position aloft.

When the halyards are squared away, check all your gear, including blocks, shackles, winches, and chair and harness. Make sure that no snapshackle or self-tailing winch is part of the system; snapshackles sometimes do, and self-tailing mechanisms sometimes don't.

A block-and-tackle with a very long tackle is one way to make it easier to get you aloft, but it involves a lot of line, and the blocks mean that you can't get very close to the top of the mast. Leading the halyard to a winch is usually preferable when you need mechanical advantage.

Whether the hoisting power is manual or machine, the most important person on deck is the one tailing the line. This individual keeps the line around a belay point and takes up the slack as it comes in. If the people or machinery doing the hauling should slip, it is the tailer who will check your fall. The tailer watches you all the way up, controlling the speed of your ascent and watching for any trouble, while simultaneously keeping everyone out from under the mast and shushing needless noise. If you have a request or a problem, you address it to the tailer, and the tailer is the only one who answers. A prestigious job.

With gear and crew set, all that remains is to test the halyards. The drill is to take up on the safety until it bears most of your weight, then bounce hard on it a few times to make sure it will hold. The shock load you're imposing here will be far heavier than any normal load. So if the halyard and blocks hold for the bounce, they'll probably hold for the haul. Ease off on the safety, have your crew take up on the primary, and bounce on that. Just in case something aloft does let go, do this out from under the mast.

Going Up—In What?

You are about to ascend to dizzying heights; what are you sitting in? The traditional plank-and-rope bosun's chair is a marvel of simplicity, economy, utility, discomfort, and danger. With the addition of the back and crotch straps shown in Figure 4-36, the plank bosun's chair becomes something like safe, and it's certainly the cheapest way aloft provided you know how to splice and seize. But a good canvas chair with wide adjustable back strap and leg straps and with handy built-in pockets is altogether a better way to go.

Figures 4-37 and 4-38 show a couple of the chairs available on today's market. The Hood is a big, padded Cadillac of a chair, as opposed to Lirakis's Ferrari. Most bosun's chairs fall somewhere between. When you go shopping for one, pay attention to fit, just as you would for clothing. Not only should the width and depth be right for your body—tight enough to keep you from sliding around, loose enough that you don't lose circulation—but the chair and its appointments should feel right, with no D-rings in your face, no hard-to reach pockets, no hard-to-adjust straps. Sit in it while it's suspended.

If you're satisfied with the cut of the chair, try to tear it apart. I mean pick it up and haul on all the load-bearing seams really hard. Try to be discreet about this, but remember that you will literally be trusting your life to this item.

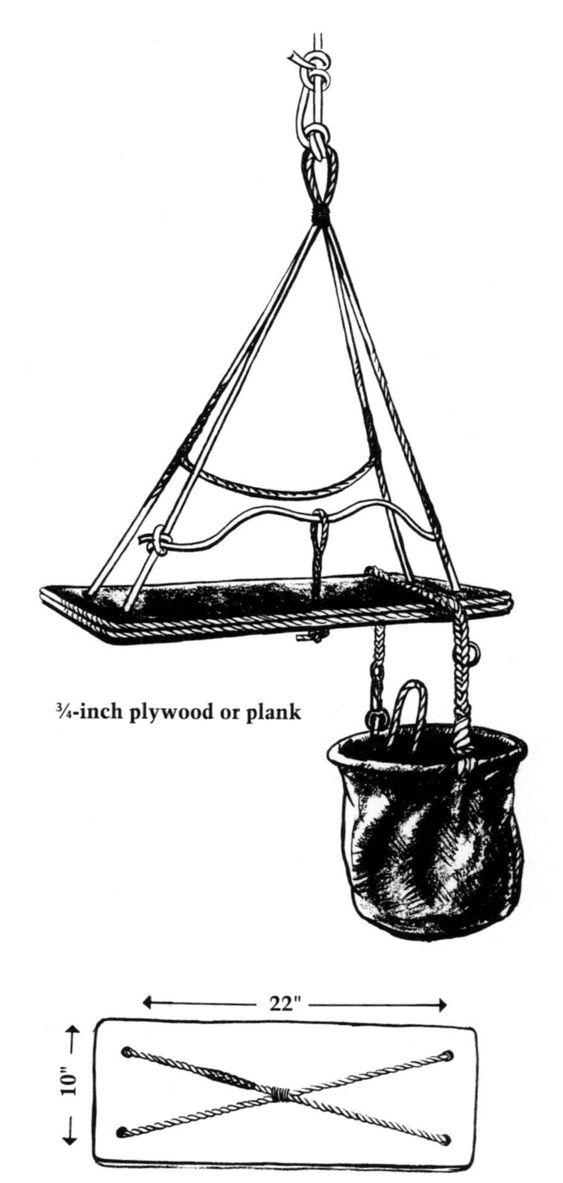

Figure 4-36. *A traditional plank chair with a few new wrinkles: Roped sides are easier on legs and mast; spliced-in back and leg straps provide security. Note in bottom view that the chair legs are crossed and seized so that one broken part will not drop the occupant. The rigging bucket has a shackle on its lanyard for hanging around mast or spreader, with an extra lanyard (bight showing) for intermediate attachment while the main lanyard is being passed.*

Figure 4-37. *Hood makes a well-designed, comfortable canvas sling chair, with large pockets, lanyard rings at front lower corners, back and leg straps, and leather-covered seat. Halyard attachment is an Anchor Hitch backed up by two half-hitches (left loose in drawing for clarity). Shackle pin is moused to prevent unscrewing. Small block on shackle is for gantline.*

Figure 4-38. *Lirakis makes an ultralight, ultrastrong, all-nylon chair. The seat is epoxy-impregnated for stiffness. An optional rigging bucket (shown) fits over the legs on one side.*

While not exactly a bosun's chair, a mountaineer's climbing harness is a handy substitute, at least for quick jobs, where you can expect to get back to deck before your legs go numb. It's extremely prudent practice to wear a harness even when you're using a chair; attach the safety halyard to the harness so you'll have two independent halyards leading to two independent seats. Reassuring. The best climber's harnesses have broad straps and built-in loops for attaching gear to (shackles, bits of spare line, tools, etc.). I particularly like R.E.I.'s Pinnacle harness for its good price, comfortable fit, and the fact that the halyard attaches to two loops for redundant safety.

En Route

Your deck crew's good, your gear is sound, you're ready to travel. Give the command to haul away and head on up, nice and easy, but concentrating like a diamond cutter. Watch out for mast fittings, wires, spreaders, and anything else you or your gear might snag. Keep an eye out for inconsiderate boobies who motor by too quickly, kicking up a wake; motion is amplified up here, and a wake that is merely annoying on deck can really thrash you around. If you have any problems, let your tailer know, but avoid unnecessary chatter.

If you can't avoid going up while the vessel is heeling or pitching, shackle your chair or harness around a tautly belayed halyard (Figure 4-39). If you only need to go up as far as the spreaders, use the forestays'l halyard for the hoist so that there's less halyard length above you to contribute to pendulum length.

In flat weather or calm, stay completely present mentally as you go up, and don't be shy about implementing extra security precautions, such as having an extra hand tail your safety so that you can hang onto the standing rigging as you go up, just in case

Figure 4-39. *To avoid becoming a human plumb bob while going aloft on a pitching vessel, shackle your chair or harness around a taut second halyard. Note that the snap shackle on the hoisting halyard in the illustration is not being used; always use a positive-lock shackle for hoisting humans. This mate is going aloft in a Lirakis mountain climber's–style harness, which can be used instead of a bosun's chair for short trips. Such harnesses provide great mobility aloft and are impossible to fall out of.*

both halyards fail, or, more likely, in case you need to stabilize yourself against rough motion. All of these precautions would have seemed excessive to old-time sailors, who thought nothing of single-halyard-and-plank-chair ascents, but bear in mind that they were in big, stable ships, hanging on heavy-duty halyards, with plenty to hang onto aloft and a professional deck crew alow. And even then they were taking chances, as the odd fall from aloft attested. Today we have smaller boats with much more severe motion. And one hopes we're smart enough not to scoff at safety procedures.

Procedures Aloft

Once you get to your work station, have the deck crew belay the primary halyard and tell you when it is fast. Then tie or shackle a short tether from your harness or chair to a fitting on the mast, or clear around the mast if you can (Figure 4-40). Now you're held in place independent of the halyards. At this point you can have the deck crew belay your safety as well. Or, if you want to be able to lower yourself, Carabiner Hitch the hauling part of the safety to harness or chair. Then tell the deck crew to ease off slowly on the primary halyard. If the Carabiner Hitch is secure, they can cast off the primary, which you can now use for a gantline (messenger line) for sending equipment up and down. If you don't need to lower yourself, leave the primary attached and use another halyard as a gantline. If you've run out of halyards, hang a block from your chair or around the mast and reeve a light halyard through it.

Sometimes, as for changing a masthead light or installing instruments, the places you need to get at are just out of reach above you. Because the halyard attaches lower on a climber's harness, you can almost always get up higher with them than with a chair. But if the job you're doing will take any length of time, you will get real uncomfortable before you're done, and might even have to come down to rest occasionally. The alternative is to go up in a chair and harness, and once in place, to stand up in the chair. This is a very dangerous procedure, but there are ways to make it much safer.

Figure 4-40. *Once aloft, secure yourself to the mast to prevent getting swung about and as insurance against halyard failure. This rigger shouldn't be up there without a safety halyard.*

First of all, have two tethers on your harness, one about 4 feet (1.2 m) long, the other about 2 feet (0.6 m). When you get aloft, pass the long one around the mast and clip it back to your harness. Then clip the 2-foot (0.6 m) tether to a tang or block fitting. Then grab the upper shrouds and haul yourself up so you can stand on the chair. A short "stirrup" hanging from the chair, with a loop for your foot, makes a convenient step. Once you're standing, you can lean back against the longer tether, lineman-style, with the short tether and the safety halyard as backups. But keep your feet firmly against the mast on either side to prevent the chair from sliding. If necessary, shorten the long tether for a snugger fit. Try this entire procedure on deck first.

Seated or standing, you're ready for work; all you need are some tools and materials.

Working Aloft

The gear you need will be coming up on a gantline. You could have brought it up with you, but why add all that weight and clutter to the exercise? If you're going to send the primary down as a gantline, stop and tie a Figure-Eight in the Bight or other loop knot in the line, about 2 to 3 feet (0.6 m to 0.9 m) from the end, before you detach the line from yourself. The knot will prevent the halyard from accidentally slipping out of the block (Figure 4-41). Now tie a Bowline with the end around its own standing part, so that the end won't blow away out of reach as it goes down. It is sometimes convenient to tie the end around a jibstay, backstay, or other piece of rigging instead. In any event, send the line down to deck. If you'll be using another, extra halyard that already has both ends on deck, have your deck crew put the loop knot in.

The first thing to come up will be your rigging bucket. This should be a stout item, preferably of canvas and with a stiff rim, with tool pockets on the inside so the tools can't snag every little thing on the way up. The bucket should be big enough to hold basic tools with enough room for paint or miscellaneous fittings—10 to 11 inch (250 mm-280 mm) diameter by a foot (300 mm) tall is a good size—and should have its own tether, so it can be hung from the mast without a halyard.

When the bucket gets up to you, tie its tether on, then detach the halyard and send it back down for the next piece of gear. If everything you need is in the bucket, leave the halyard attached, the other end belayed on deck.

The sure mark of a pro aloft is a profusion of lanyards. Lanyards for all the tools, a lanyard for eyeglasses, a lanyard for each piece of hardware you'll be working with, and two lanyards for heavy objects, plus

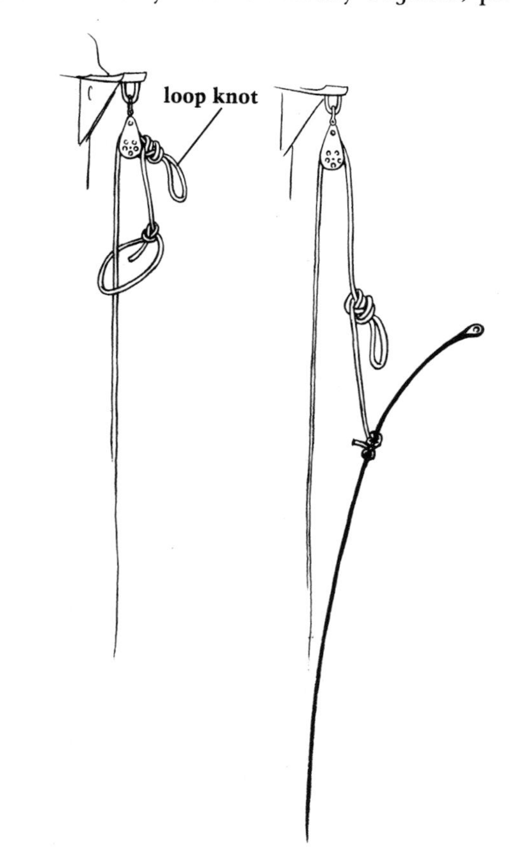

Figure 4-41. *Preparations for sending a line down from aloft are shown on the left. When sending items up, the deck crew ensures that it will not be necessary to untie the line before installing the object. Here (right), for example, a stay is coming up with the gantline Rolling Hitched to its standing part, not its end.*

BUTTERFLY KNOT—IMPROVED METHOD

The Alpine Butterfly Knot, also known as the Lineman's Loop, is a strong, secure, easily untied knot, suitable for hanging purchases from, looping over posts, or for any job where you want to make a loop without having to pull the end through. This usually means the knot is in the middle of a long line, but you can also make it at the *end* of a line if an Eyesplice-with-shackle makes it awkward to tie a Bowline.

For a long time, this knot's sole drawback was an awkward method of tying. It was particularly difficult to do in braided rope, because the method involved twisting, which makes braid ornery. The old method was also fairly difficult for most people to learn. The accompanying method addresses this drawback, proceeding in a falling-downhill fashion to an inevitable conclusion, *sans* twist.

To tie: Make three loose turns on the hand. Put the turn closest to your thumb over the middle turn. **(1)** Put what used to be the middle turn over the other two turns **(2)**, and then back under them **(3)**, to form the loop. Draw up after taking your hand out.

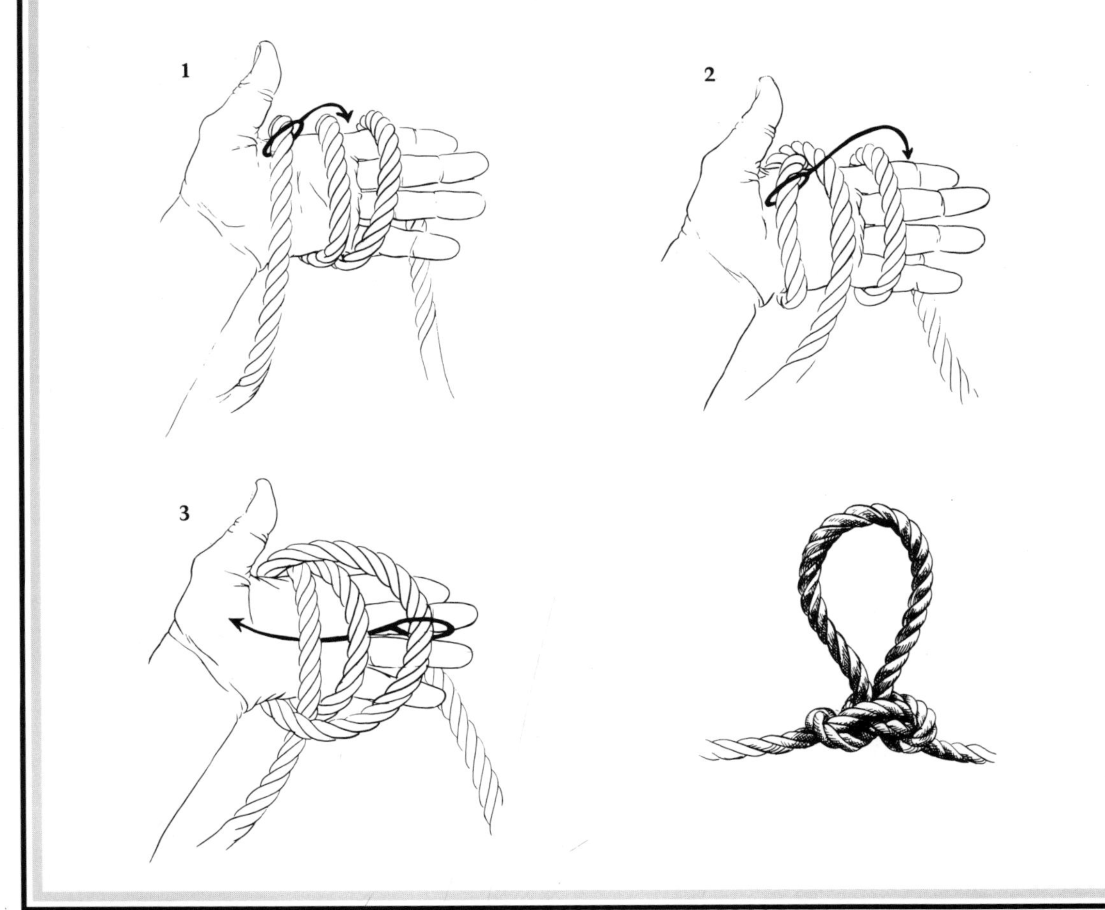

spare lanyards for any extra requirements that might come up, with each lanyard scaled to the weight of its lanyardee.

There will be times when you say, "I can't deal with all these strings!" But if the lanyards are getting in your way, you're probably trying to work too fast. It's a special world aloft; clearing and stowing tools and lanyards during each step of a job is a necessary ritual, one that will prevent hard, expensive objects from crashing down on crew or deck, bouncing overboard. Slow down, work on organization, and the strings won't be so intimidating.

MICHELLE, MA BELL

Take this idea aloft with you. The crescent-wrench head is welded to the marlingspike, and both this tool and the knife are secured to the sheath with lanyards. But here's the crowning touch: The "lanyards" are household telephone extension wires, belayed to tools with Knute Hitches. You get plenty of working lanyard scope without trailing long bights of twine through the rigging. I call this configuration Michelle, with apologies to the Beatles.

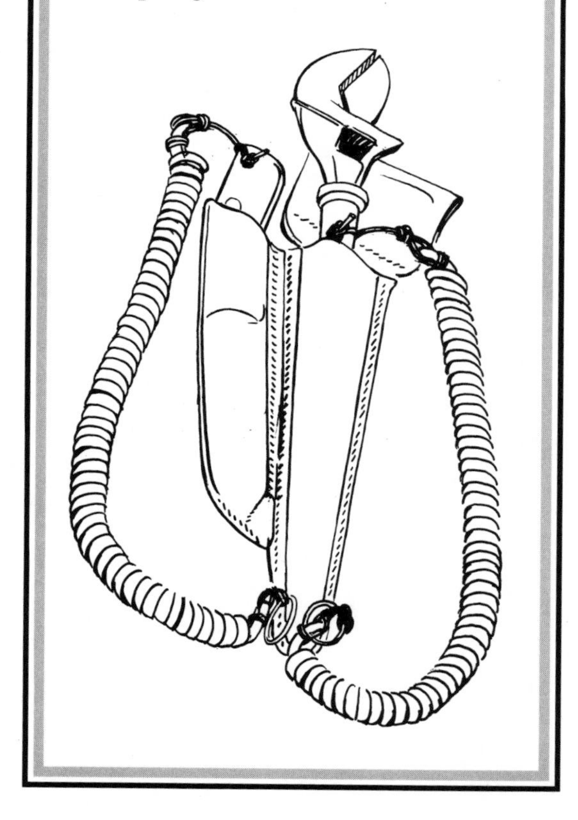

Some specific confusion-reducing tips:

(1) Distribution. Rigging buckets are always overcrowded; relieve the clutter by keeping some items, such as tape, seizing wire, and especially electrical tools, in the pockets of your chair. And wear a knife, spike, and pair of Vise-Grips on your belt or the belt of your safety harness. The gear loops on the harness are also a good place to keep spare shackles and lanyards.

(2) Lanyard sharing. Seize small rings to your rigging bucket just below the rim (Figure 4-42). Attach a lanyard

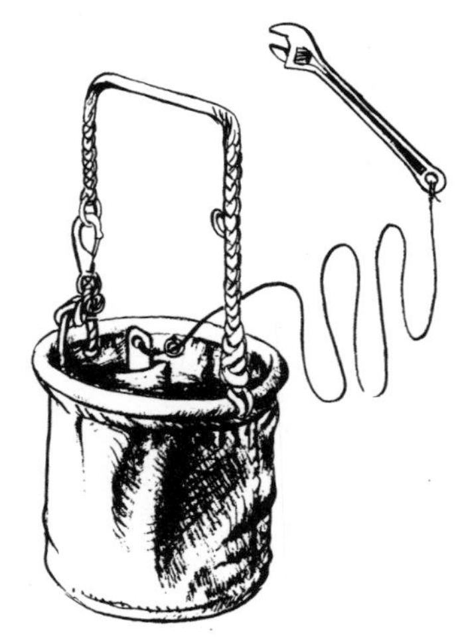

Figure 4-42. *To reduce the number of lanyards you must deal with, thread a single lanyard through a ring seized to the rigging bucket and attach ends to tools that you won't use simultaneously (in this case, adjustable wrench and putty knife).*

to a tool, reeve it through the ring, and attach it to a second tool, one you won't be using at the same time as the first. Likely pairs include a crescent wrench and chisel, hammer and file, hacksaw and screwdriver, etc. You'll be dealing with half as many lanyards, in complete safety. You can seize rings onto your chair, too, or just run lanyards through the halyard-attachment eyes. Or build a sheath that will hold both knife and spike—so their lanyards can't get tangled between separate sheaths—and seize a ring to the bottom of the spike sheath.

(3) Separating jobs. Say you're installing a radar aloft. You need to drill holes, run wiring, mount the bracket, and finally mount the radar. That's four jobs with four partly or entirely different tool requirements. Separate the job into two or more phases before you leave deck, laying out tools and gear in the order you're likely to need them. Then the deck crew can shuttle things up and down in the rigging bucket, instead of overstuffing it and leaving you to struggle with it aloft.

(4) Maximize versatility. While job-specific consciousness is important aloft, so is resourcefulness and versatility. Welding the head of a crescent wrench to your marlingspike (Figure 4-43) enhances the worth of that already priceless tool; half-round files are a two-in-one blessing; the positive-lock on Sears ratchet handles prevents dropped sockets; and an electric drill with a variable clutch can work for both making holes and setting screws.

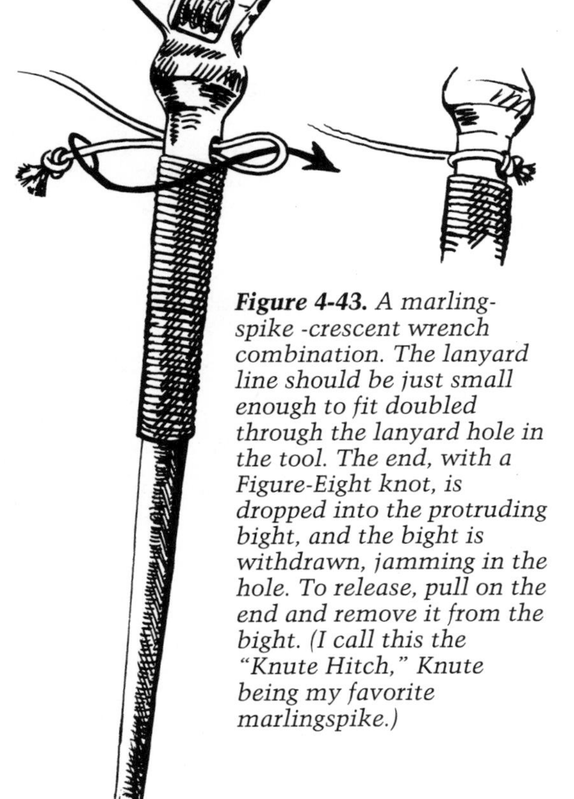

Figure 4-43. *A marlingspike -crescent wrench combination. The lanyard line should be just small enough to fit doubled through the lanyard hole in the tool. The end, with a Figure-Eight knot, is dropped into the protruding bight, and the bight is withdrawn, jamming in the hole. To release, pull on the end and remove it from the bight. (I call this the "Knute Hitch," Knute being my favorite marlingspike.)*

Miscellany

- Extremes of motion aside, the single greatest potential for annoyance and hazard aloft comes from how the deck crew attaches the gantline to the gear they send up. Let's say you're replacing a shroud. The hole in the end of the new terminal is the obvious place to tie the halyard. But when the wire gets to you, you must remove the halyard before you can install the wire. This involves either the bother of securing a separate lanyard or the hazard of casting off the halyard and trusting that you won't drop the wire before you get the clevis pin in place. A far better practice is to Rolling Hitch the gantline onto the wire 2 feet (0.6 m) or so below the terminal. The halyard will be out of your way, and you can get the clevis in before casting the halyard off. If the halyard is too large or stiff to grip the wire securely, hitch on a smaller line and tie the halyard to that.

- If you are sending up a tool, don't tie the halyard to its lanyard. Again, the tool must be secure before the halyard is untied.
- For sending up sharp-cornered objects, avoid having the line touch the corners. If this is impossible, pad the corners first.
- Have a Polaroid camera handy when you go up, to get a portrait of details aloft that you can examine and analyze before any future work. Hold a ruler or section of tape measure in the picture for scale.
- Whenever possible, avoid going up in bad weather. Not just because it's uncomfortable and dangerous, but because communications become difficult—it's hard to outshout a storm—and because you just can't do as good a job.
- Minimize future trips aloft by making every trip an opportunity to survey the rig (see "Surveys" in this chapter). Have an assortment of clevis pins, cotter pins, screws, bolts, shackles, tape, etc. handy, so you can deal with small problems immediately.
- Wear white-soled or scuff-proof dark-soled shoes to avoid marked masts.
- Wear long pants to avoid chafed and bruised legs.
- Spinnaker halyard blocks are convenient to work from, since they're up high and they swivel. But like all swivel blocks, they're more fragile

HUGH LANE'S LOADED BAT

Conventional-shaped mallets are okay for conventional situations, but try to pull one out of a crowded rigging bucket and you'll likely pull a few other items out with it. If you put it in head-up, it is liable to fall out by itself, after tangling its lanyard on other items.

As an alternative, get a fish bat—available at most any sporting goods store (anglers use them to, uh, subdue fish)—drill its end out and fill the hole with lead. Melting the lead and pouring it in works best, and lead is easy to melt. Just provide plenty of ventilation for the fumes, and chisel the hole out a bit to make it cone- shaped, so the lead can't come out when it solidifies.

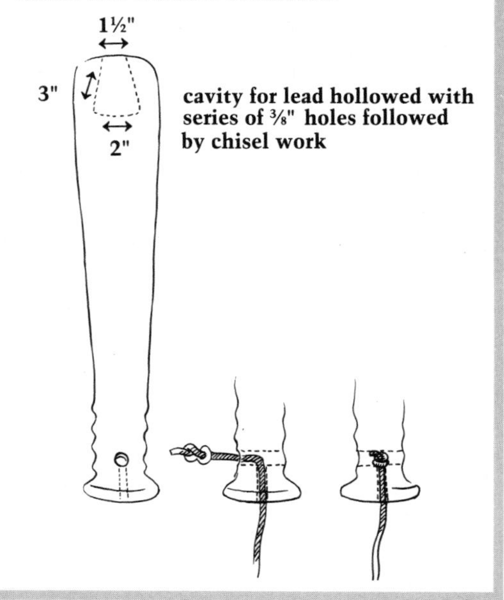

than nonswivelling blocks. Use a safety halyard.

- The deck crew will appreciate any help you give them in getting you up, but avoid pulling yourself up so quickly that you put slack into the main halyard; sudden slack can cause the halyard to "wrap" on the winch, or jump off a loose sheave above and jam between sheave and mortise.
- If a vessel has ratlines, make things easier for everyone by climbing as far as you can on those, then switching to halyards.

MAST STEPS

Mast steps are a popular way to get aloft, "as easy as climbing a ladder," as the sales literature says. But did you ever try to climb a wet, cold, awkwardly shaped ladder that was waving back and forth in the air? For all but flat calm conditions, mast steps are no treat, and even in flat calm you need to have a safety line attached to you and tended on deck.

It makes much more sense to have an efficient bosun's chair routine set up, one that enables you to go up in any conditions and to stay up there without having to hang onto a "ladder." And without the weight, windage, and expense.

But there is one place—about four feet down from the masthead—where mast steps are a really good idea. Just a pair of them at this height gives you a place to stand when you need to get at the very top of the mast, higher than a halyard can take you. Of course, you want to be sure you're tied to the mast before you do this, so there's no danger of pitching out of your chair.

- With mast steps, a safety line on you, well-tailed, is essential. Instead of having steps all the way up, consider installing just two, near the top of the mast, for a place to stand when working at the top.

Mountaineers and sailors have a long history of information sharing. Two of the most valuable ideas from the mountains are leg-and-hip-encircling safety harnesses and cam-grip rope-climbing devices (Figure 4-44). A good harness is far superior to an ordinary safety belt in the event of a fall, because it distributes so much better the force of fetching up against the safety line. A harness is not exactly comfortable to hang in, but it is sufficiently painless to be used in lieu of a bosun's chair. Climbers' supply stores carry a wide assortment of harnesses. Lirakis's, the

Figure 4-44. *A mast climber (taken from an illustration in Lirakis catalog): The "Ropewalker's" cam-grip action gives you a stairway to the spars. Release one side and slide it up, raising the corresponding leg, then engage the cam and straighten that leg while sliding the other side up. Note that one walker is belayed to the safety belt. For a faster ascent, use two halyards.*

only harness I know of designed specifically for sailors, allows the wearer to assume a semi-seated position for comfort.

Cam-grip devices used in conjunction with one or two belayed halyards allow you to get aloft under your own steam, an invaluable ability in shorthanded vessels. Again, Lirakis and climbers' stores are your sources. It is most advisable to practice with these devices at low altitudes before walking up a mast.

A rope or webbing "diaper" and Prusik knots (Figures 4- 45 and 4-46) are low- or

Figure 4-45. *Prusik knots are an inexpensive equivalent to ropewalkers. Make a 15- to 18-inch circumference grommet in line smaller than the halyard. Ring Hitch the grommet to the halyard, then pass the bight through once more. Draw up securely, and hitch on foot pendants and safety line. Pull outward then downward before each step to keep knots drawn up.*

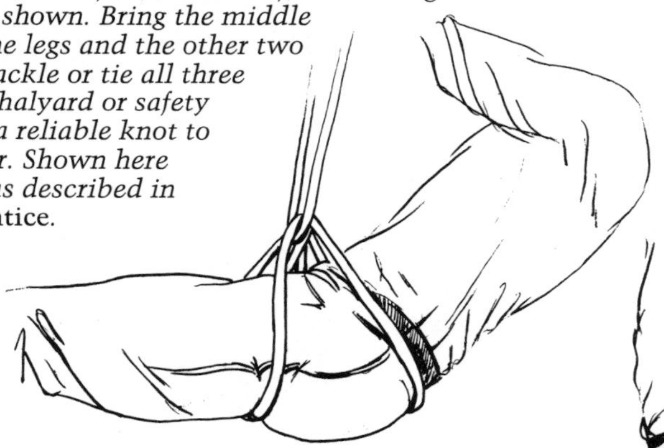

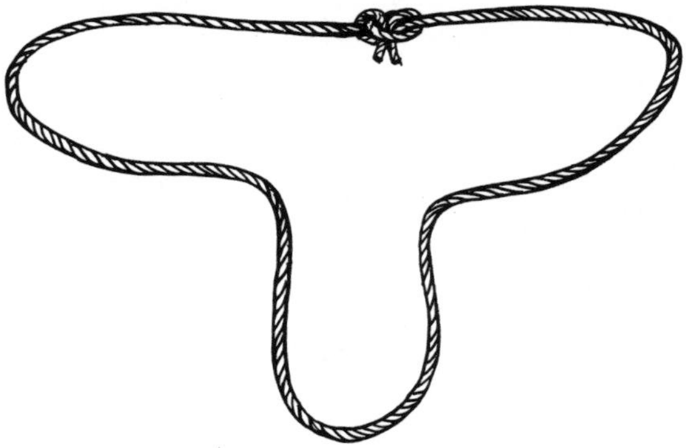

Figure 4-46. *Not quite as uncomfortable as it looks, a rope harness is good for emergencies or limited budgets. To make, tie the ends of a 7- to 9-foot piece of line together (the thicker and softer, the better), and arrange it in three bights as shown. Bring the middle bight up between the legs and the other two outside the legs. Shackle or tie all three together and to the halyard or safety line. Be sure to use a reliable knot to tie the ends together. Shown here is a Straight Bend, as described in* The Rigger's Apprentice.

Figure 4-47. *Self-lowering, method A: Pull a bight of extra halyard through the chair and drop it behind your back. Catch the bight with your feet, and bring up in front of chair. Take out slack to form hitch. Transfer strain gradually to extra halyard, then cast off first halyard. To lower, feed slack into hitch.*

Figure 4-48. *Self-lowering, method B: Before leaving deck, tie a loop knot in an extra halyard and seize a carabiner into the loop. Once aloft, pull a bight of the spare halyard forward through chair and hook into carabiner. Pick up a bight from back of chair and hook into carabiner, then hitch a third bight below the carabiner. To lower, carefully cast off hitches, then feed slack through the carabiner. Replace hitches at each step.*

no-cost equivalents of harness and cam grips. Like the plank-and-rope bosun's chair, they are not as comfortable, convenient, or foolproof as their manufactured cousins, but they are adequate for many jobs and excellent in an emergency.

For a primer on their use, see the excellent booklet, "Ropes, Knots, and Slings for Climbers," by Walt Wheelock (La Siesta Press, 1985; Box 406, Glendale, CA 91209).

Coming Down

When the work is done, the pressure is off, and all that remains is to ease on back to deck. But this is just the time when many accidents happen. Keep caution at a high level until you're past this job-ending transition point. Get to deck first, then relax.

STAYSAILS

It's called a staysail because it's attached to a stay. This is ironic, since a staysail is inclined to do anything but stay. It flaps and flops around when tacking, and is always demanding to be fussed with, even when you've settled on a course. It takes muscling to get it where it needs to go, and even after it's furled it's been known to go racing up the stay on its own if caught by a puff of wind. The time-honored method of reducing a staysail's intractability—lashing a boom to its foot so that it becomes self-tacking—also reduces its efficiency. And that can mean the difference between a daysail and a day-and-a-half sail, or between escaping a lee shore and calling your insurance agent.

HEADSAIL PREFEEDING

With foil-mounted stays'ls mount two prefeeders at the base of the stay, one just below the foil, one on a short lanyard at the deck. The double pre-feeders make for much smoother, quicker sail hoisting.

The Basic Self-Tending Staysail

In the quest for sailing efficiency without corresponding muscular effort, people have concentrated on improved mechanical advantage (read: larger, more expensive winches) or easier sail handling (read: fragile, expensive roller-furling devices). More on these two approaches later. Right now I'd like to take another look at self-tending staysails, because only a self-tending sail can relieve you of sailing's most arduous burden: tacking. And with intelligent design, loss of efficiency need not be great.

First let's look at the traditional design

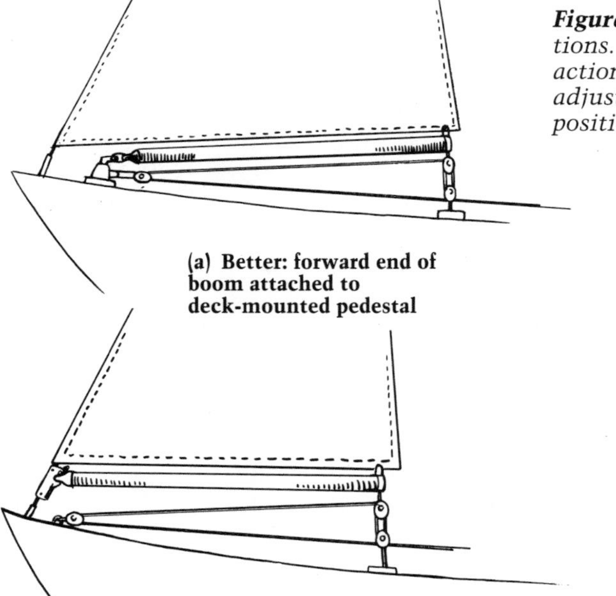

Figure 4-49. Self-tending staysail configurations. A curved traveler provides more vanging action, especially if used in conjunction with adjustable cars on the traveler to control position of traveler sheet lead.

(Figure 4-49). A block-and-tackle sheet slides on a traveler, for trim and some vanging action. The fall of the sheet runs forward under the boom to a turning block, then aft, for a fair sheet lead at any boom angle. You'll frequently see the forward end of the boom attached to the stay the sail is hanked to (Figure 4-49). This results in a thrust load on the stay, which is hard on the wire and its turnbuckle. Worse yet, it lowers sail efficiency, because the sail has the same shape whether it's sheeted in hard or eased when off the wind.

For more efficiency, try the first arrangement shown, in which the forward end of the boom swivels on a pedestal mounted on deck, aft of the base of the stay. When the sail is sheeted in, the boom stretches the foot of the sail out for a flatter shape. Off the wind, the difference between boom length and sail foot length causes the sail to become fuller for more efficient reaching and running.

Variations

Mr. Angleman of Angleman Ketch fame came up with a nifty elaboration on the length-differential theme: attaching the forward end of the boom to a heavy-duty track (Figure 4-50). The turning block at the forward end is so mounted that it pulls the track car aft when the sail is sheeted in and lets it slide forward when the sheet is eased, thus maximizing sail shape variation. As an added feature, the boom slides forward clear to the base of the stay when the sail is lowered, eliminating the need for luff jacklines.

A pedestal, used with or without a track, can help optimize sail shape but still leaves

ADJUSTABLE TACK PENDANTS

Most stays'ls are attached to deck with short tack pendants. The pendants are just long enough so that the foot of the sail will clear the bow pulpit or is high enough to see under. And most stays'l sheets are fixed to sliding cars to adjust the sheet lead, for optimal sail shape at different wind angles or when reefed. A less labor- and money-intensive option is to lead the pendant through a stout block at the deck and then aft, as for a downhaul. Then you can adjust sheeting angle by adjusting halyard and pendant. Use heavy line for the pendant, to reduce the effects of chafe.

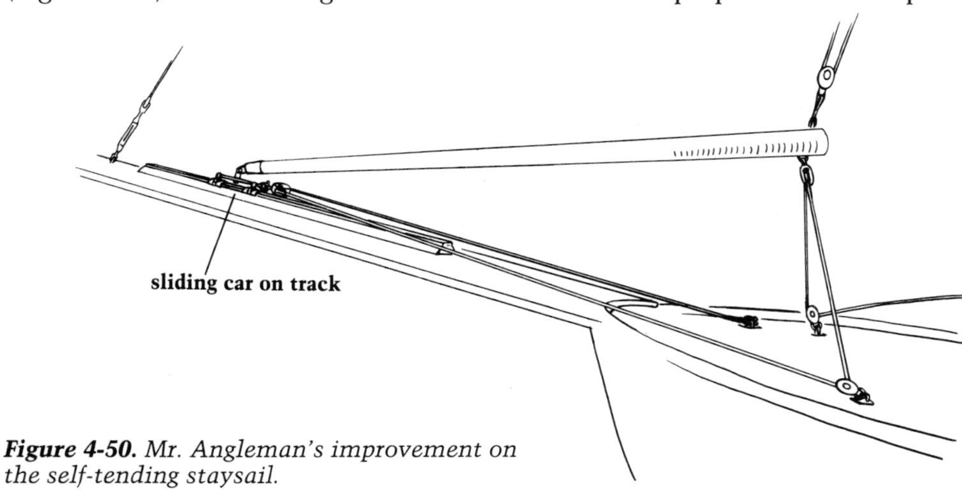

Figure 4-50. *Mr. Angleman's improvement on the self-tending staysail.*

self-tending rigs weak in the efficient sheeting department. As the boom moves outboard when the sheet is eased, it also rises, and most tracks provide no more than a modicum of vanging action, so that sail shape suffers from an excessively open leech. A rising boom and baggy leech also make an accidental jibe much more likely—self-tending staysails are notorious foredeck depopulators.

One intricate and expensive way to deal with this is to install a curved traveler with locator blocks to position the sheet for different points of sailing (Figure 4-49). This works well, and can be used in conjunction with a sort of secondary sheet that varies the distance of the clew from the boom end. But self-tending staysails, a feature of shorthanded vessels, are prized because the sailors of those vessels disdain intricacy and expense every bit as much as inefficiency. Is there no simpler, cheaper way to efficient sheeting?

Some modern self-tending designs are based on the assumption that a small staysail isn't much good off the wind anyway—it's too flat and is often blanketed by the main—so, focus instead to optimize close-reaching and beating efficiency. Tillotson-Pearson fits a boomless staysail to some of its boats, the sheet riding on a U-shaped traveler. There's no boom to deal with, and some degree of sail shape variation can be had by tightening or easing the sheet.

The Bierig Cambersail incorporates a pivoting wishbone boom inside the sail, which simultaneously bears the luff/clew compression load and holds the sail in an

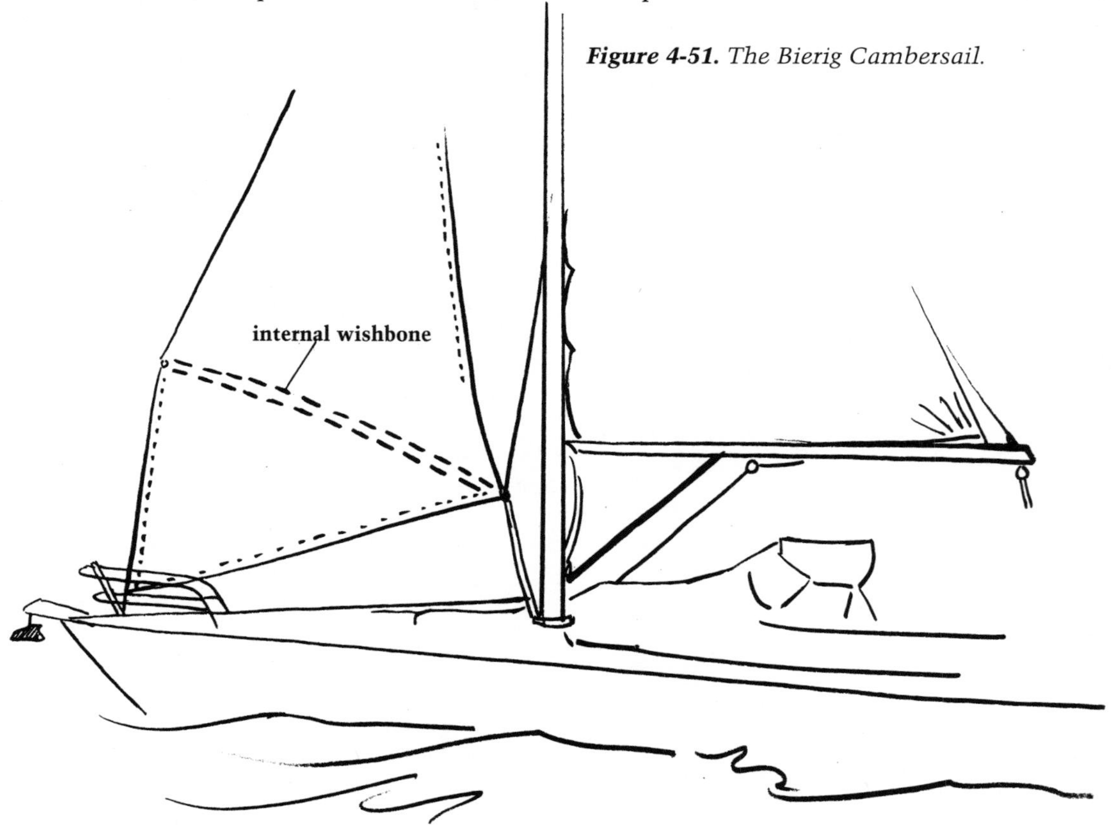

Figure 4-51. *The Bierig Cambersail.*

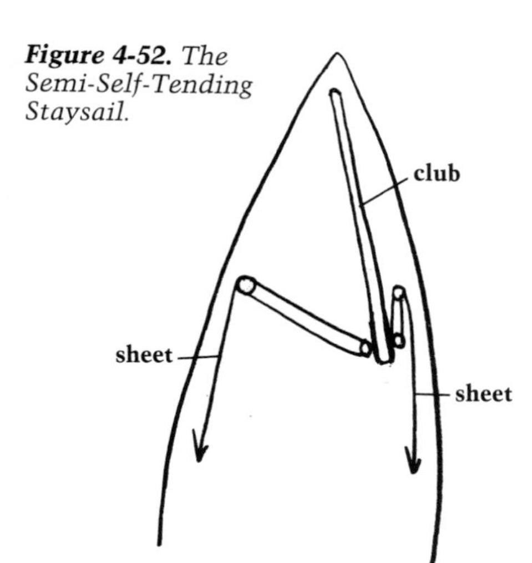

Figure 4-52. The Semi-Self-Tending Staysail.

airfoil shape, one optimized for on-the-wind sailing. The boom automatically flips to the leeward side when you tack. This sail, used with a forward-curving traveler, would have good offwind usefulness, but then again we're getting into intricacy.

So here's one more option: the Semi-Self-Tending Staysail (Figure 4-52). The club is controlled by two separate sheets, in the manner of a regular staysail. The deck blocks for the sheets are mounted well outboard and forward. The sheets can be a multipart purchase, or get their mechanical advantage from a winch aft. For maximum sailing efficiency, the procedure is to use the weather sheet as a lateral boom locator and the leeward sheet as a vang (vertical locator) or preventer. When you tack, ease the vang to let the boom swing across the deck. The vang now becomes a sheet. It's a simple matter to mark the sheets to indicate optimum settings for beating, reaching, or running angles. When the tack is complete, haul in on the old sheet to make it a vang/preventer.

If you're short-tacking or just feeling lazy, leave both sheets slack so each will function on alternate tacks. Not quite as efficient, but completely self-tending.

STAYS'L TACKLE

A sheet configuration with a 2:1 advantage. A short pendant, or "lizard," ends in a block. The line rove through it is dead-ended to an eyebolt on the rail and passes aft through a fairlead, also on the rail.

This is the traditional arrangement for gaffers and square-riggers with relatively small staysails and large crews. On shorthanded boats, a small winch back by the belay can be used to set up the sheet when going to weather.

If your boat is a little slow in stays, leave the vang side belayed as you tack. This automatically "backs" the sail to help you through the wind. You can also back the sail in order to heave to or just to pull the boom out of the way when the sail is down and furled. And of course, the preventer is a wonderful crew-saver, making the foredeck a safe place when broad-reaching and running.

The Semi-Self-Tending Staysail is a little more work than other systems, but is more efficient, versatile, and safe. There's no traveler to invest in, and no traveler to clutter up the deck. And if you put this rare system on your boat soon, you'll enjoy the added advantage of being the only one in the harbor with it. For a while.

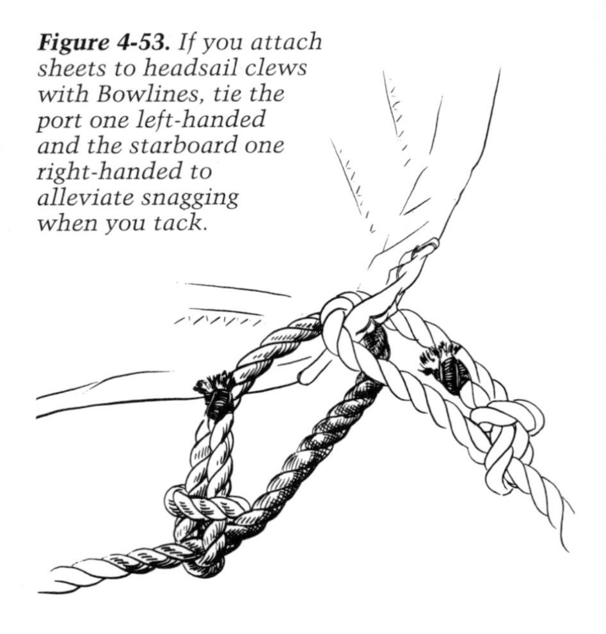

Figure 4-53. *If you attach sheets to headsail clews with Bowlines, tie the port one left-handed and the starboard one right-handed to alleviate snagging when you tack.*

Tackers

Boomless stays'ls, the ones you have to tack yourself, are a lot more work than their boomed brethren. Not only are the forces on them higher—no spar to take compression along the foot—but these sails are also usually larger than boomed stays'ls. Force on the sheets is related to sail area, so large sails need stronger sheets. And that's why you need big winches to trim big sails. This concentration of forces has implications outside the subject of sheets per se (see Chapter 5, for example), but it's sheets we'll deal with here.

First and foremost, you have to attach the sheet to the sail. Most people tie on the sheets with Bowlines, and that's okay except that Bowlines frequently hang up on shrouds and stays in mid-tack. To prevent this, tie the port sheet left-handed and the starboard sheet right-handed (Figure 4-53).

Another problem with Bowlines in these days of slippery, stiff, synthetic lines is that Bowlines can come untied. A locking tuck will fix this.

But even if you take the above measures, Bowlines are by no means the ideal sheet-attachment knot. Though supremely convenient, they weaken the line they're tied in by about 45 percent, and they're just plain bulky and clumsy-looking out there on the corner of your sail.

Eyesplices, the strongest and most compact of all knots, are ideal sheet ends for braided or three-stand line. For braid only, the Brummel Splice (Figures 4-54 and 4-55) is a way to make both sheets from a single piece of line in minutes.

If you use the same set of sheets for more than one stays'l, or if you have a reefable stays'l, splices are out—you need a detachable attachment. One alternative is to splice each sheet to a shackle, or both sheets to one shackle. Expedient, but shackles can come undone, are costly, and are hard. As in hit-you-in-the-face-when-the-sail-is-flogging hard. No, what you want is a soft, strong, se-

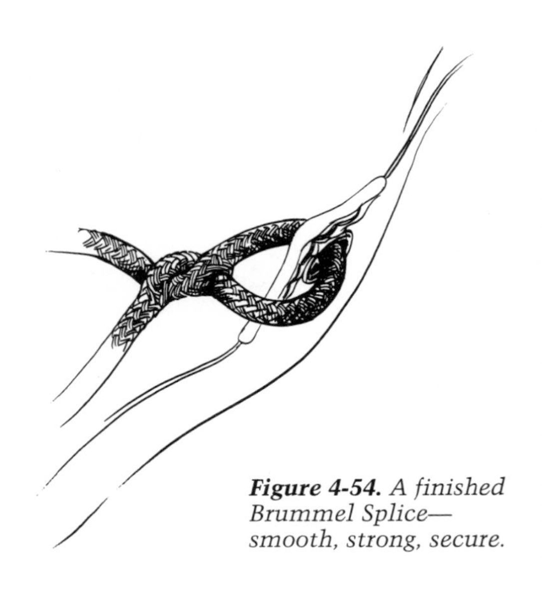

Figure 4-54. *A finished Brummel Splice—smooth, strong, secure.*

cure, inexpensive, and easily detached attachment.

The Bind series (Figures 4-56–4-58) all work on a jamming principle similar to the "Knute Hitch" (see Figure 4-43). They're secure, given the right size clew ring and line diameter. When I first tried these binding variations I was skeptical about their ability to hold under heavy loads and flogging. Since then I've seen them do fine on some

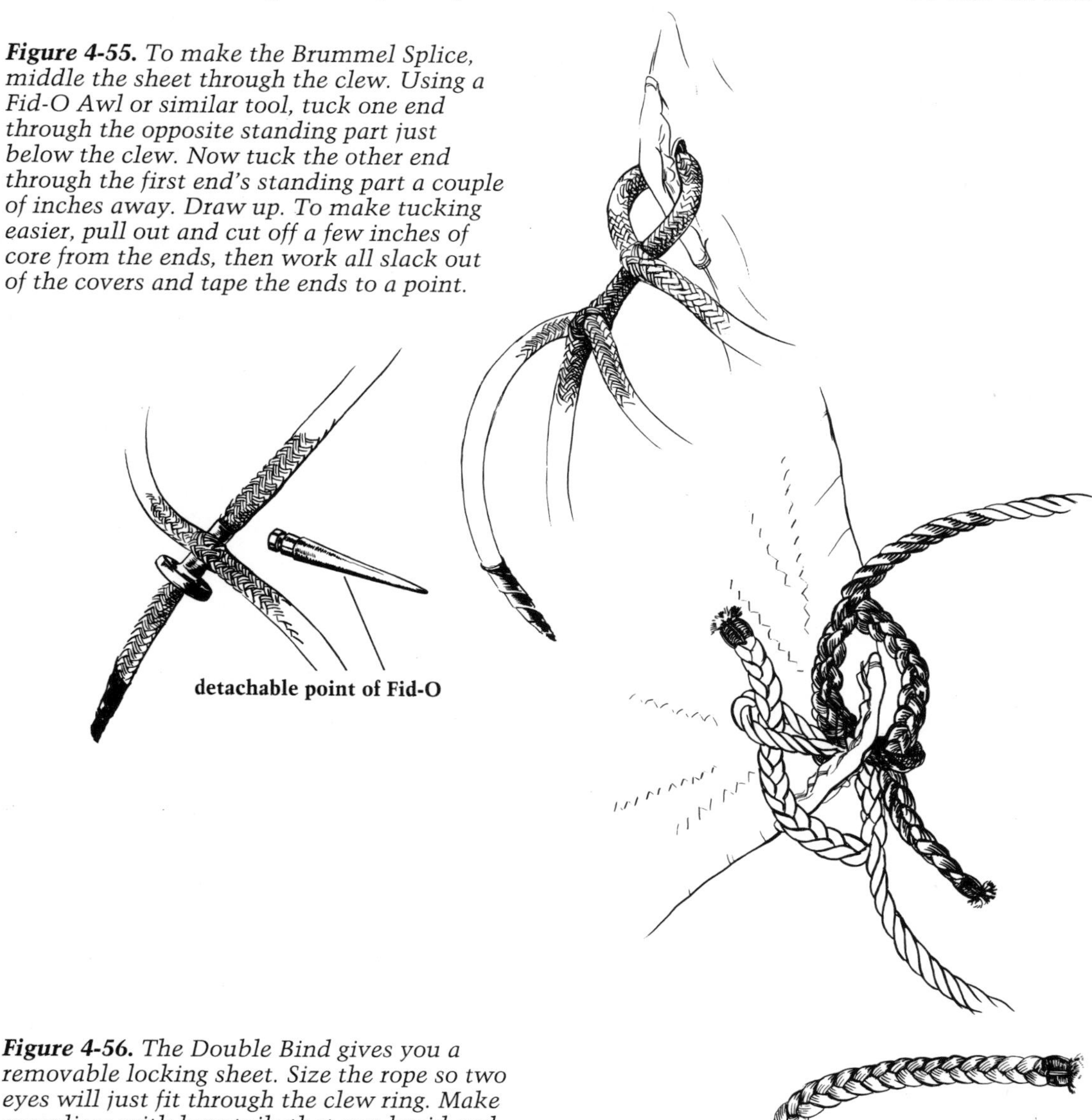

Figure 4-55. *To make the Brummel Splice, middle the sheet through the clew. Using a Fid-O Awl or similar tool, tuck one end through the opposite standing part just below the clew. Now tuck the other end through the first end's standing part a couple of inches away. Draw up. To make tucking easier, pull out and cut off a few inches of core from the ends, then work all slack out of the covers and tape the ends to a point.*

Figure 4-56. *The Double Bind gives you a removable locking sheet. Size the rope so two eyes will just fit through the clew ring. Make eyesplices with long tails that you braid and whip. Pass the eyes through the clew and the tails through the eyes, then snug the eyes against the ring. The result is quicker, more secure, stronger, and less likely to snag on rigging than Bowlines.*

Figure 4-57. *The Single Bind. When the sail's clew ring is small enough and the sheet diameter is large enough, a single eye can be locked in place with a tail braided from the long ends of the eyesplice. Then the second sheet can be "side-spliced"—it's started just like an eyesplice—to the first sheet. The Single Bind is a particularly compact, snag-free configuration.*

large, hard-sailed boats. And they come apart quickly when you want them to, making for quick headsail changes.

WINCHES

A modern sail plan comprises a few large, very powerful sails. The intent is to produce greater efficiency and less complexity than the traditional approach of more and smaller sails. But the modern sail plan concentrates forces to such an extent that a large, expensive machine—the winch—is the only practicable way to make things work. Largely because of market pressures, boat manufacturers tend to fit winches that are smaller (and thus cheaper) than you and I might consider ideal. Smaller winches are less powerful, so the price we pay for lower cost is increased physical effort.

In determining adequate winch size, work generated by the sails is the obvious

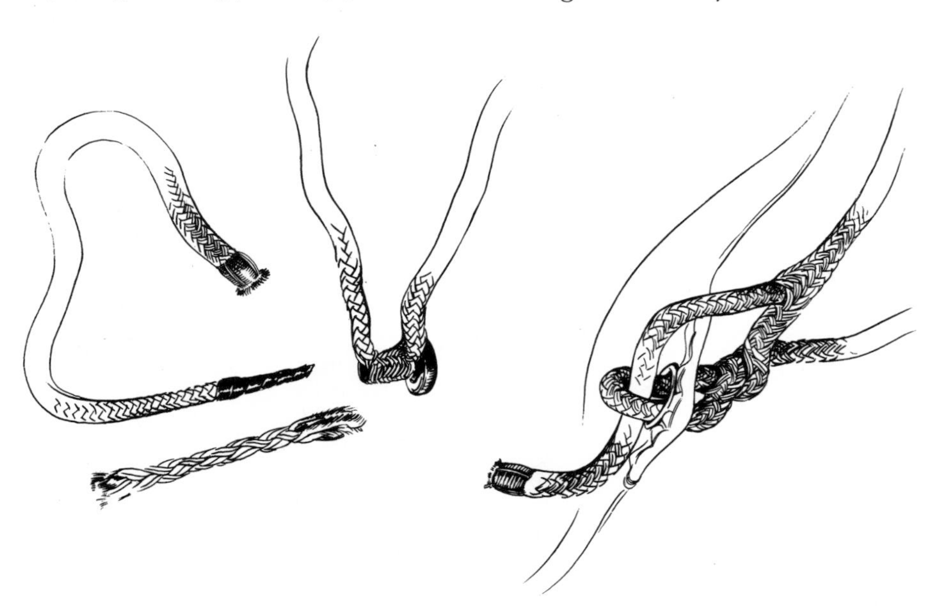

Figure 4-58. *The "Toggled Brummel" turns a Brummel Splice into a Single Bind. Side-splice a short length of rope to the standing part immediately below the Brummel Eye. (This consists of merely removing a short piece of core from the toggle, taping the end into a taper, and inserting it as shown.) Stitch through the rope to secure the spliced piece; don't worry, it won't have to bear much load.*

consideration. But work generated in the winch itself is nearly as significant. Like blocks, winches lose power to internal friction, typically 40 to 50 percent (!) in low gear.

What does this power loss mean in the real world? Say you have a winch rated at a mechanical advantage of 40:1 in low gear. A 35-pound (15.9 kgs) load exerted at the end of a standard 10-inch (254 mm) handle on that winch yields a theoretical force of 1,400 pounds (35 x 40) (636 kgs), but with friction, you'll be lucky to deliver 850 pounds (386 kgs). The jaws on a self-tailing cap can kick in another 10 percent in friction, and turning blocks can easily contribute another 5 percent or so.

How big, then, does a winch have to be to overcome both the pull of the sails and its own drag? The answer starts with how much effort is acceptable for you. For most people, 35 pounds (15.9 kgs) is a comfortable maximum sustained load. Most production boats have maximum loads of more like 45 pounds. That's why it's so hard to sheet that genoa in.

Bear in mind that we're talking about loads that you'll only encounter going to weather in a stiff breeze—relatively infrequently encountered situation, but one in which boat motion, fatigue, and discomfort all conspire to render you least capable of concerted physical effort. It's a situation in which you're most appreciative of adequate mechanical advantage, and hang the extra cost. By investing in a worst-case-scenario power level, you also get extra-easy sail handling in lighter airs. A final bonus is that the larger drum size means more surface area, and thus more gripping friction on the rope for every turn you make around the drum. So fewer turns to put on and remove, and better control when easing slack around the drum.

Workload Formula

To calculate desirable sheet winch power, figure your foretriangle area (the area bounded by the forestay, foredeck, and mast), multiply it by 6 (29 if you figure your foretriangle area in square meters) (a number that accounts for friction and sail force), and divide the answer by 35 (your desired maximum force in pounds), or 15.9 kgs for kilograms. The answer you get will be the rated mechanical advantage in second gear (low gear) of your optimally powerful winch. For example, start with a foretriangle of 300 square feet (28 sq. m.). 300 x 6 = 1,800. Divide that by 35 and you get 51.4; you want a winch with about 50:1 advantage. If you felt comfortable with more or less than 35 pounds (15.9 kgs), your optimum winch would be somewhat less or more powerful.

If you want to know what the maximum handle load on your current winch is likely to be, start with the same Sail Area x 6 (or 29) number, then divide by the rated mechanical advantage in second gear (this information is available through chandlers and manufacturers). The result will be the maximum handle load in pounds (or kgs). If we stay with our 300 square foot (28 sq. m.) foretriangle and assume a winch with a typical 40:1 advantage, we get 1,800 ÷ 40 = a typical 45-pound load (795 ÷ 40 = 19.9 kgs). Too high.

Efficiency Alternatives

If the cost of a big enough winch is too high, or if you're driving a race boat and the extra weight is a consideration, there are five other ways to get more from your winches:

(1) Brawn. Keep a very large, muscular, and willing individual around to do your winching. This is the traditional option for racing craft.

(2) Handle Leverage. A winch is a form

of lever, with leverage from internal gearing compounded by leverage from the handle. A 12-inch (305 mm) handle will provide 20 percent more leverage than the 10-inch (254 mm) handle your winch is probably fitted with now. This advantage is somewhat qualified by the slowness and awkwardness of swinging the handle through a wider arc, but many people hardly notice the difference, and love the ease. Also consider getting a two-hand handle, either 10 (254 mm) or 12 inches (305 mm) long, so you can make better use of the leverage you have, getting the strength of both arms into the effort.

(3) Compound Advantage. By combining a winch with a block and tackle, you compound your mechanical advantage. So a 40:1 winch hooked to a 4:1 block and tackle yields 160:1, minus friction. For quick, coarse take-up at low loads, you can use the block and tackle alone, hooking up the winch for power and refinement. This setup is the rule for mainsheets but is rarely applied to staysails; blocks hanging from sail clews can be real crew-killing deck floggers. On large traditional boats, with clews well above deck, it's still the viable option that it's always been.

Another old practice is to put a block on the head of a sail, for a 2:1 advantage to be compounded by the halyard winch. With a 50 percent lower load on the winch, you can use a much smaller, cheaper winch. This generates savings that offset even the long-term costs of the 50 percent longer halyard.

(4) Fairleads, big blocks, and lubrication. By using a minimum number of large, high-quality, strategically placed turning blocks, you reduce friction. By being one of the minuscule minority of sailors who strip down and lubricate their winches on a regular basis (at least once a year), you reduce friction by a lot more.

(5) Design. Staysails are any boat's most significant edge to weather, but their size and significance have been greatly exaggerated by racing rules. So if you are daunted by the prospect of shelling out the bucks for a comfortably powerful winch, consider making your sails smaller instead of making your winches bigger. This can mean getting smaller, more efficient staysails, or reefing or changing down more deeply and sooner than you currently do, or increasing the size and power of your main, particularly with full-length battens, among other options. Skillful trimming of skillfully made sails can compensate for the loss in headsail area. So if your sails are wearing out now anyway, consider a design change.

Canny use of design principles, combined with some or all of the efficiency-enhancing procedures mentioned above, will reduce the onerousness of winch work, leaving you freer to enjoy sailing itself.

SURVEY AND MAINTENANCE

A good rig has a designed-in safety factor: a degree of overbuilt toughness that will allow its components to deteriorate or even fail without precipitating a dismasting. A survey is an opportunity to maintain that safety factor, to spot and correct flaws before they're serious. A survey is also an opportunity to minimize expense and labor by assessing lubrication, adjustments, improvements, and other maintenance needs.

Attitude

Before starting, stand back for a minute and get into "survey mode," a state in which you see and feel the rig as a balanced, integrated whole. Take in the details of running and standing rigging, and feel how they interrelate with mast, sails, and hull. Entering this frame of mind is going to do you at least as much good as the usual procedure—starting with a list of Things to Look For. By envisioning the whole rig you'll be inclined to notice if something is missing, or could lead better, or is worn.

Next, stir in some general ideas to give your gestalt a little focus. The most succinct survey rules I know of are these four from yacht designer Eva Holman:

(1) If it is fastened, it will try to undo itself.
(2) If it touches something, it will try to chafe itself or that other something to death.
(3) If it is slack, it will try to snag something.
(4) If it is metal, it will try to corrode itself or its neighbor.

Details

Armed with an informed attitude, you can now make up a list, enumerating as many details as you can think of. This list will be more complete in the state of mind you're now in than it would be if you'd begun in classic Western analytical mode.

For example, things that will try to unfasten themselves can range from the bolts securing chainplates to the pin restraining your windvane, with spreader bolts, link plates, toggles, screws, antenna wiring ties, and sheave pins in between. It can include swages, Sta-Loks, and other wire terminals; sail track, bolts, and lashings; welds and glue joints; and on and on. You can even get compound fastenings: say, a cotter pin holding a clevis pin. Is the cotter secure and in good condition? If it should fall or get pulled out, is the clevis head downward so it too will fall out, or head upmost, so it will stay in place for a while, giving you a chance to spot the missing cotter?

As you make your list, feel free to let your categories overlap; you'll get a more complete list and a clearer idea of how rig components interrelate. For instance, some spreader fastenings secure the spreader to the wire, and some secure the spreader base to the mast. But good fastenings are only as strong as what they're fastened to. In this case, wire sometimes will suffer chafe or accelerated fatigue where it passes over spreader tips; check the wire's condition and the suitability of the tips as well as the condition of the fastenings. Then see if the spreader angles up, as it should, to bisect the angle formed by the shroud. If it doesn't, the wire is always trying to push the tip down, straining the outboard fastenings.

For Rule 2, the Certainty of Chafe, look not just for gouges and tears but also for shiny spots—something's been rubbing there. And avoid the trap of looking only where you expect chafe to be. I knew a boat

DRIFTER HANK

By putting at least one hank at the top of a drifter or other free-flying sail, and hanking it to the stay, you'll greatly reduce halyard chafe and the chances of getting a jammed halyard. Be sure the sailmaker sews in a generous reinforcement patch.

SURVEYING A HOOK

Hooks are unlikely to break, but they can straighten under heavy loads. The interior curve of a hook should be the radius of a circle. If there's any evidence of distortion, junk that hook.

on which the crew always made off the running backstays well forward when not in use, to avoid chafing the mainsail. Instead, the runners chafed on the after edge of the lower spreaders. Moderate pressure combined with the movement of the vessel was all it took for the spreaders and wires to saw into each other.

Sometimes chafe is hidden, so get in the habit of ducking and squinting into unlikely spaces, like the underside of standing rigging terminals. I once saw a rig in which a sheerpole threaded through the upper jaws of the turnbuckles chafed most of the way through the eyesplices on the lower ends of the shrouds. All it took to spot it was a cursory crouched-down look, but for years no one crouched.

Running rigging chafe is usually obvious because the line is always literally passing through your hands. Sheaves, stays, cleats, hawses, and stoppers tend to bear repeatedly at the same point; get in the habit of checking known chafe points so you can end-for-end or adjust the length of the line before chafe becomes too severe. To adjust length, cut off a short length at the hauling end, then reattach. This moves fresh line onto the chafing area.

But checking and shifting do not lessen chafe. To do that, you need to analyze what is causing it in the first place, and act accordingly. If you have good-quality rope clutches, for example, chafe should not be a serious problem unless the rope size is small compared with the loads exerted on it. Likewise, chafed-through rope or wire strands on a halyard can be caused by a too-small sheave (see "Blocks," earlier in this chapter) and mainsail chafe can be caused by the cloth bearing on an after lower shroud. In the latter case, it's not practicable to increase the wire size, so instead pad it with service, leathering, or baggywrinkle, or have a chafe patch sewn onto the sail.

Chafe can also be caused by things being where they don't belong. Staysail sheets delight in bearing on stanchions, shrouds, and anything else between clew and winch. But a little thought and intelligent use of turning blocks will eliminate this problem.

Rule 3 leads us out of the relatively simple world of fastenings and friction and into the subtler realm of ballistics. An unsecured running backstay, for example, will chafe as it flops around, but is liable to cause even more problems by grabbing at stanchions, turning blocks, gear lashed to cabintops, and you. Surveys are opportunities to take action to prevent snags.

Most often, it is jibsheets that snag, and usually in mid-tack. When doing a survey,

CLEAT DETAILS

In a storm, cleats for mooring lines cannot be too large or too smooth. Use files and sandpaper to remove anything like a rough edge from your cleats. While you're at it, smooth out your chocks and hawses. Think wide radius. In a blow, lead mooring lines around winches first, then to cleats, or belay each line to two cleats if they're available. And in a storm, no matter how you belay your lines or to what, double them up. Cheap insurance.

Figure 4-59. *A strut on a Dorade vent alleviates headsail sheet snags.*

flop the sheets around intentionally to see how close they come to snaggable cleats, vents, spinnaker poles, bitts, etc. If a snag seems likely, modify or relocate the potential snagger. A simple strut added to a Dorade vent, for instance, can prevent an inadvertent vent launch (Figure 4-59).

More than running rigging can be slack. If you've attached antenna wire to a shroud with those little electrical ties, watch out for their degrading in sunlight. When they let go, the wiring will start waving around, eager to catch the odd line or sail. Likewise, the tape that secures spreader boots can come undone, exposing a little opening in the boot that halyards just love to crawl into. Worst of all are masthead sheaves that are slack in their mortises, leaving room for a halyard to wedge itself between the sheave and the mortise wall.

Corrosion is always a problem in a saltwater environment, particularly when you mix antagonistic materials such as aluminum and stainless or stainless and carbon fiber; these materials are on different points on the galvanic scale, and when joined by the conductive medium of salt water they set up an electrical potential. The resulting activity corrodes whichever of the two materials is least noble (lower on the galvanic scale)—aluminum in both these examples. While not as serious with rigging as with permanently immersed items such as hull fastenings, galvanic corrosion can over time weaken aluminum and carbon fiber spars and clog such machinery as winches and blocks.

The first thing to do is to take note of places where dissimilar materials are in contact. Then go about isolating them with some form of nonconductive bedding. This can be as low-tech as parceling and serving to isolate galvanized steel wire from bronze thimbles, or it may involve 3M #5200, Alumelast, or other polyurethane or plastic compounds to isolate aluminum or stainless fasteners. When surveying an aluminum spar, pull a few fasteners and examine the screws and the holes they came from. If you see a white powdery substance on either, you've got galvanic corrosion. If the fasteners shear off when you try to pull them, you've got serious galvanic corrosion. The most likely places for galvanic corrosion to occur are at the mast step, under sail tracks, and under winches.

Stainless steel, the dominant material in rigging today, is susceptible to its own special form of decay: crevice corrosion, also known as oxygen starvation. Stainless steel contains significant amounts of chromium. When exposed to the atmosphere the surface oxidizes slightly and a thin film of chromium oxide forms, preventing any further oxidation. If exposed to water, salt or fresh, without the presence of air, this film will not form and the metal will corrode. If the water in question is salt water, the process is accelerated.

You risk oxygen starvation anytime you cover stainless, as when applying spreader boots, shroud rollers, or service. The trick is

TOGGLED GALVANIZED TURNBUCKLES

Most turnbuckles are "jaw-and-jaw"—that is, both end fittings are jaws with clevis pins through them. But galvanized turnbuckles are also available in "jaw-and-eye" configurations. A shackle has a better bearing on the eye than it does on the clevis pin of a jaw, so use jaw-and-eye turnbuckles when you use shackles for toggles.

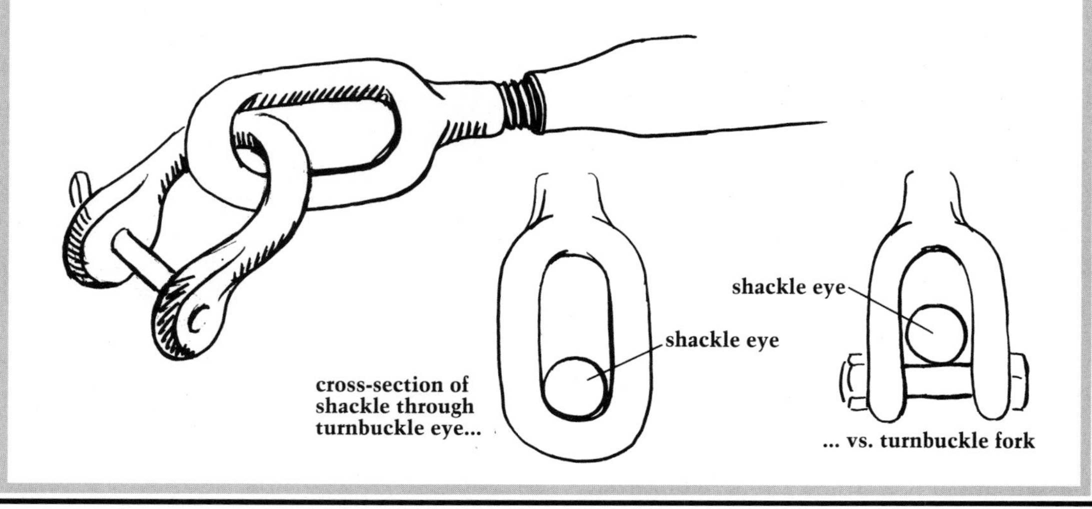

to exclude both water and air. When serving, some anhydrous lanolin covered with proper parceling and service works fine. Lanolin or mineral oil under shroud rollers is also good. Just rinsing stainless with fresh water whenever you can will lessen the corrosive effects of salt water.

When surveying for stainless corrosion, don't be distracted by stains. Contrary to what the name implies, the stuff does stain, mostly from bits of mild steel scraped off the extruding dies when the wire is formed. But do look closely, preferably with a magnifying glass, for any sign of pitting in the metal—the surface will seem to have teeny-tiny craters in it. Any significant pitting is cause for replacement. When in doubt, proof-test, running the wire up to 50 percent of its rated strength on a testing machine. If nothing breaks, it was a relatively cheap way to check. If it breaks, you'll feel very prudent for checking. Proof-testing is a good idea after five to ten years in northern climates, or three to five in the tropics, even without evidence of pitting, just to make sure the wire

SKIP GREEN'S HANK ROTATION METHOD

Hanks on staysails wear most severely at the lower end of the sail, and the next most severe wear is at the top. To prolong hank life, use extra-large hanks at the top and bottom, and switch these two hanks when the load on the lower one becomes severe. As an added step, also switch the second-lowest and second-highest hanks; they wear faster than the rest, but not fast enough to warrant oversizing.

hasn't fatigued (see below) or suffered other damage.

Sometimes metal just plain rusts. Stainless steel rusts more slowly, but tropical climates will get to it in just a few years. Galvanized steel left untended can dissolve in a matter of months. Any survey of metal must be a survey for rust.

The most-often-unlooked-for place for galvanized wire to rust is at the throat of the splice, at the pointy end of the thimble. Even competent splicers will sometimes neglect to "diaper" this spot adequately. The result will be that salt water will splash up inside and rust the throat, while the rest of the wire stays like new under its service.

A wire seizing made with galvanized seizing wire can rust from the inside out, even though brightly painted on its surface, if the wire isn't rustproofed with thinned tar or other wire preservative first. This is one reason to use stainless steel seizing wire.

Another hidden rust spot is under any item that is lashed to served wire. Unless the service is covered with a chafe-resisting layer of leather or canvas, the lashing can cut through the service and expose the wire.

Swages are much trickier to inspect for corrosion than a served splice; about the only way to tell anything's happening inside a swage is to have it crack from internal pressure or fatigue. Eva Holman recommends tapping swages with a tool handle to determine their condition. With a little practice you can learn to tell the hollow sound of a corroded swage from the live ring of a solid one.

To Holman's four surveying rules I would add a fifth: Stainless steel fatigues. No, this is not an all-metal army uniform. It refers to the characteristic of alloyed steels of hardening and becoming brittle with age. The more heavily a piece of alloy is stressed relative to its ultimate strength, the faster it will fatigue. Therefore, you'd survey lightly rigged race boats for fatigue more carefully and frequently than heavily rigged cruisers.

Fatigue reveals itself with cracks. Sometimes small, "gee-I'm-glad-I-spotted-that" cracks, sometimes "Oh-my-God-I-could-drive-a-truck-into-that-thing-and-it's-holding-up-the-jibstay" cracks. Obviously it pays to invest in high-quality stainless and to make it plenty heavy, to delay the onset of fatigue. Bronze is nearly impervious to fatigue, which is why it is so often used in toggles, turnbuckles, tangs, and chainplates. Galvanized steel is likewise just about fatigue-proof, so if you can keep it from rusting, it will outlast stainless. If your survey reveals broken yarns in a halyard wire, and the sheave is adequate-sized and the wire is fairly new, consider using a galvanized wire halyard. It will require only periodic oiling (Marvel Mystery Oil is great), is stronger than stainless, and stretches less. Alternatively, you can increase the size of the stainless wire, but this will also involve increasing the sheave size. Can of worms.

When rigging wire fatigues, its strands will begin breaking. Note that a single broken yarn in 1 x 19 wire reduces strength by more than 5 percent. Wires will usually break first at the lower ends of standing rigging, where corrosion and fatigue work together. But check both ends and all the wire between, just in case. Fatigue can be reduced by increasing wire size, but again this

CHOOSING A FOUNDRY

Have bronze castings done by a foundry specializing in bronze; they'll know which type of bronze is best-suited to your application, and will be pickier about ingredients and proportions than a general-purpose foundry.

Figure 4-60. *A cracked swage.*

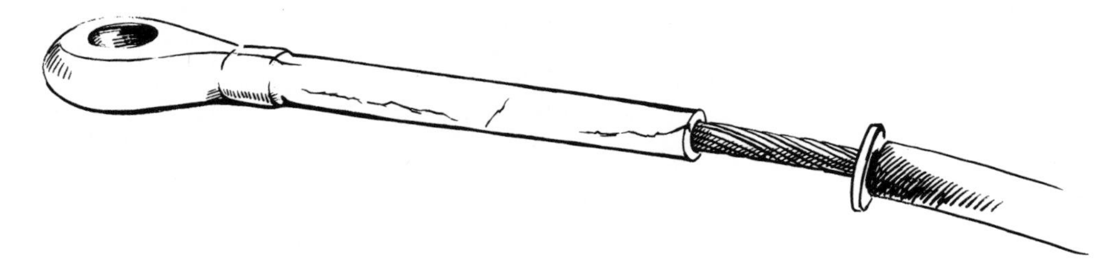

is not always practicable, especially for racers, as it increases weight aloft. It's usually better to use an appropriately sized wire (see Chapter 7) and to employ other fatigue-reducing strategies. The easiest one is the addition of toggles. Put one at either end of each turnbuckle or buy turnbuckles with built-in toggles. Add another toggle at the wire's upper end, particularly on stays with sails hanked to them, as these are most heavily worked. And keep your rigging snugly tuned so that sailing motions won't cause your mast to bang around, shock-loading your wires.

When swages fatigue, they'll crack, too. Again, this can also be caused by internal corrosion—the corroded wire expands, trying to split the swage apart (Figure 4-60). Cracked swages can survive for years or days. Replace any wire that has a cracked swage on it immediately, unless you enjoy that sort of gamble. A horizontal crack is always more dangerous than a vertical one. Use a magnifying glass or dye penetrant to spot fine cracks. Check the eye as well as the barrel of the fitting. To maximize swage and wire life, consider inverting each wire when you figure it has reached its half-life, before cracks appear. That way each end will spend some time in the clean air aloft. And when swages are new, seal them against corrosion by melting wax down into the terminal.

Some swages are made by a rotary swager, which hammers the fitting rapidly from all angles, making a smooth-finished surface. If you see a lengthwise ridge on the barrel of the swage, it was formed by passing it between the dies of a Kearney swager. Kearney swages are far more likely to crack, and frequently end up with a disquieting banana shape. Don't use them.

Sta-Lok and Norseman fittings are the best mechanical terminals—right up there with splices in terms of trustworthiness. They're screwed onto the wire, which means there are no hammer-or die-induced stresses. And they're reusable, so when you re-rig you only need to buy wire, not terminals.

The above is by no means a complete list of things to look for, but it gives you an idea of how free-ranging and inclusive a survey mentality must be. To give you an idea of how this might translate into reality, a sample survey follows. It's a bit of a flaw collage, excerpted from several vessels. If the number and severity of flaws seem high, bear two points in mind: (1) A moderately run-down rig and one in good condition will have roughly the same number of notations; as you fix big problems, you start noting smaller ones. (2) Assuming the mast is still standing, the list of things that are OK is always longer still.

Survey of a 34-foot Cutter

Starting at the top:

1. *Topping lift block has unmoused shackle. Shackle pin is upside down.*

   Here's one you can fix right

away, assuming you brought along some nippers and a length of seizing wire.

2. *Recommend spare halyard at masthead.*

An extra halyard is good as a safety when going aloft (see "Living Aloft" in this chapter) or as a backup in case you lose one of the regular halyards. It should be on a stout swivel block so it can substitute for jib, main, or even spinnaker halyard. Affix the block to the side of the mast and reeve it with a length of flag halyard line. This way you avoid weight and windage aloft from a full-size rope, as well as UV degradation of that rope. When the time comes, you can easily seize a real rope to the end of the light stuff and pull it through. Meanwhile, you've got a flag halyard at the masthead.

3. *Jib halyard sheave not turning.*

The sheave sides probably just need a light sanding. If main and jib sheaves share the same axle pin, secure both sheaves with lanyards before driving the pin out (or see the accompanying "Tapped Sheave" essay). Since you're probably hanging from the main or jib halyard at the moment, you'll also want to switch to an alternate means of suspension before driving that pin out. The procedure is to get your deck crew to haul you up just as high as the halyards allow. Then tie two short

TAPPED SHEAVE

It's very tricky to remove and replace masthead sheaves without dropping them—they're difficult or impossible to get a lanyard around. Cruiser Steve Dashew has a solution: Drill and tap a small (1/8 to 3/16 inch or 3.2 mm to 4.8 mm) hole in the bottom of the sheave groove. A hole this small won't bother the wire or rope halyard. When the time comes for sheave removal, just thread a bolt or machine screw into the hole. Tie a lanyard to the bolt and one to the sheave pin, then remove the sheave.

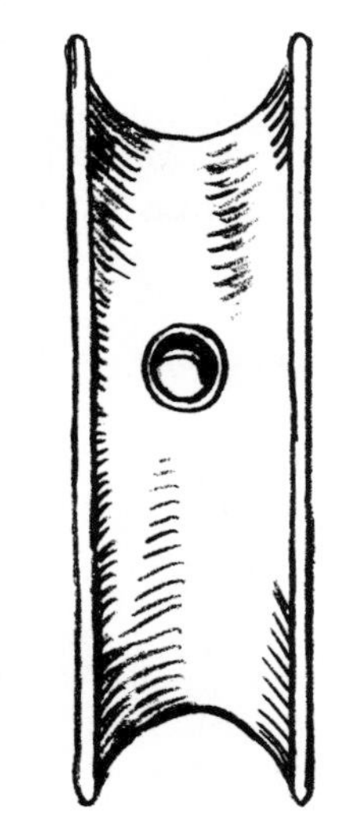

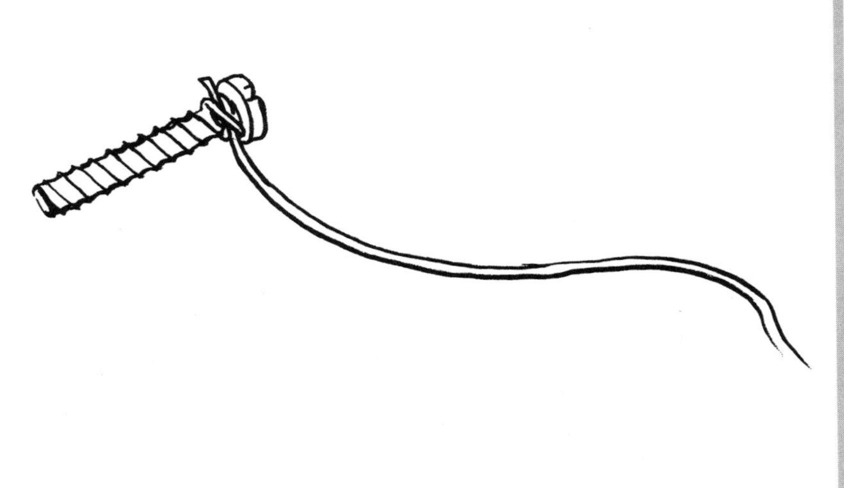

lengths of stout rope to the chair (or, if you're wearing a harness, one to the chair and one to the harness). Secure these lines to the masthead, hitching around lugs, shackles, or whatever is up there. If there are spare halyards, you can hitch those to the chair instead. When your replacements are tight and secure, have the deck crew slack away on the other halyards. Detach them from the chair and tie them to a convenient shroud or stay. Drive the pin out—constrictoring a lanyard onto it as it comes—and pull the sheaves out. Check for wear on the pin, the sheave bushings, and the mast mortise. If they look okay, sand the sides of mortise and sheaves with fine sandpaper. Reassemble. Check for a fair halyard lead, slop in the mortise, proper alignment of the pin, and chafe on the halyard. If there's more wrong with the sheave or pin than the sanding will fix, you'll need to reassemble and come back up when you've got a replacement, unless you already have one aboard.

4. *Clevis pin in port upper shroud tang is too thin or too long.*

   At some time in the past, someone needed to replace this clevis and didn't have exactly the right size. So he went with the nearest thing he had to a fit. When a clevis pin is too small, the load from the tang and wire terminal bear at a single point on the pin. This "point loading" causes accelerated wear and fatigue. And of course, a too-small-diameter pin is a weak link. If the pin is too long, it provides that much more weight, windage, and cost. It also might jam into the mast, damaging at least the paint.

   If you don't have the proper size clevis replacement, make a note on size and fix it on another trip up. If you do have the right size, you're faced with the disquieting task of detaching a shroud from a mast you're sitting at the top of. As long as this isn't the weather shroud of a boat driving to weather, and as long as this isn't one of those criminally spindly racing masts, this need not be a dangerous action. First have your deck crew slack away some on shrouds at your level, plus a bit on jib and backstay. This will prevent any sudden distortion of the mast when the shroud you're working on is detached. Next, as a precautionary measure, rig a halyard as a temporary shroud, belaying both ends at the rail. It probably is unnecessary, but it can't hurt. Put a lanyard onto the standing part of the mis-pinned shroud, just below the upper terminal. Make the lanyard off to the masthead. All set? Have your deck crew loosen the shroud until you can easily take the pressure off the old clevis. Have everyone stand from under. Remove cotter, then clevis, and stow them in a closable pocket. Install the new clevis and cotter. Have the deck crew retighten all wires.

5. *No toggle on upper end of jibstay.*

   No toggle with you or aboard, so make a note of the size required and of the space between the sides of the tang, so you will know if the right size toggle will fit. Maybe the space is too small, and that's why there's no toggle here now. If so, you'll have to do something creative, like modifying the tang. Check with a yacht designer first.

6. *Halyard for roller-furling jib is wire with rope tail. Rope tail will not fit through wire-only sheave. Result is that jib cannot be lowered all the way to deck.*

   Sheesh. Any more like this and you'll be thinking of pulling the stick out so you can work on it more quickly and easily on the ground. What happened was that whoever rigged this jib did so on the ground. They made up a halyard of the right length, ran the wire end through the masthead sheave, then Nicopressed an eye in it for the sail, then shackled the eye to the sail, then stepped the mast with the sail hanging in place. Had they tried lowering the sail to check the layout, they would have discovered their mistake. And it was a very dangerous mistake: Picture yourself in a rising wind with a jammed roller-furler mechanism, trying to lower a monstrous, flapping jib and discovering that it only comes halfway down. "Shotgun reefing"—blowing a few holes in the sail—might be your only recourse. Anyway, this is more of a problem than you can solve right now. Measure for an appropriate sheave and halyard and inspect the head of the roller-furling unit. They usually need an occasional rinse and/or a squirt of WD-40 or similar lubricant. Check manufacturer's maintenance recommendations.

7. *Moving down from the masthead, all is well until we get to the spreaders, which are horizontal instead of bisecting the angle formed by shrouds.*

   This is the single most common flaw in rigging. If the spreader is angled properly, it functions as a pure compression member. If it is horizontal, the shroud will act to push its outboard end down. This leads at least to an excessive buckling load on the spreader. If the end seizing slips, the spreader could collapse altogether. Dismasting. Besides, horizontal spreaders look dowdy, lifeless.

   To fix this, cast off the outboard seizings and tap the ends up to the proper angle. Have one of your deck crew with a good eye get well in front of the boat to help you with this. If you're on the ground, use a tape measure from the masthead to get the angles identical. Finish by seizing the spreader ends securely to the wires.

8. *Horizontal crack in starboard after lower shroud swage.*

   Oops. Measure for a replacement. Then ask yourself: How old is this rig? Is this wire heavy enough? If the wire is new and adequately sized, this might just be a fluke. But it's much more likely that all the wires are fatigued or badly swaged—you just found the first one to show signs. At least consider a new, good-quality gang.

9. *Spinnaker pole track has machine screw as stop at upper end; screw is bent.*

   This stop prevents the car that the butt of the spinnaker pole rides in from coming out of the top end of the track. A machine screw is a quick and dirty substitute for a real stop, an item available, cheap, at any chandlery. Get one.

10. *You're back on deck (whew), but not done yet.*

   Go to the bow and make sure there's enough thread on the outside of the jibstay turnbuckle barrel that you can add that toggle aloft and still tension the stay. It could be that the turnbuckle is tightened down too far for this to happen, and that's why there's no toggle aloft. You just might be able to shorten the wire by the length of the terminal, apply a new terminal, add the toggle, and have a perfect-length stay. Measure carefully. It might also be that you'll have to replace the entire, expensive stay. But with no toggle aloft, you're going to have to do this relatively soon anyway.

11. *Sticky stays'l lead block car.*

    Sandpaper and rinsing will probably do it, though cars and tracks sometimes get sufficiently dinged to require some artful filing.

12. *Mainsail halyard cleat angled backward.*

    You've had this boat for years and always known there was some reason why this was an awkward belay. The cleat should be angled so that the halyard touches the lower end first.

13. *Nick in mainsheet near standing end. Line badly twisted.*

    End-for-end the line. Try to figure out what nicked it. Resolve either to coil by figure-eight or alternate-hitch methods, or to have a halyard bag made up and mounted by the mainsheet winch. Coiling braided line by the conventional clockwise method results in twists that don't come out.

Sailing Is Surveying

The best sailors notice things: wind patterns on the water, how an engine sounds, the shape of a sail, the shape of a cloud. By noticing, they have a greater reference base, and can act quickly and appropriately when they need to. What we call "surveying" is really just a formal exercise in noticing things. So get formal for your surveys, but consider living every day by another of Ms. Holman's maxims: "Feel, rattle, pull, knock, and touch absolutely everything."

Chapter 5
Angles

"The square of the hypotenuse is equal to the sum of the squares of the two sides."
– Pythagoras

Angles are to rigs as genes are to bodies: They determine the shape, size, and proportions of the finished structure. The angle a shroud makes relative to its mast, for instance, determines the ratio of compressive and lateral force which that shroud exerts and therefore how strong both mast and shroud need to be. Very steep shrouds can exert compressive loads that could crumple any mast, so in practical terms angles limit rig design options. Rig design involves working out realistic relationships among load, angles, tension, and compression.

Happily, most sailing rigs are simple things, understandable with simple formulas based on proportions. And if the word "formula" intimidates you, you can reach the same understanding by drawing funny little pictures.

A LIFELINE

Consider a deck lifeline, say a piece of ⅛-inch (3 mm) 1 x 19 stainless steel wire with a breaking strength rated at 2,100 pounds (955 kgs). It is secured at the cockpit at one end and at the foredeck at the other, and is long enough that if you pick it up at the middle, exerting 150 pounds of force, it will deflect about 12 inches (305 mm) over its 30-foot (9.1 m) run. By clipping the shackle of a safety harness around this wire, a crew-

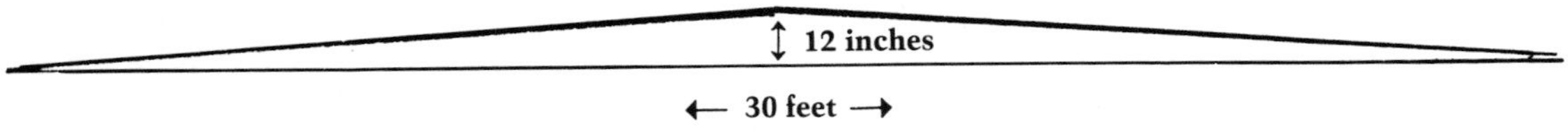

12 inches

← 30 feet →

Figure 5-1. *A load of 150 pounds applied at the middle of a lifeline run of 30 feet causes a deflection of 12 inches. The tension on the lifeline can be calculated from this observation using the formula in the text or the diagrammatic representation in Figure 5-2.*

Figure 5-2. *To calculate tension on a line or wire graphically, draw a vertical line of any convenient length to represent the deflecting load at the midpoint (in our case, 150 pounds). Now draw in the deflected legs, starting from the top and bottom of the vertical; in our example we get the angles of deflection from Figure 5-1. (We could walk them over to Figure 5-2 with parallel rules if desired.) Line ac or bc divided by line ab and multiplied by 150 pounds yields an approximation of the tension on each end of the lifeline.*

member can walk fore and aft, secure from the danger of going overboard. If the lifeline is strong enough.

That chilling "if" might make you receptive to a helpful formula. If so, here's one: Load x Length of one leg ÷ Deflection = Tension on both legs combined.

In this case that's 150 pounds x 15 feet $\frac{12}{32}$-inch ÷ 1 foot = 2,255 pounds. This is the combined tension, so the load on each leg is 1,127.5 pounds (531.5 kgs).

A pictorial representation of the same formula starts with the load (150 pounds) represented by a vertical line of arbitrary length. Lines parallel to the two sides of the lifeline are drawn from the top and bottom of this vertical line (Figure 5-2). Their length where they intersect divided by the length of the load line yields a proportion which, when multiplied by the vertical load, gives the load on each end.

Same answer by either method. In practice the formula is more precise, and the picture (called a stress diagram) is handier for showing the effects of changes in configuration.

But let's come back to that lifeline. Our calculations were based on a static, sustained load of 150 pounds (68 kgs), the weight of a lean crewmember. Under those conditions we have a factor of safety of less than 2—marginal at best. Now consider that in real life, that 150-pound (68 kgs) load will come on abruptly when the crewmember falls or is washed across the deck. In this "shock load" circumstance, the momentum from the load can easily double or triple the load arrived at by our calculations. It's also worth considering that some members of the crew might weigh more than 150 pounds (68 kgs), and that the wire might not be 100 percent efficient, and that the fasteners that anchor the lifeline are mostly subjected to a shearing force, which they cannot withstand as stoutly as they can an upward pull. In other words, the formula or diagram is just a starting point.

Taking this into account, it's clear that this lifeline is ironically named. We need to increase wire strength, reduce tension on the wire, or both.

Let's start by reducing the tension. Lengthening the wire until it deflects 18 inches (457 mm) results in a load on each leg of about 750 pounds (341 kgs) (Figure 5-3). By increasing the deflection, you reduce the leverage the load exerts on the wire's ends. You can keep on increasing deflection until the two sides are nearly parallel. Here there's minimal load on the wire, no shear on the deck fastenings—and your crewmember is 200 feet (61 m) to leeward. So

IN PRAISE OF THE LOGGER'S TAPE

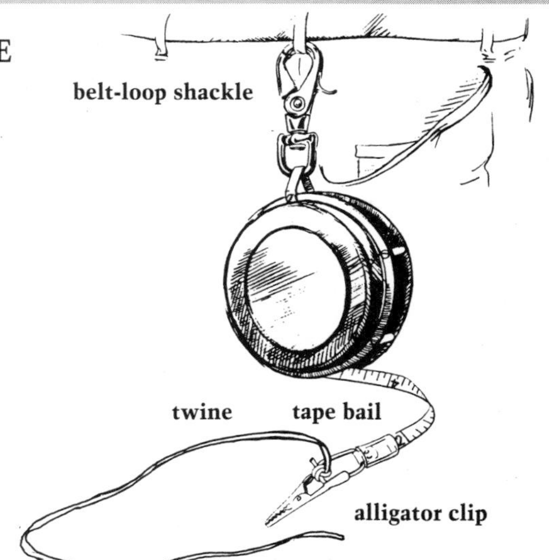

There's more to measuring a log than seeing how long and wide it is. More often than not it involves fighting your way over difficult terrain while lugging heavy, awkward tools, without even the luxury of a helper to hold the dumb end of the tape for you. And loggers, paid by the board foot, have no time for cranking conventional long tapes up with those silly little handles.

That's why loggers use logger's tapes, often called Spenser Tapes after their original and still preeminent manufacturer. A logger's tape clips to a belt loop, so it's always out of the way but instantly usable. The tape blade is narrow, flexible, reads on both sides, and comes in 50- and 75-foot lengths. Repair kits are available, so an errant saw cut won't necessitate buying a new blade. After multiple cuts and kinks, clever little refill kits defang the powerful spring mechanism for safe, easy reloading.

Logger's tapes are a measurement revelation for sailmakers, carpenters, riggers, and any other trade involving large-scale layout, because they leave both hands free to hold tools, pens, or clipboards, and because they can be used without a helper. To accomplish the latter, loggers customarily attach a horseshoe nail or the like to the tape bail. They lightly tap the nail into one end of the log, walk away, letting the tape unspool itself, get the measurement, then retrieve the tape with a light tug. The nail pulls out and the tape winds itself up.

A nail works well for carpenters and sailmakers, but is uncool procedure around varnished wooden spars and impossible around aluminum ones. But an alligator clip and piece of twine make an excellent rigger's alternative. Remove the bit of insulation from the clip, then use the point of an awl to open up the little tube where the wiring is supposed to fit. With a pair of pliers, crimp the tube on around the side of the tape bail. Hitch a 2- to 3-foot length of twine onto the end of the bail. You can now measure anything. You can hitch the string onto the standing part of a halyard, above the shackle and splice, for a true deck-to-sheave measurement (always tie the other end of the halyard through the shackle bail as a downhaul). To measure a shroud, pass the twine through the turnbuckle barrel and back to the clip. Adjust twine length until the tape bail rests on the middle of the shroud's lower clevis pin. Then go aloft, both hands free, to the other end. Or you can work from the top down, passing the twine around the mast and back to the clip. On deck or ashore you can hitch tape and the item to be measured to the same cleat, post, stanchion, etc. Adjust the string length to your "O" point, and walk away. By clipping the alligator jaws onto just enough of the string to hold securely, you can tug the end loose when you're done, and watch the tape snake its rapid way back to your side.

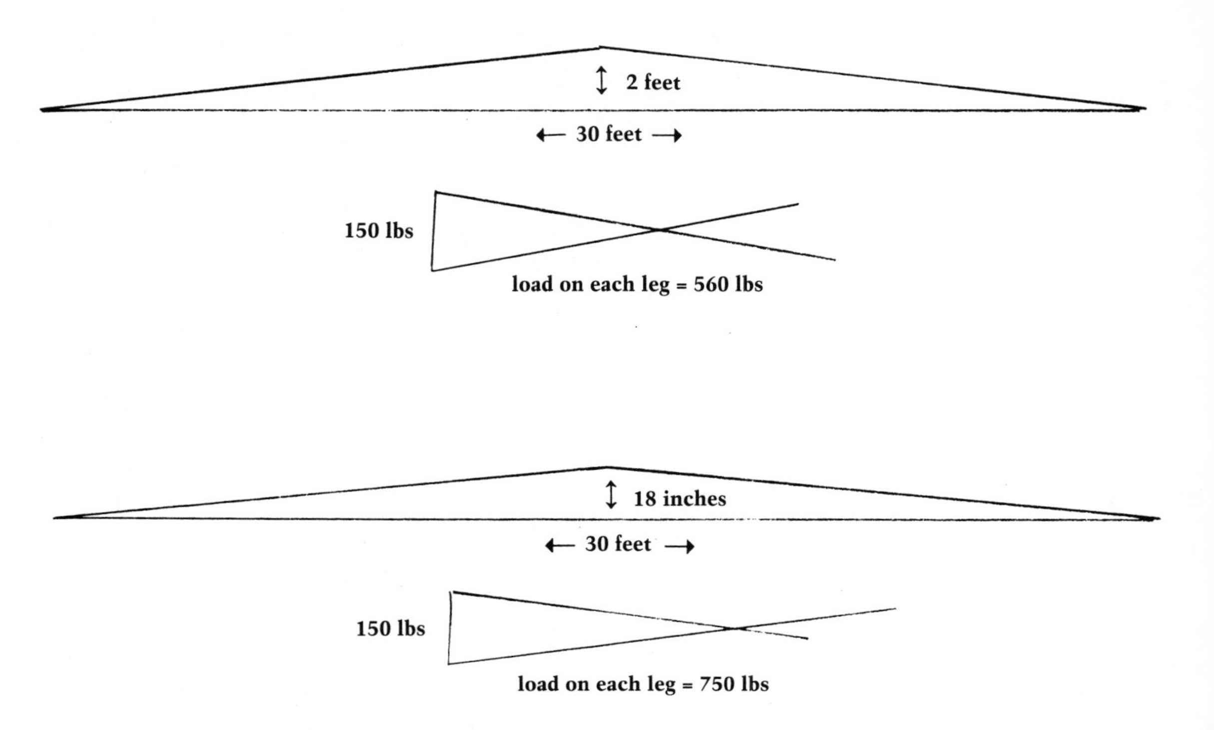

Figure 5-3. *Increasing deflection in the lifeline (decreasing the resting tension) reduces the leverage a load exerts on the wire's ends.*

you back up, working out a compromise between minimum tension and minimum deflection. Playing with tether and lifeline configuration can make the compromise less painful.

After you've settled on deflection, choose a wire size that will provide you with a comfortably massive safety factor. In our case, a combination of short tether, 2-foot deflection, and ¼-inch wire (7 x 7 construction, because it resists damage from kinking) gives us a 10:1 safety factor. Now that's a lifeline.

SLINGS AND COMPRESSION

The strain on a two-legged sling can be calculated using the same formula or diagram as for the lifeline. But neither will tell us the compression load on the object the sling will lift—remember, the ends are being pulled toward each other. This wasn't an issue with the lifeline; the compression load wasn't going to buckle the deck of the boat. But the sling might be picking up a pallet or crate that could be crushed by excessive compression. So here's a related formula: Compression = Load x One-half the horizontal span of the load ÷ Vertical length of the sling. And in Figure 5-4 you can see how to represent the compression pictorially. Marvelous.

So far we've let the load dictate things. But let's now say that you have an existing two-legged sling of ½-inch (13 mm) 7 x 19 wire, rated strength 22,800 pounds (10,364 kgs). You want to put a spreader bar between the legs of the sling (Figure 5-4).

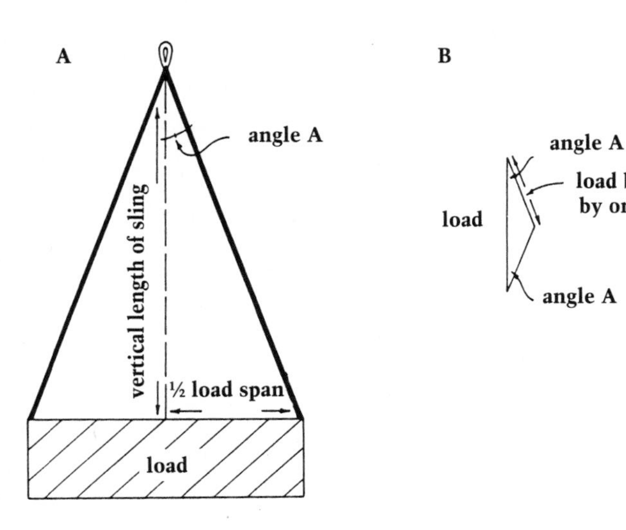

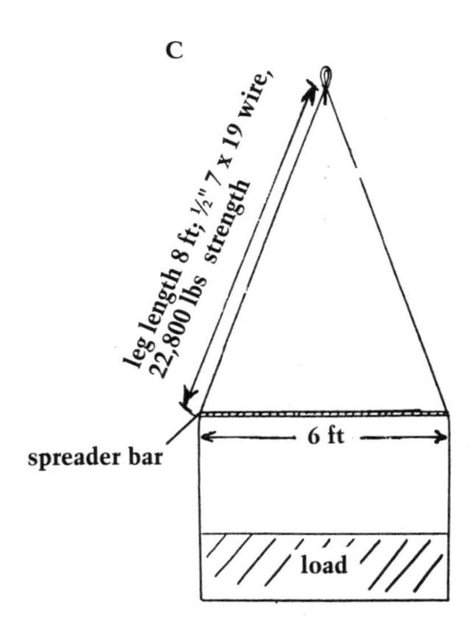

Figure 5-4. A load hangs from a simple sling in ***A****. If the load is 10,000 pounds, one-half the span of the load is 3 feet, and the vertical measure of the sling is 7.42 feet, then the compression force is 10,000 x 3 ÷ 7.42 = 4,043 pounds, or 2,022 pounds pressing inward at either end of the load.* ***B*** *shows a pictorial method to solve for the load borne by each leg of the sling—in this case about 56 percent of the supported load, or 5,600 pounds. In* ***C*** *we've added a 6-foot spreader bar to protect a load from compression. If the load were heavy enough to strain the 7 x 19 wire to its rated strength of 22,800 pounds, the spreader would have to withstand 17,100 pounds of compression.* ***D*** *shows this graphically.*

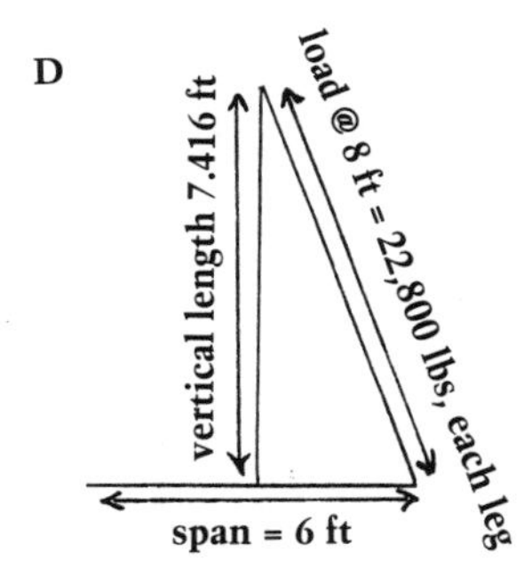

What is the maximum load that can come on this spreader bar before the wire breaks? If you know the answer, you'll know how strong to make the spreader bar. You could take the long way around, first determining what kind of load would put a 22,800-pound (10,364 kgs) strain on each leg of the sling, then calculating the compression load. But it's quicker just to juggle the formula again: Total Compression = Tension x Width ÷ Length of one leg. In this example, that's 22,800 pounds x 6 feet ÷ 8 feet = 17,100 pounds (10,364 kgs x 1.83 m ÷ 2.44 m = 7,773 kgs) that bar has to take, half that amount pushing on it from each side.

And once again, for the graphically minded, a little picture tells the same story.

Use this "prescriptive" method whenever you have some of a system's loads worked out and want to fit other components in.

SLINGS ON HOOK

Extremely wide-angle sling legs are bad practice because of excessive loading and also because it's very easy for one leg of the sling to slip off the bill of the hook.

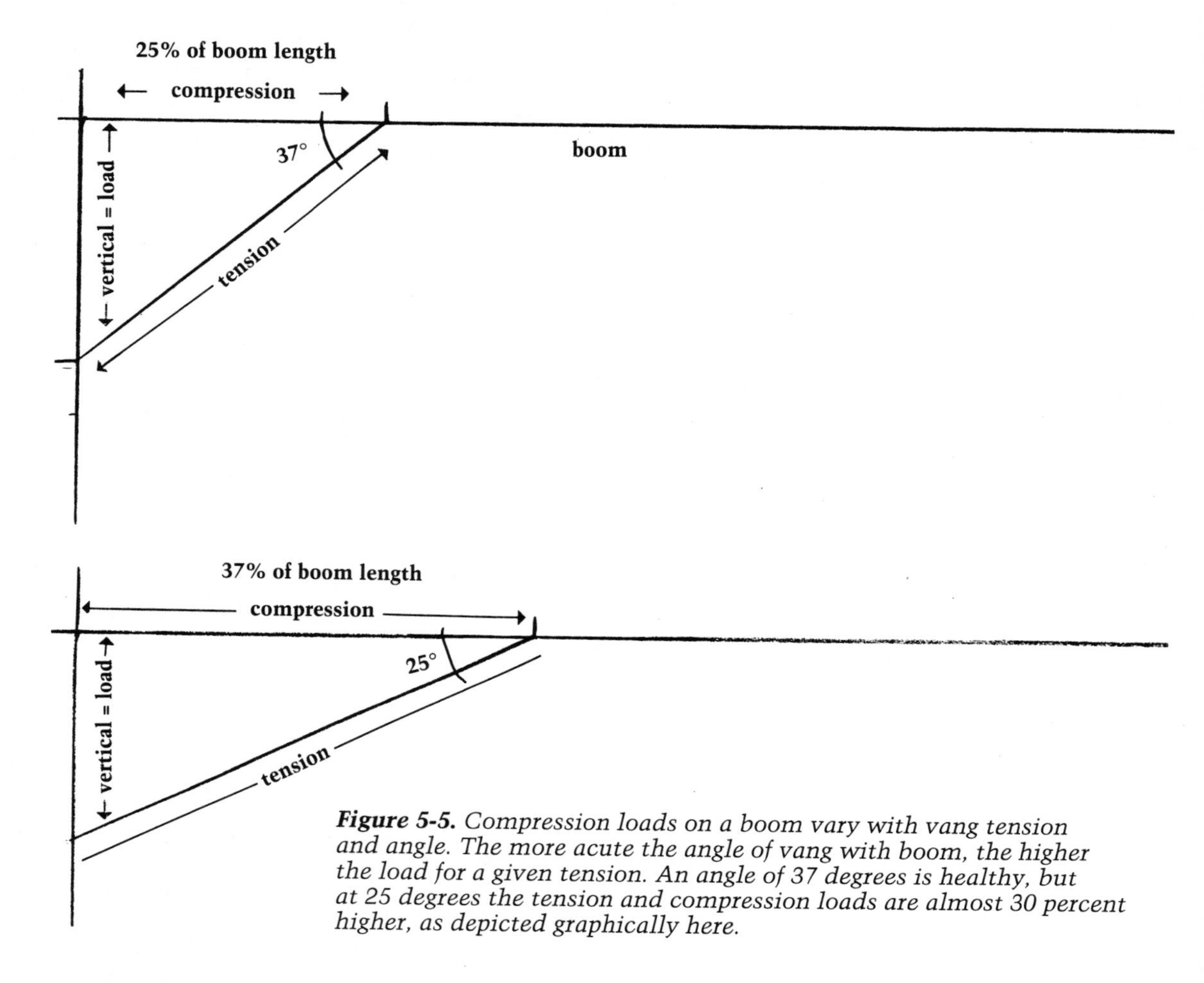

Figure 5-5. *Compression loads on a boom vary with vang tension and angle. The more acute the angle of vang with boom, the higher the load for a given tension. An angle of 37 degrees is healthy, but at 25 degrees the tension and compression loads are almost 30 percent higher, as depicted graphically here.*

BOOM VANG

A boom vang (Figure 5-5), like a sling, imposes a compression load on the object it's attached to, the load varying with angle.

Older-style boom vangs run from the boom down to deck instead of from boom to base of mast. They impose little or no compression load on the boom but must be cast off and reset with each gybe or tack. So although they control sail shape efficiently, they are labor-intensive, useful today primarily aboard infrequently tacked cruising vessels.

The modern vang swings with the boom, but as with so many modern conveniences, it brings with it complications. To start with, it must attach to the boom at least 25 percent of the boom's length aft of the gooseneck in order to have sufficient leverage for a downward pull. But as you move the attachment point aft, you narrow the angle of vang to boom, increasing both tension on vang and compression on boom relative to sail load. A high cabintop, low gooseneck, or extra-long boom can also conspire to narrow vanging angle. You'll be very lucky to get a 45-degree angle and will probably have to settle for more like 30 degrees. If you can't get even that much, you might be better off either raising the gooseneck or learning to live with a boom-to-deck vang.

No matter what angle you get or what kind of vang you use, bear in mind that vangs are hard on booms, deflecting the midsection by downward pull as well as compression. Therefore, vanged booms need to be appreciably stiffer than vangless ones, particularly when the angle is tight. So before installing a vang, get boom dimensions (depth, width at several points, length, wall thickness, sheet attachment point, and projected vang angle), sail area, and vessel size and displacement figures. Take them to a sparmaker to see if your boom will be up to the loads you want to add to it.

WILL IT FIT?

When laying out a new rig, pay close attention to hardware compatibility. Do the thimbles fit into the turnbuckles, tangs, and shackles? Are all clevis pin holes the same size, and do they line up? If you're installing a jib foil or other rollerish fitting, is the wire size compatible? Are all components of comparable strength? Rigging is a lot more than length.

MASTS, SPREADERS, AND UNSUPPORTED LENGTH

Spreaders are essential for most rigs, since without them the angle of shrouds to mast would be extremely narrow. Spreaders widen this angle, interjecting themselves as compression struts to siphon off some of the load. As Figure 5-6 shows, the longer the spreader, the lower the load on mast and rigging. Long spreaders would seem to be desirable. But there are, as usual, complications:

- As the load on mast and rigging goes down, the load on the spreader goes up. So you have to make the spreader stronger, and thus fatter and heavier,

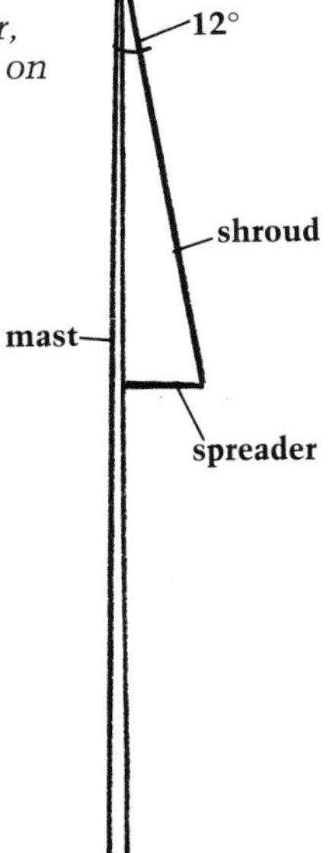

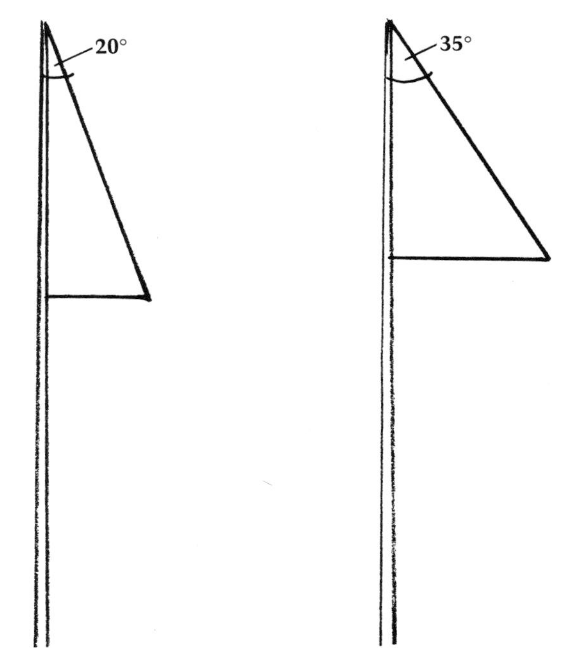

Figure 5-6. *The longer the spreader, the lower the load on mast and rigging.*

Figure 5-7. *It's much easier to buckle* ***A*** *than* ***B*** *because the stiffness of a column under compression varies with the square of the unsupported length.* ***B*** *is about one-half as long as* ***A*** *and thus about four times stiffer.*

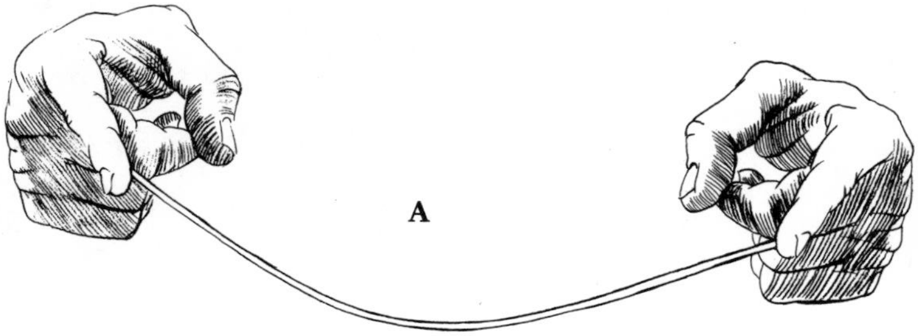

A

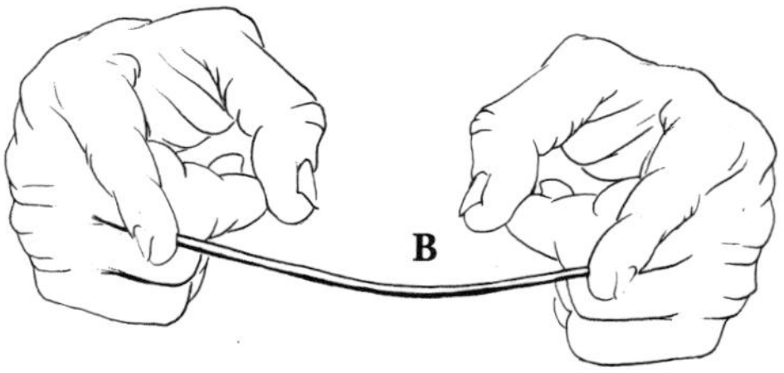

B

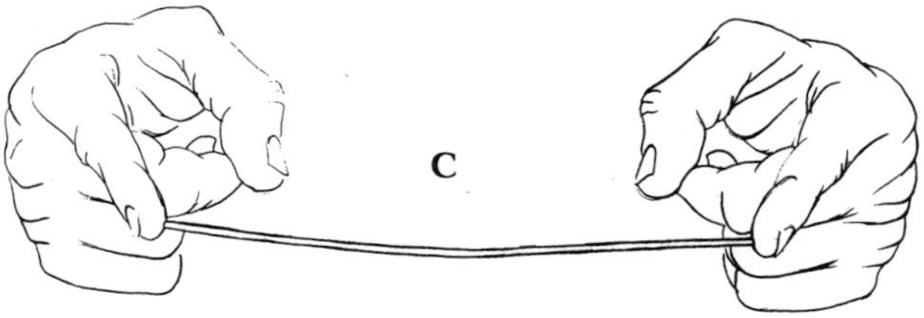

C

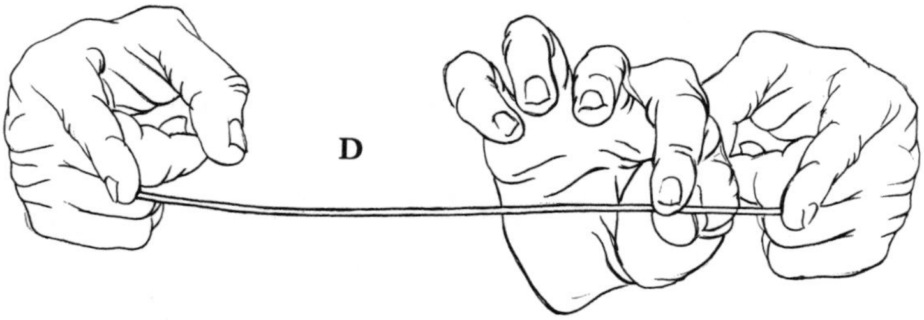

D

Figure 5-7 (cont.). *The unsupported lengths of* ***C*** *and* ***D*** *are the same, but* ***D*** *is stiffer because the third hand pinching near one end acts like mast partners, stiffening the section above it and causing it to bend less easily, and higher up.*

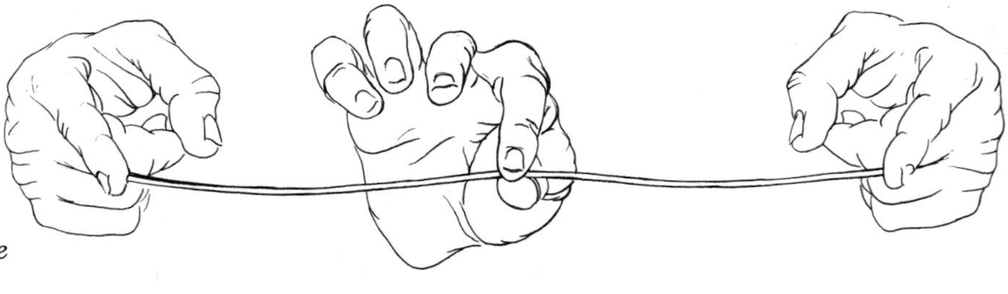

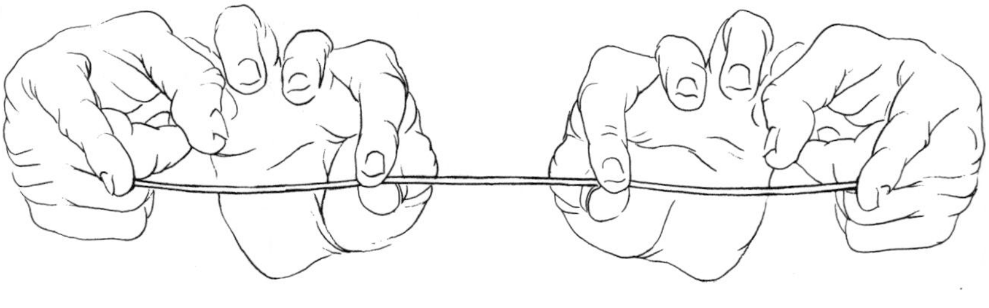

Figure 5-7 (cont.). *Here the pinching fingers simulate the effect of single and double spreaders. With double spreaders the unsupported lengths are shorter, so a noodle—or mast—of the same overall length is stiffer than it would be with no spreaders.*

to compensate. This can cancel out weight savings in mast and rig.

- A longer spreader is inherently more inclined to buckle than a shorter one under the same load. The stiffness of any compression member is relative to its "unsupported length"—the length of the section not braced by shrouds, brackets, or other deflection-preventing devices.

To demonstrate the relationship between unsupported length and stiffness, get out a piece of uncooked spaghetti and brace it between your palms (Figure 5-7). It is very easy to deflect and then break the spaghetti by moving your hands together. Now take one of the resulting shorter pieces and repeat the procedure. This time it takes more force to deflect and break the spaghetti. Finally, have a friend hold the middle of a short piece between thumb and forefinger. This shortens the unsupported length, not the overall length. Push on the ends. Now you're in danger of poking a hole in your hands before the noodle will break.

To move back to rigging, a mast is usually a moderately stiff item with a long overall length. But it is broken into two or more relatively short unsupported lengths by shrouds, stays, spreaders, and partners (see Chapters 6 and 7). Spreaders, on the other hand, have a short overall length but

LUCY BELLE'S BOBSTAY

For vessels with twin bobstays, the usual procedure is to install a separate fitting for each on the stem. Instead consider the configuration used on the Friendship sloop *Lucy Belle*: a single piece of wire is seized around a thimble, the two legs measured to suit. Advantages: One fitting instead of two; two splices and a seizing instead of four splices; and a more easily sealed and maintained lower eye, particularly when the bobstay is galvanized wire.

Of course, the seizing must be of excellent quality, since it must hold if either leg fails.

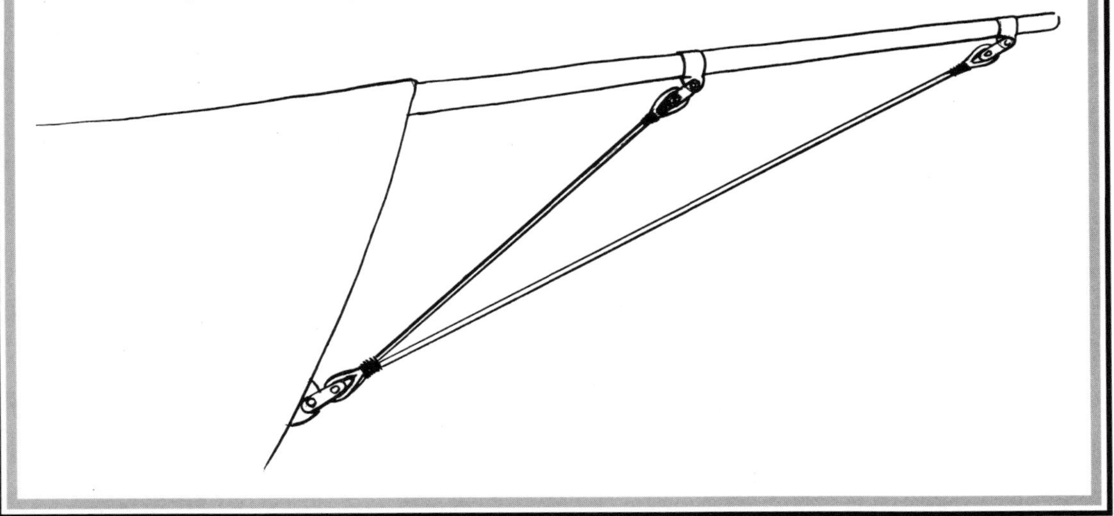

no intermediate supports. Overall length is the same as unsupported length. Spreaders therefore can only be made to resist deflection by being made stiffer and heavier. And the longer they get, the stiffer they have to be for the same load. So there is a balance between shroud angle and spreader stiffness which tends to limit spreader length (the option being massive scantlings of objectionable weight and windage).

- The maximum practical length for spreaders is in any event about half the vessel's beam at the mast. Otherwise you risk snagging docks or other people's spreaders in close quarters. And very few boats have even half-

WIRE STRENGTHS

Wire rope strength varies according to manufacturing procedures, and is graded according to strength. Each grade is about 15 percent stronger than the next weaker. Thus Plow Steel is 15 percent stronger than Mild Plow, and Improved Plow Steel is 15 percent stronger than Plow Steel.

Most stainless steel alloys have strengths comparable to improved Plow Steel. The major exception is type 316 which, although highly corrosion-resistant, is closer in strength to Plow Steel.

VOYAGER'S BOBSTAY

It is prudent to make bobstays extra-stout, both to take heavy jibstay loads, and to hold up under likelihood of corrosion and collision. But an extra-stout bobstay will often simply not fit. Especially when served, its diameter is too much for stem fittings, shackles, and turnbuckle jaws.

One clean solution, used on the schooner *Voyager*, was to make a wire rope grommet, seized around thimbles at either end. The wire is thin enough to fit into even narrow spaces, yet the combined strength of the two legs of the grommet is equivalent to that of a massively fat conventional bobstay.

In addition, the seized configuration makes for a built-in safety feature: If one leg fails due to collision or corrosion, there's a backup leg to keep the 'sprit in place until you can effect repairs.

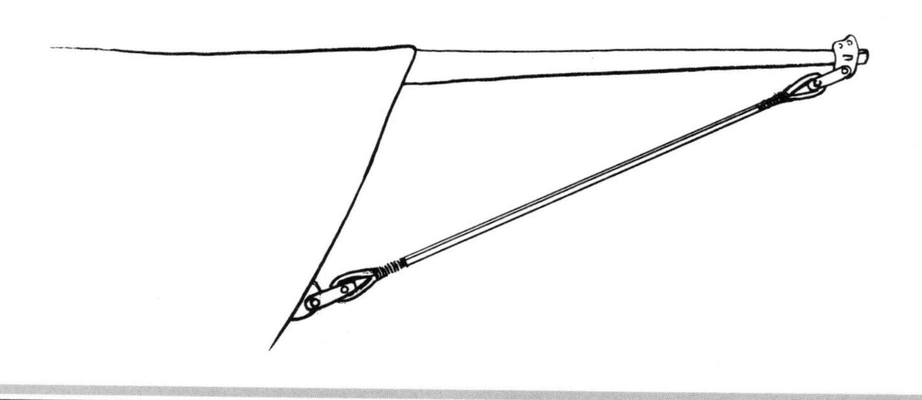

> beam spreaders, because spreader length limits how closely you can trim your headsails. Which is why all racers and not a few cruisers sport dinky little spreaders, increased mast and rig loads be damned.

With very tall rigs, the game of long-or-short spreaders becomes even more complex. A big height-to-beam ratio means that even a very long set of spreaders won't produce an acceptable staying angle. And the great unsupported lengths above and below these spreaders would necessitate an extremely heavy mast section to resist deflection. That's why designers go with multiple sets of spreaders. The topmost set gives the upper shrouds a good lead to the masthead, and because they're up high, they can do this and still be stubby enough not to poke a hole in the jib. Lower sets of spreaders are waystations for the upper shrouds on their way to the top, as well as angle-providers for intermediate shrouds. Because all these spreaders break the mast into shorter unsupported lengths, the mast section can be significantly smaller or thinner-walled (Chapters 6 and 7).

Multiple spreaders are no placebo. On moderate-height rigs, they're just so much extra clutter and expense. And the virtues of multiple spreaders are regularly abused by race-crazy sailors. They want really, really short spreaders for that proverbial close sheeting angle, so they can end up with mast compression and rig tension loads even higher than what they'd get with a single set of moderate-length spreaders.

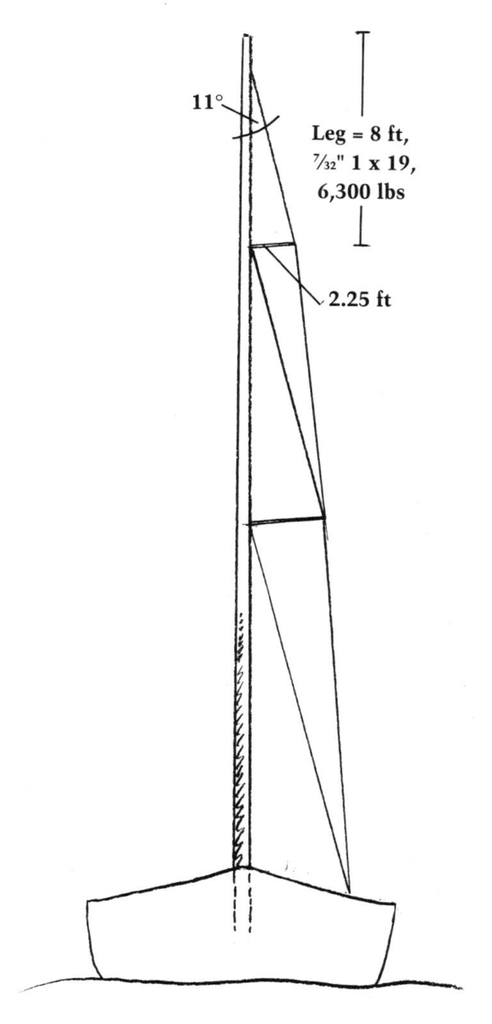

Figure 5-8. *A double-spreader rig. To calculate the maximum load that would bear on the upper spreader, solve graphically as in Figure 5-4, or use the formula, Total Compression = Tension x Width ÷ Length of one leg. Substituting, we get 6,300 x 4.5 ÷ 8 = 3,544 lbs. of compression load, or 1,772 lbs. on each spreader.*

As you can see, once you start playing with angles, you start playing with a lot of other things, too. And all must interrelate, or none will. Where, then, do you start? What factors do you consider first? In sailing vessels rig forces relate directly to the stability characteristics of the hull, but as we'll see in a subsequent chapter, hull characteristics can be largely determined by the design of the rig. Design is a matter of gestalt, arrived at by attentiveness to the interplay of forces. So the double-spreader rig in Figure 5-8 isn't just something to stick on a hull to make it look "modern" (although that has, sadly, happened more than once). Rather, it's a reflection of system awareness.

To get back to puzzles, let's find the load on the upper spreaders in Figure 5-8. They're 2 feet 3 inches (0.69 m) long, and the upper shroud is 7/32-inch (6 mm) 1 x 19 stainless with a rated strength of 6,300 pounds (2,864 kgs). The angle of shroud to mast is 11 degrees.

To solve, use the prescriptive diagram detailed under "Slings," earlier in the chapter. This matches spreader and wire strength, with the assumption that wire strength has been calculated previously to match the loads the hull and sails will impose.

Proportion. Amazing stuff. It can even reveal the loads in much more complex systems, like for instance . . .

A CRANE

A simple crane, as shown in Figure 5-9, has compression loads on boom, mast, and the ground or deck beneath the mast. Then there are varying amounts of tension on the halyard, topping lift, and guys. And unlike the fixed forces on a sailing vessel's standing rigging, all these loads change as the boom is raised and lowered. So crane design must take into account a range of possible configurations. It might sound intricate, and the stress diagram does have more pieces to it, but the relationships are the same as for simpler structures.

No matter what the level of complexity, rigging tends to be invisible to most people's eyes. If they notice it at all, it is to comment on "all those lines," and they seldom see beyond to the relationships that make it work. The next time you go to playing with rope or wire, pay appropriate attention to materials, intended use, and durability, but first and foremost, look to the most invisible and essential angles.

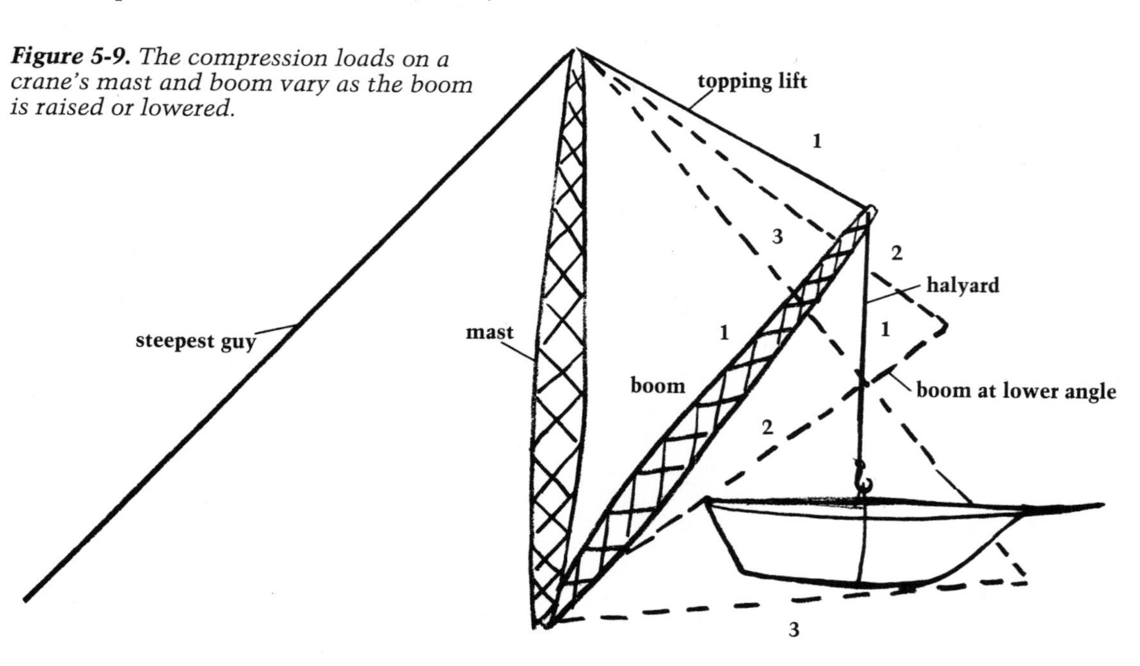

Figure 5-9. *The compression loads on a crane's mast and boom vary as the boom is raised or lowered.*

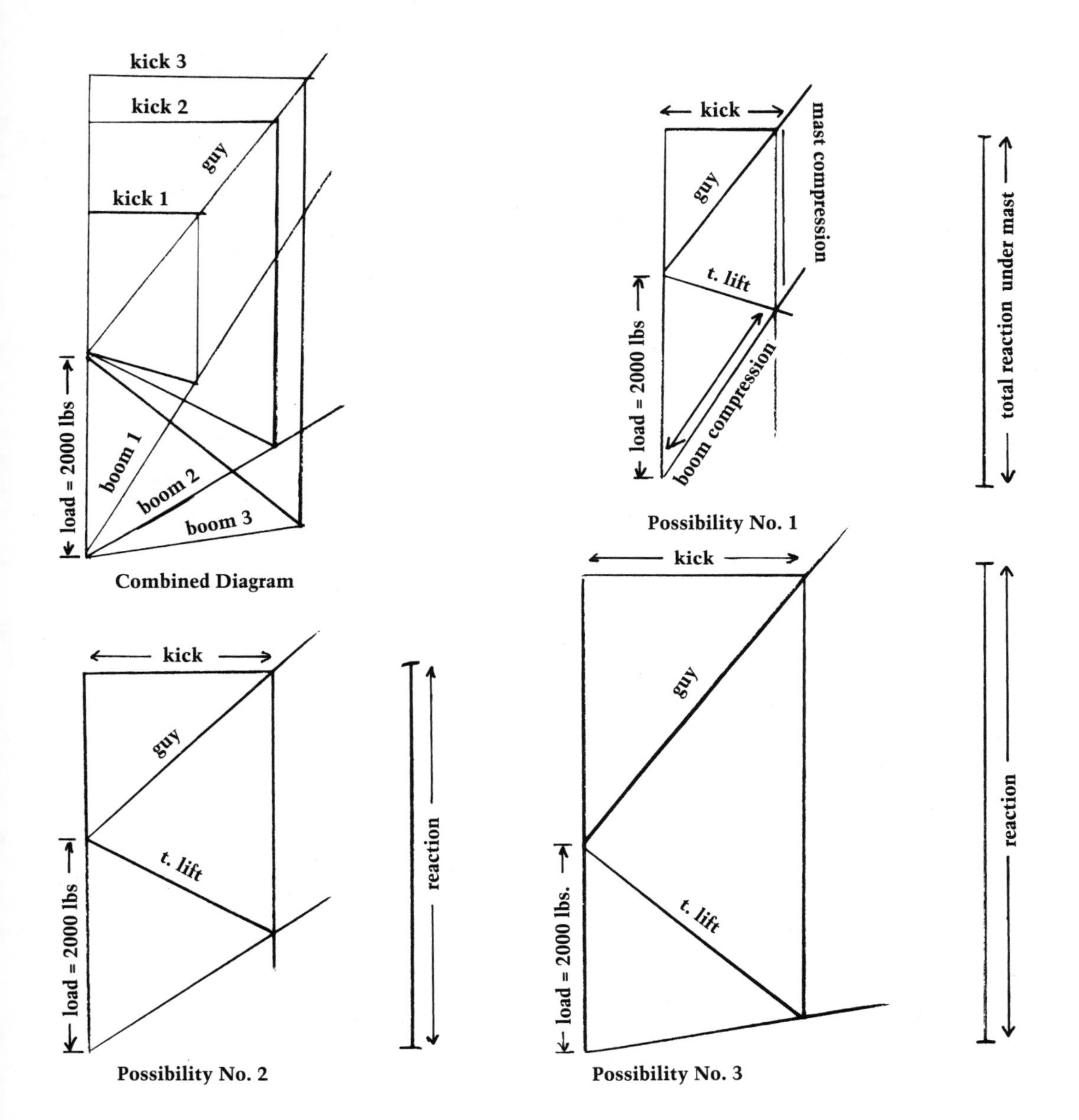

Figure 5-10. *A graphic representation of the forces resulting from the three boom positions shown in Figure 5-9. The three possibilities are sketched separately, then combined in upper left diagram. If the suspended load is 2,000 pounds, the forces work out as follows:*

	Possibility 1	*Possibility 2*	*Possibility 3*
Boom	*1,840 lbs*	*1,880 lbs*	*2,040 lbs*
Topping Lift	*1,000*	*1,760*	*2,760*
Mast	*1,740*	*2,900*	*4,520*
Guy	*1,640*	*2,620*	*3,340*
Kick	*920*	*1,540*	*2,000*
Reaction	*3,300*	*4,120*	*4,660*

Chapter 6

Mizzens

"The elaborations of elegance are at least as fascinating, and more various, more democratic, more healthy, more practical—though less glamorous—than the elaborations of power."

– Wendell Berry

In this sloop-happy world, mizzenmasts don't get a lot of respect. Ketches and yawls generally don't go to weather as well as their single-masted cousins, and so are viewed by many sailors as inefficient. That is, by those whose sole definition of "efficient" is "able to tack through 70 degrees."

But a mizzen can be more than just an extra mast. It can be evidence that designer and owner have decided that versatility and comforting redundancy offset a loss of absolute weatherliness. That the expense and complexity of an added mast is offset by reduced size, expense, and labor- intensiveness of the mainmast. That any inconvenience and clutter—the mizzen of a ketch does sit right in the boat's busiest work area—can be more than offset by a center of effort lower than that of a comparable sloop, by less sharply focused hull stresses, by a more versatile sail plan, and by increased power on a reach. This last reason is why so many of the vessels in the most recent Whitbread Round-the-World race were ketch-rigged.

Because small (under 35 feet or 11 meters) sloops and cutters already have relatively easily handled sails, mizzens are most appropriate on larger vessels. Crew laziness or non- agility or a particularly large sailplan might justify a mizzen on smaller vessels.

Regardless of vessel size, a mizzen always presents a challenge in rig design: How do you stay it adequately without interfering with the main? With few exceptions (see *Sundeer* section) there isn't room between the mizzenmast and the main boom for a forestay. There often isn't even room for much of an angle on the forward-leading mizzen shrouds. And because the mizzen is so far aft, there's also rarely room for a backstay. Designers have risen to these and other mizzen challenges with varying degrees of success. What follows is a spectrum of configurations, analyzed for interrelationship.

CUTTY SARK

A gaff ketch looks archaic to modern eyes, and it is the least weatherly rig around, but it's also simple, strong, low-cost, and very powerful when reaching. The mizzen on *Cutty Sark,* a 44-foot Angleman design, has a bulletproof simplicity. And because it is so simple, it is easier to see principles that are buried in more complex, latter-day mizzens.

The standing rigging here consists of four shrouds (Figure 6-1), all going to the masthead. Because the mast is relatively short and the vessel relatively beamy aft, the shrouds have a wide staying angle, so there's no need for spreaders (see "Angles"). The shrouds lead well forward and well aft, staying the mast fore-and-aft as well as laterally. When the mizzen stays'l is set, particularly if the breeze or chop is up, there is also a pair of running backstays to reinforce the aft-leading shrouds. There's enough space between the runners so that, on the wind, they can both be left set up and still have room to tack the mizzen back and forth.

Note that the chainplates are all the way outboard; there's no need to move them in, since the only staysail on this mast—the mizzen staysail—is an off-the-wind sail, and, unlike the main staysails, is never sheeted in

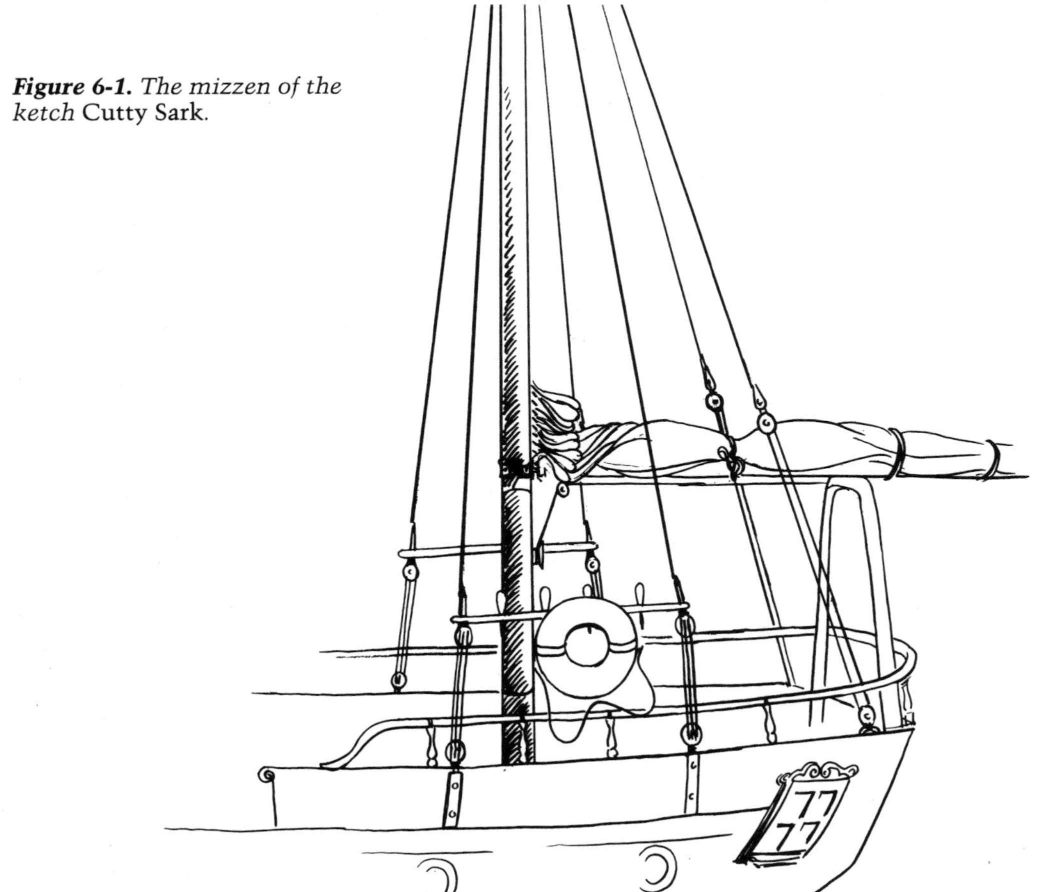

Figure 6-1. *The mizzen of the ketch* Cutty Sark.

so far that the shrouds might be in the way. And because the stays'l here is never sheeted in tight, it never puts huge compression loads on the mast (see "Mains"), so the mast and its rigging can be lighter than a comparable-size main. Lessened compression load also means that *Cutty Sark*'s mizzen can have a long, unsupported section of mast between tangs and deck. The stick will flex some, but never to the point where lower shrouds are necessary.

JENNY IVES

If *Cutty Sark*'s mizzen were taller, or thinner, or had a narrower staying angle, the compression loads relative to the mast's stiffness would be greater, and that ultra-simple

MIZZENMAST SCANTLINGS

Because mizzensails are much smaller in area than mainsails, and because mizzen staysails, set off the wind, do not impose genoa-grade compression loads, and because mizzens are usually furled when the wind picks up to prevent weather helm, mizzens aren't exposed to the level of forces that mains are. Accordingly, the standard formula for mizzenmast scantlings, while still based on RM_{30} (see Chapter 7 for mast scantling formulas), uses a much lower constant. The formula is:

$$\frac{RM_{30} \times 0.5}{\tfrac{1}{2}\text{ Beam (at chainplate)}}$$

Because the staysail is low-load, the same formula is used for both mast and rigging. A recommended safety factor of only 1.5 to 2 further reflects most mizzens' easy life.

The formula works well for most boats. Even if you lower the main in a storm and sail under staysail and reefed mizzen, the latter sail, maybe 20 to 30 percent the size of the main, just isn't big enough to generate a maximum righting moment load. An exception would be vessels like *Sundeer*, in which mizzensail area is uncommonly large (around 40 percent).

WELDING ROD COTTERS

Stainless steel welding rod (flux-free) makes a great substitute for cotter pins in turnbuckles. To install, bend a length of rod into a C-shape. The height of the "C" matches the distance between the cotter holes in the turnbuckle threads. Insert the rod and bend the ends down to finish.

Ask for 1/16- or 3/32-inch (1.6 mm or 2.4 mm) diameter (depending on turnbuckle size) heliarc welding rod, type 304 or 316.

Source: T. Tracy
Better Boat, Vol. 9, pg. 13

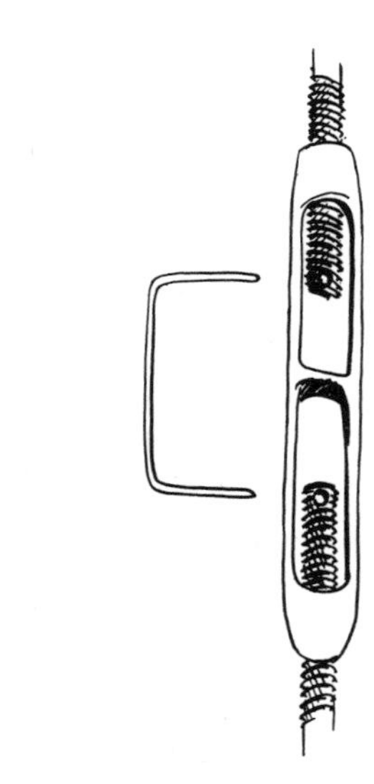

Figure 6-2. Jenny Ives*'s mizzen has a long unsupported length below the spreaders and is deck-stepped. The mast stiffness is adequate, except that as built the mast had no aft-leading lower shrouds; aft staying was provided by the aft-led upper shrouds. The spreaders, along with the pull of the jumper strut, bowed the mast forward in the middle. Aft-led lowers, sharing chainplates with the uppers, solved the problem.*

standing rigging wouldn't work; the stick would be inclined to "pump" in its mid-section. The Bermudian mizzen on the ketch *Jenny Ives* is taller and lighter with a narrower staying angle, so it needs a different standing rigging configuration (Figure 6-2).

Since strain is localized at the masthead, a forward- facing "jumper stay" running over a strut stabilizes the upper section of the mast, preventing the head from sagging aft under the pull of the sail. To widen the staying angle, the upper shrouds run over spreaders. To prevent the mast's whipping forward in a chop, the spreaders are swept aft, so the uppers pull aft as well as laterally, eliminating the need for running backstays.

Unfortunately, the compression load on the spreaders is trying to make the mast buckle forward in the middle. The lower shrouds, which have a slight forward lead, stabilize the lower section of the mast laterally, but exacerbate the spreader-induced forward bow. What to do? We could move those lower shrouds aft, but, oops, this is a deck-stepped mast. It needs forward-leading lowers to stay up.

To make things even worse, the bottom end of the jumper stay also pulls forward as

MAST WIRING GOOSENECK

Instead of routing mast wiring out through the butt of the mast (or down through the compression post of a deck- stepped mast), install a stainless steel or bronze gooseneck on deck next to the mast, and lead the wiring out next to the gooseneck.

This keeps the wiring out of the bilges and makes it much easier to install, maintain, and repair. The gooseneck can be made from heavy-duty marine plumbing fixtures.

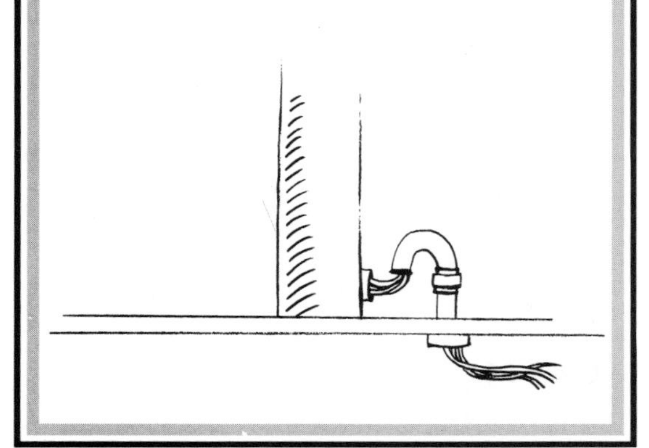

well as up at the same height as the upper shroud spreaders and lower shrouds. The grand effect of all this is to make the mizzen frighteningly mobile in any kind of wind or sea.

The lid for this particular can of worms is an aft- leading lower shroud, going to the same (overbuilt) chainplate that the uppers lead to. Most of the stress is still on the forward legs, to hold the mast stable forward; sufficient tension is applied to the aft legs to keep the mast stable fore-and-aft.

CIRRUS

The N.G. Herreshoff yawl *Cirrus* is from an era—the 1920s—when rigs and hulls were changing from low-aspect designs like *Cutty Sark* and *Jenny Ives* to the sleeker, spindlier rigs of our own era. The mast is relatively tall and set in a skinny boat, so there are higher compression loads on a narrow mast. Accordingly, the mast is supported at more points—with lower shrouds, upper shrouds, and a pair of lateral jumper stays called "diamond stays." The distance between the staying points is called "panel length;" the higher the compression loads and the lighter the mast, the shorter the panel lengths need to be to prevent buckling (see "Mains" for mast scantling details).

Short panel lengths are characteristic of modern masts, but *Cirrus* also has some old-timey details (Figure 6-3): Although her lower shrouds go to tangs, her uppers and the upper ends of her diamond stays attach with mast-encircling "soft eyes" set on carved bolsters. Another unusual archaic detail is that the main backstay and aft-leading mizzen lowers are attached by a curved tang; essentially, the mizzen is trapped by a split main backstay. In the interests of easier handling, *Cirrus* was converted from sloop to yawl early in her career, and this novel fitting was one result. It eliminates the problem of the mizzen's being in the way of the main backstay, and simultaneously provides fore-and-aft support for the lower section of the mizzen. This is an effective configuration for a racer like *Cirrus*, but it's a no-no for a cruising vessel, precisely because it ties the two masts firmly together. In the event of

MINIMIZING HALYARD FATIGUE

To avoid accelerated wire fatigue, never let a splice, nicopress sleeve, or other terminal get within 2 inches of a sheave or fairlead.

a rigging failure, it's "as the main goes, so goes the mizzen."

As a final note, there are running backstays here, because, unlike *Jenny Ives*'s aft-swept upper shroud spreaders, the diamond stays supporting *Cirrus*'s masthead provide lateral staying only.

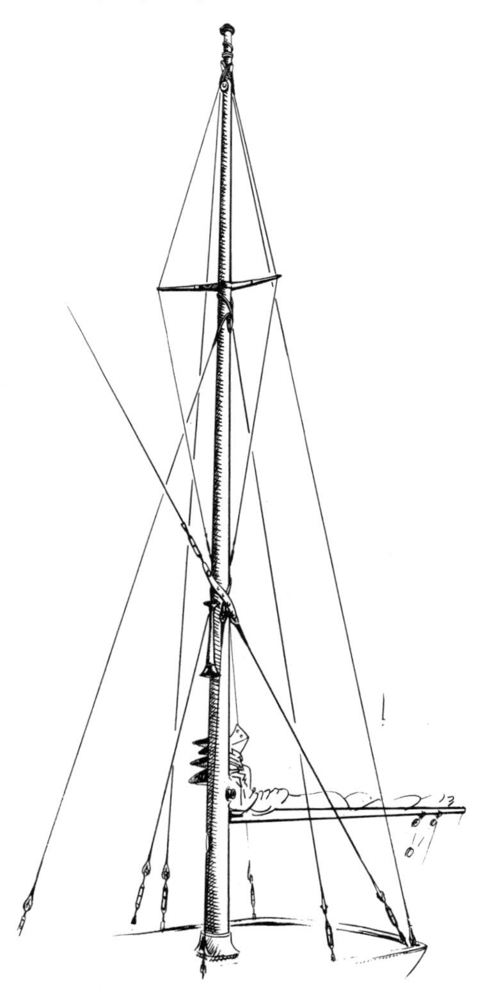

Figure 6-3. Cirrus*'s mizzen is early Bermudian with traditional and contemporary details. The upper shrouds and diamond stays terminate in soft eyes at the mast, while the lowers end in tangs. The main backstay terminates at the mizzen, and a custom fitting takes its load directly to the aft-leading mizzen lower shrouds.*

CONCORDIA YAWL *PORTUNUS*

If *Cirrus* has an innovative change-of-era rig, the Concordia yawl (Figure 6-4) represents an early mature version of Bermudian mizzen design. The soft eyes are gone, replaced by tangs all around. The main back-

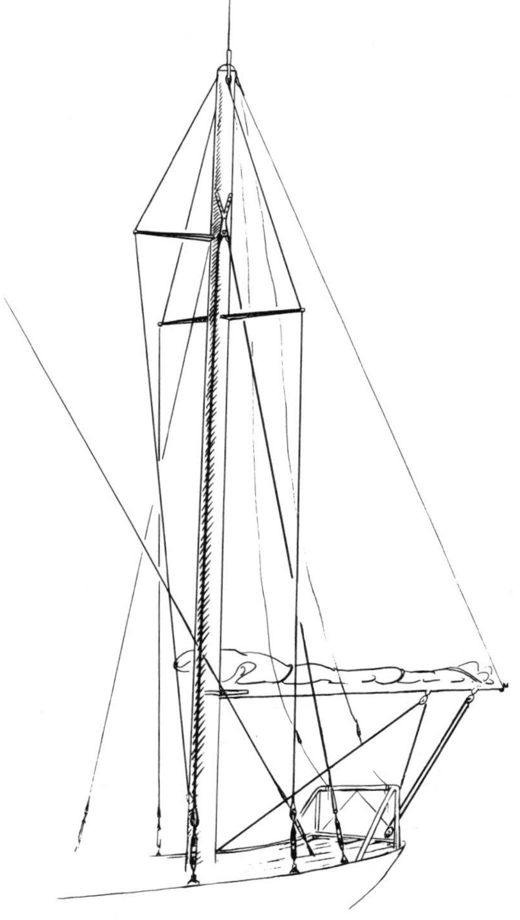

Figure 6-4. *The mizzen of* Portunus *has a long unsupported lower section, but avoids Sojourner's problems (Figure 6-6) by dint of heavy scantlings, an oval cross section, and being stepped on the keel. In addition, the jumper strut leads to the base of the mast instead of the middle, where it would pull the mast forward. The running backstays can be made off to the quarters in heavy weather to prevent the mast's upper section from "whipping."*

stay angles off to one side, ju-u-st enough to miss the mizzen, so the two masts are independently stayed. There's a jumper stay, a la *Jenny Ives*, but here the lower end of the stay comes clear down to the base of the mast, so it doesn't buckle the middle. This has the added advantages of lessening compression on the jumper strut and getting the turnbuckle down where it's easy to adjust.

Mast shape has changed, too. The previous masts were round, but *Portunus* has a rectangle, with the long side running fore-and-aft. This makes the mast stiffer in this plane, where most of the buckling forces are. And the mast is big and stiff enough in both planes at the lower end that the long unsupported panel below the spreaders is no problem. Now the lower shrouds can go way up, to stabilize things in the way of the jumper strut. The upper shrouds hold the masthead laterally, while their spreaders stabilize the mast below the lower shrouds. With so much fore-and-aft stiffness, the running backs are practically vestigial, set up far less often than for the preceding boats.

By playing rig configuration and mast size and construction off against each other, the Concordia's designer, Raymond Hunt, was able to come up with a mast that could take a lot of load, yet be quite simple.

NABOB II

So much for simplicity. *Nabob II*'s mizzen is infested with rigging, radar, baggywrinkle, antennae—it looks like the winner of a design competition for Most Occupied Mast. But once you look at the gear piece by piece, in the context of the vessel it fits, you'll see that it's thoughtfully configured.

Nabob II is a beefy, 55-foot (on deck) Spaulding Dunbar ketch. Her entire purpose in life is to take people to far- off lands in utmost comfort and security. Europe, Africa, the South Pacific, Far East—*Nabob*'s been there.

So let's start with the upper shrouds (Figure 6-5). They're easiest to find. There are

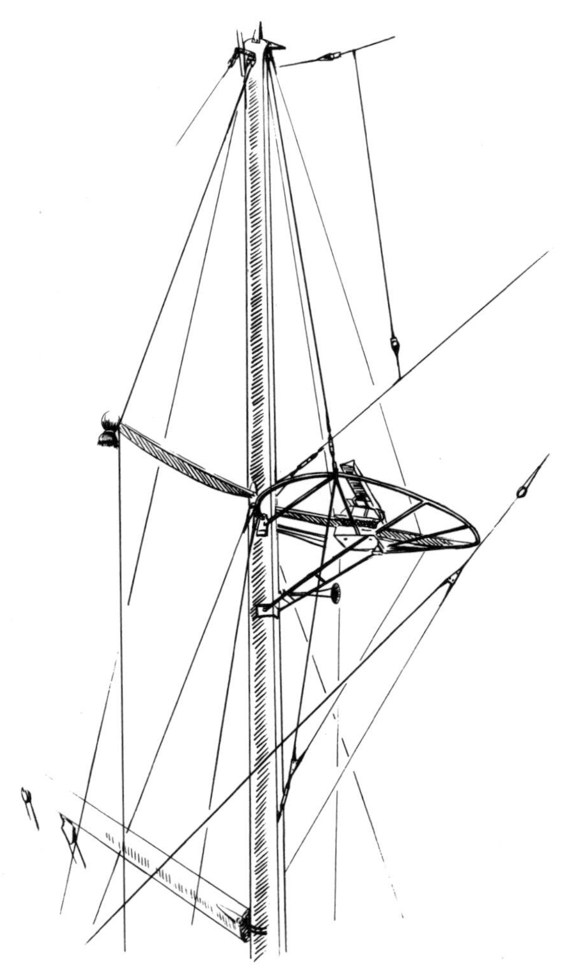

Figure 6-5. Nabob*'s mizzen is a heavily and redundantly rigged mast suited for serious cruising. The two sets of lower shrouds are complemented by two sets of uppers, one leading over the spreaders and the other led aft to serve both as backup and backstays. The radar bracket doubles as a strut for the jumper stays. A lower springstay (just above the radar) holds the mast forward, while an upper springstay functions as an antenna. The vertical wire between the springstays is also an antenna leg.*

COMPASSES AND EYEGLASSES

Metal eyeglass frames can throw off readings on "hockey puck"-style compasses.

two sets, one going over wide, well-padded spreaders for upper-panel lateral support, and another leading aft as backstays; this second set could have been running backs led right to the stern for a more favorable angle, but that would have meant more strings in a sometimes-shorthanded boat. Instead, the spreaderless uppers angle far enough aft for good support, yet are far enough forward that the mizzen boom can swing out on a broad reach.

Two lower shrouds per side fix the lower panel in all directions. Notice how the wire ends go to different heights; they're independent of each other as redundancy in case one should break, and they spread the load over a wider area of mast than if they both went to the same tang.

The big news on this mast is the jumper stay/radar housing setup. The double jumpers, angled diagonally forward, hold the masthead forward, prevent the middle of the mast from buckling forward, and help stay the masthead laterally.

WIND AND TIEDOWN STRAPS

Anytime a trailer tiedown strap is over space—as between the gunwales of an open boat—it will flutter and hum in the wind, greatly shortening strap life. To prevent this give the strap a half twist where it crosses open space.

Just above the radar is a springstay leading to the mainmast. This does tie the masts together, but it's a calculated tradeoff. The compression load from the radar/jumpers wants to push the middle of the mast aft, and those aft-leading uppers don't lead far enough aft to counteract with sufficient compression-bow forward. The springstay takes the excess jumper compression and delivers it to the mainmast. The aft-leading lower shrouds assist the aft-leading uppers in preventing the entire mast being pulled forward by the springstay. Tricky, but it gets the job done.

There's also an upper springstay at the masthead, but this is solely a place to hang an antenna. Same with the vertical wire connecting upper and lower springstays. The owner is into electronics.

That leaves the main backstay. It has an antenna on it, too (note insulator above "Y"). But it's of interest to us because it splits into two legs just below the radar. This is the standard contemporary method of getting the main backstay past the mizzenmast. It leaves an uncluttered area directly behind the mizzen, allowing for an aft cabin on this boat, and splits the backstay load between the sides of the hull. You'll sometimes see two entirely independent backstays in lieu of this arrangement. They make for more weight and windage, but can be a comforting redundancy on a cruising boat. Twin backstays are essential for vessels with twin jibstays, to keep both jibstays taut.

SUN AND TIEDOWN STRAPS

Braided tiedown straps are susceptible to damage from sunlight. As with all regularly exposed plastic items, inspect them regularly for signs of discoloration and brittleness.

SOJOURNER TRUTH

This beautiful Morgan Giles–designed ketch is plagued by a mizzenmast so limber it makes the one on *Jenny Ives* look positively docile.

The lower shrouds attach high up (Figure 6-6), a la Concordia, but unlike the Concordia, the spreaders for the upper shrouds are also up high, leaving a long unsupported panel to deck. Worse yet, the bottom of the jumper stay attaches to the middle of that unsupported panel, encouraging it to buckle. Even so, these details would not be so much of a problem if the mast were simply stiffer. But it isn't, and being wooden, it can't readily be stiffened, like aluminum, by the addition of an inner sleeve.

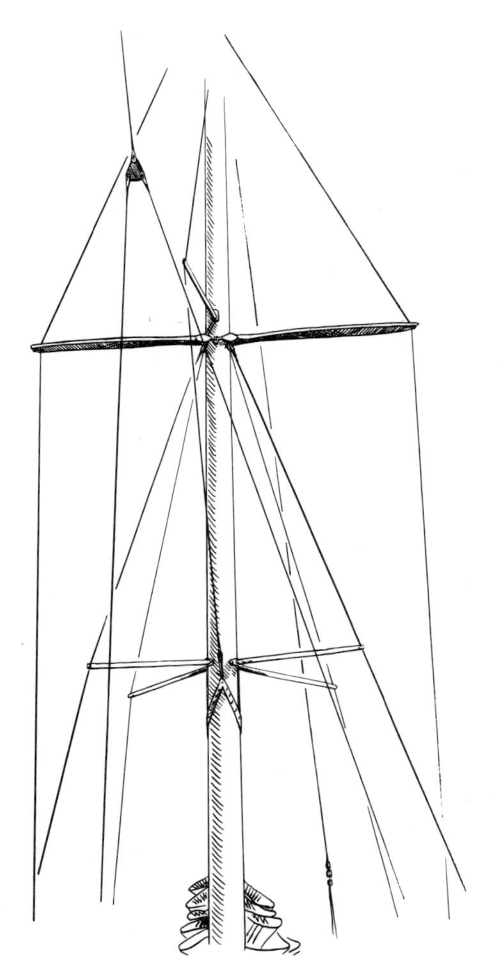

Figure 6-6. *Sojourner Truth's mizzenmast has a light section, a long unsupported length below the spreaders, and is deck- stepped. The result is an excessively flexible lower section. "X-spreaders" on the lower shrouds effectively shorten the unsupported length. Together with careful tuning, this stabilizes the mast.*

It might help to relocate those spreaders down below the lower shrouds, but this stick is such a noodle that that might not be enough; it would be a shame to undergo a project like that and not fix the problem. And so, although it's a bit unorthodox, this mast has been fitted with "X-spreaders," four struts affixed in the way of the jumper stay's lower end, and braced against the four lower shrouds. It helps some.

It's hard to be sure, but I would guess that someone, somewhere along the way, neglected to take into account that this mast is deck-stepped. A keel-stepped mast, because it's supported at deck level by the partners, is stiffer than a deck-stepped mast of the same scantlings. In essence, the partners take up some of the compression as a lateral thrust load, while the rest of the compression is delivered to the step on the keel. A deck-stepped mast, bereft of this load-sharing setup, must be stiffer—much stiffer, by about 50 percent—in order not to buckle under the same loads. *Sojourner Truth*'s mizzen acts like a deck-stepped mast designed to keel-stepped specs. For further details on calculating mast loads, see "Mains."

SUNDEER

Yacht designer and world cruiser Steve Dashew brings mizzens into the New Age. There's a high-aspect, double- spreader, intentionally "bendy" rig on his evolutionary new ketch *Sundeer* (Figure 6-7). And there's

Figure 6-7. *Steve Dashew's* Sundeer.

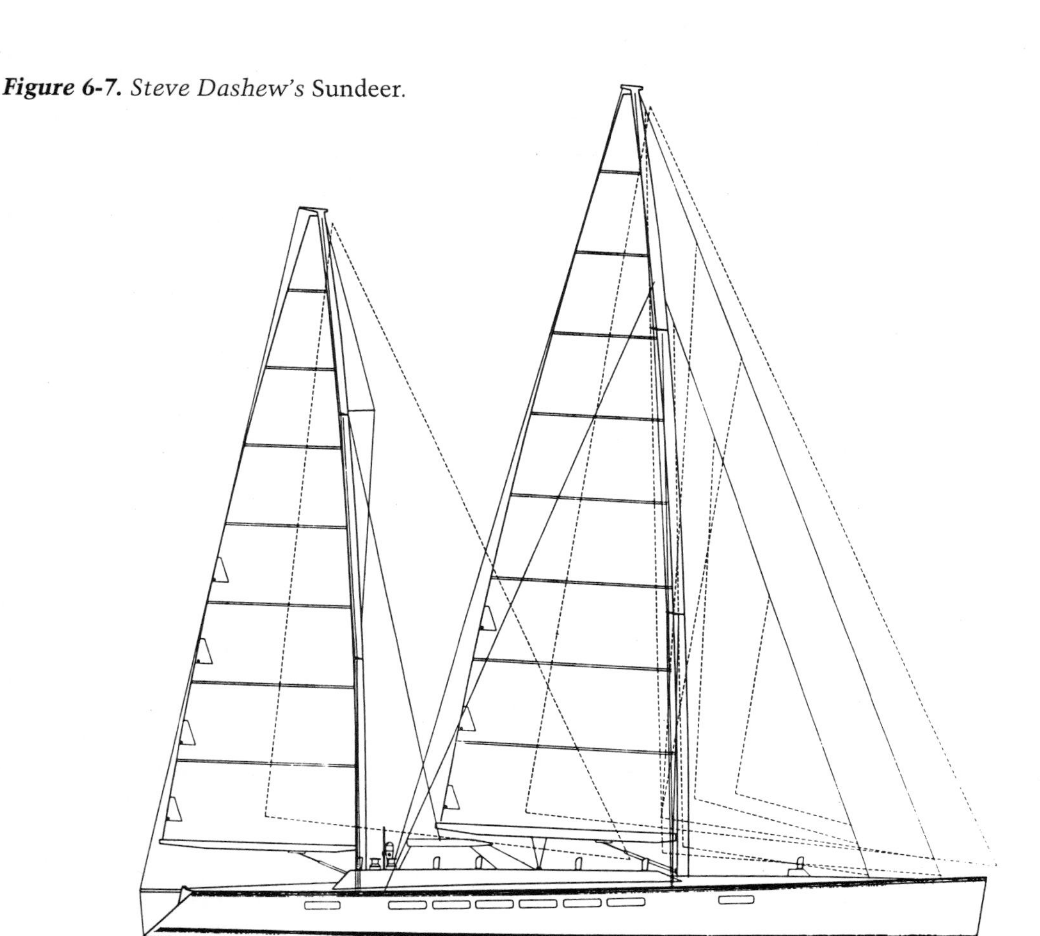

even a forestay and backstay. Since these details are more commonly associated with mainmasts, I'll be discussing their significance in the next chapter, "Mains."

Modern details aside, this mizzen has a lot in common with the ones mentioned previously. Like them, it's a place to hang a staysail for reaching power, makes for a lower center of effort than a sloop of comparable sail area, and is part of a versatile, easily handled sail plan. But there are two other important mizzen virtues that *Sundeer* in particular exemplifies. One, mentioned briefly at the beginning of this essay, is the mizzen's helpmate relationship with the main: sloop proponents talk about a split rig's "inefficiency," then usually go on to how having a mizzen means you have to buy a whole extra mast, sails, and rigging. They admit only grudgingly that a ketch or yawl might be easier to handle or more versatile. And they never mention that the main on a ketch can be much smaller and cheaper than it would be if it had to absorb the mizzen's sail area. Nor do they take into account that the mizzen prolongs the main's life by reducing the intensity of the cyclic loading that leads to metal fatigue. In *Sundeer* the mizzen is over half the size of the mainsail. This is a big mizzen (20 to 40 percent of main is more typical) for a ketch, but any appreciable mizzen is a lot more than an extra mast stuck in the back of the boat.

The other mizzen virtue has to do with

the relationship of the mizzen to the hull. By distributing stress over a wider area, a split rig is kinder to its hull than a monomast. With many boats, this distribution advantage is qualified, since mizzens, at least on ketches, are often reefed or lowered first when the wind comes up, leaving the main to deal with heavy weather. This can be because main and staysails provide more drive than mizzen and staysails, but mostly it's because on many vessels, weather helm increases sharply with increased heel (see "Seaworthy"). Mizzens, being so far aft, only exacerbate weather helm, so down they come. But this is a design flaw in hull, not sail. A balanced hull like *Sundeer*'s does not suffer hull-induced weather helm as it heels.

And on *Sundeer*, Dashew has gone a step further, intentionally matching hull and sail plan so that there is always a great deal of weather helm, all of it mizzen-induced. On most vessels this would result in a hard-to-steer boat, but *Sundeer* has a large balanced spade rudder, so the helm always feels neutral. Why do this? Because a big, properly shaped balanced rudder can provide lift, just like a keel. If it can provide enough lift, you can make the keel smaller and still go to weather well. So *Sundeer*'s rudder is helping the keel, just as the mizzen is helping the main. The net result is that this 67-foot (20.4 m) O.D. ketch draws only 6 feet (1.8 m) loaded, yet will outpoint many sloops, especially in a breeze, when speed gives the rudder more lift. Balanced spade rudders are generally frowned on by cruisers as fragile,

NICOPRESS CRIMP SEQUENCE

When crimping sleeves which require three crimps (1/8 inch, 5/32 inch, 1/4 inch, 5/16 inch, 9/16 inch, and 5/8 inch) crimp the middle first, then the thimble end, then the standing part end. This lets the sleeve expand away from the middle, preventing internal stresses. For sleeves requiring four crimps (3/16 inch and 7/32 inch) start at the standing part side of center and work toward the thimble. Finish with the standing part end.

vulnerable things, but *Sundeer*'s is built around an 8 inch (203 mm) diameter (!) rudder shaft, and has a sacrificial "crushable" bottom; it's extremely unlikely that even a violent grounding would cause significant damage.

It is unusual to have rig and hull so creatively interlinked, but it's possible to optimize the performance of any split rig relative to the hull it sits in. On some boats this might involve flatter- or fuller-cut sails, adding a bowsprit, changing mast rake, etc. A qualified rigger or yacht designer can help you with particulars. Meanwhile, I hope this section has given you enough information to extrapolate from, whether it's for a configuration that will allow you to disconnect a springstay, or to let you see force relationships more clearly, or just as an introduction to the next chapter: Mains.

Chapter 7
Mains

"Nothin' too strong ever broke."
– Maine Proverb

Mainmasts are more than just great big mizzens. The loads they bear are of a whole other order of intensity and complexity. Loading from staysails is particularly significant; unlike mizzen staysails, the ones on the main are meant to be trimmed in hard when the boat goes to weather, and this adds tremendously to the compression load on the mast.

Nevertheless, design is simply a matter of balancing a few factors: hull, mast, and standing rigging. This chapter will take a close look at those factors, using formulas to crystallize the general information in "Angles" and "Mizzens," to raise that information to the level of design.

THE HULL

The wind is always trying to knock the hull over, the hull is always fighting to remain upright, and the rig is caught in the middle. Accordingly, the strength of the rig must be scaled to stand up to the amount of knockover-resisting force the hull generates. Think of the rig as a big lever stuck into the hull; you don't want that lever to break.

STUCK MAINSAIL

If your mainsail halyard has jammed with the sail up, and a squall is approaching, let go the outhaul and lash the sail to the mast by wrapping the main halyard around the mast, "Maypole" style inside the lower shrouds.

The hull also acts as a lever, and it gets its power from ballast and buoyancy. The more ballast it has, and the lower it is mounted, the more the ballast will help lever the boat upright. Likewise, the fuller and more buoyant the hull, the greater amount of leverage buoyancy will exert.

Leverage is a matter of force applied over distance—the farther from the fulcrum one exerts a given amount of force, the greater the effect it has. So we'll measure our forces in "foot-pounds," to translate our particular forces and distances into measurable effects. The name for the amount of foot-pounds a lever can exert is a "moment." The moment that a hull exerts in trying to stay upright is a

"Transverse Righting Moment" (R.M.). This righting moment varies with each degree of heel, depending upon the shape of the hull, the distribution and amount of ballast, the weight of construction materials, etc. You can plot these shifting moments on a stability curve (Figures 7-1 and 7-2). This curve is as distinctive as a fingerprint, and can tell you a lot about the performance and safety of a hull (see "Seaworthy"). It's also a fundamental mast design tool, since it can tell you the loads the mast and rigging will face.

Righting Moment starts at 0 with the hull upright, then climbs in a nearly straight line to at least 30 degrees, sometimes up to 40 degrees. After that, R.M. increases more slowly, to its maximum, and then begins to decrease. Since maximum R.M. indicates the maximum sustained load the rig will have to bear, you'd think that designers would use this figure as the basis for mast scantlings. But this is not so; they find the R.M. for 30 or 40 degrees, depending on the formula they're using, then multiply by a safety factor to take maximum R.M. into account. Why? For one thing, it's very easy to find R.M. at 1 degree of heel, then extrapolate along that nearly straight line to find RM_{30} or RM_{40}, then work in the safety factor. It's much harder to calculate an entire stability curve to get maximum R.M. directly. For another thing, almost all of a vessel's sailing is done within 30 degrees of heel; you can use that as a benchmark, then safety factor generously or stingily, depending on how safe and solid for cruising or skinny and scary for racing you want your rig to be.

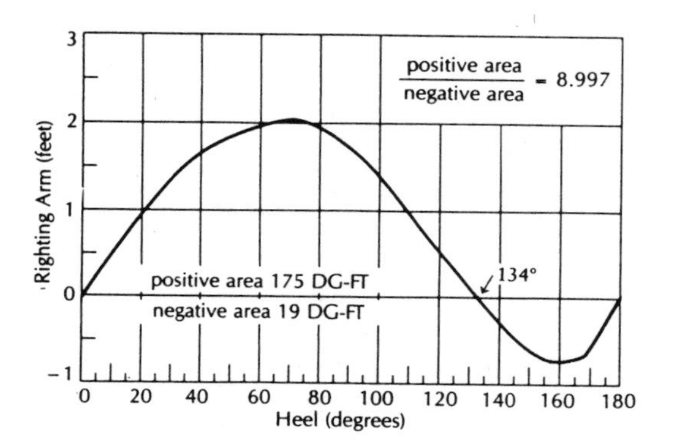

Figure 7-1. *Static stability curve for an Ohlson 38A as calculated under the International Measurement System. The righting arm is simply the righting moment of the boat divided by its displacement. Above the baseline the curve shows positive stability; below the line, negative stability. The larger the negative stability, the more disinclined is the boat to right herself from an inverted position after capsize. The degree of heel at which the curve crosses the baseline from positive to negative righting moment is the angle of heel at which the boat will capsize—134 degrees in this case. The ratio of the positive area of the curve to its negative area—here 8.997—is a measure of seaworthiness. Most boats measured by the IMS will capsize at about 120 degrees and have a ratio of around 4—minimal numbers for medium-size offshore sailing yachts; smaller seagoing boats need higher numbers. (From Sea Sense, 3rd Edition, by Richard Henderson. International Marine, 1991.)*

A FORMULA

To put what we have so far into numbers, the transverse load on the rigging = RM_{30} (Righting Moment at 30 degrees of heel) x 1.5, where 1.5 is our safety factor. If our RM_{30} was 50,000 foot-pounds (6,925 meter-kilograms), then RM_{30} x 1.5 would result in 75,000-foot pounds (10,388 meter-kilograms). But remember, foot-pounds is a convenient way of describing force over distance, whereas when we go to buy the wire, we need to know the load on it in plain old pounds. To make this translation, we divide by the length of the lever arm of the hull, which is to say one-half the vessel's beam at the chainplates. My assistant, Erin Sage, says this is like putting foot-pounds into a colander and straining out the feet.

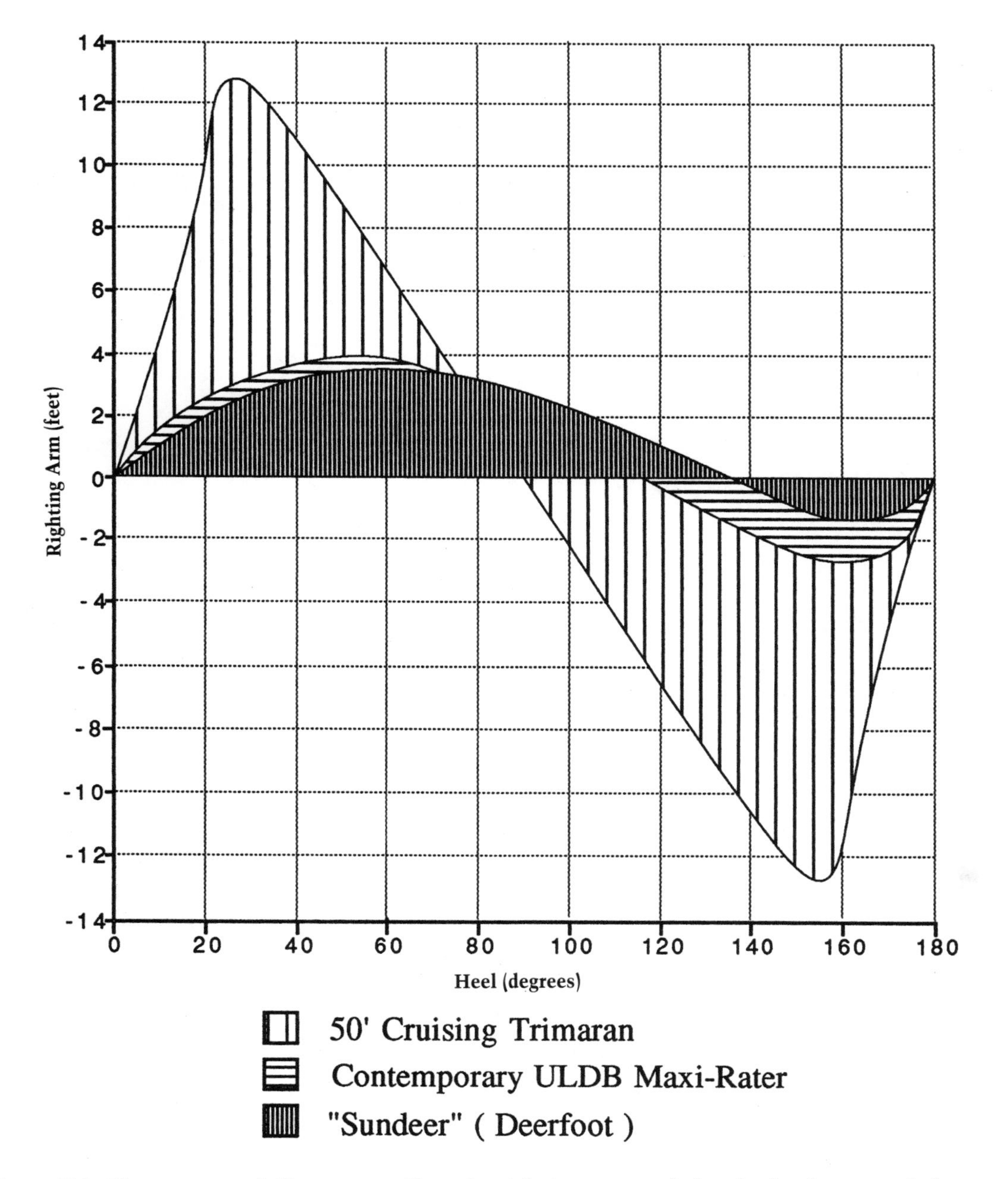

***Figure* 7-2.** *Three more stability curves. Here the righting-arm scale has had to be expanded to accommodate the much higher stabilities, both positive and negative, of a multihull. Note Sundeer's exceptionally high capsize angle (135 degrees) and ratio of positive to negative stability; even if she does capsize, she'll right herself immediately. The exceptionally high initial stability of a multihull dictates sturdy rig scantlings. (Courtesy Jeff Van Peski)*

MAST PULPIT/PINRAIL

Here's a great idea for any boat big enough to accommodate it. A sheerpole is bolted to the mast pulpit legs, and halyard ends are taken to rings on the sheerpole. Hauling parts can be secured to the cam cleat, and the sheerpole is also a handy spot to hang extra lines and coils. Belaying pins and cleats are further options. This is an excellent way to reduce cockpit clutter while adding security for crewmembers forward.

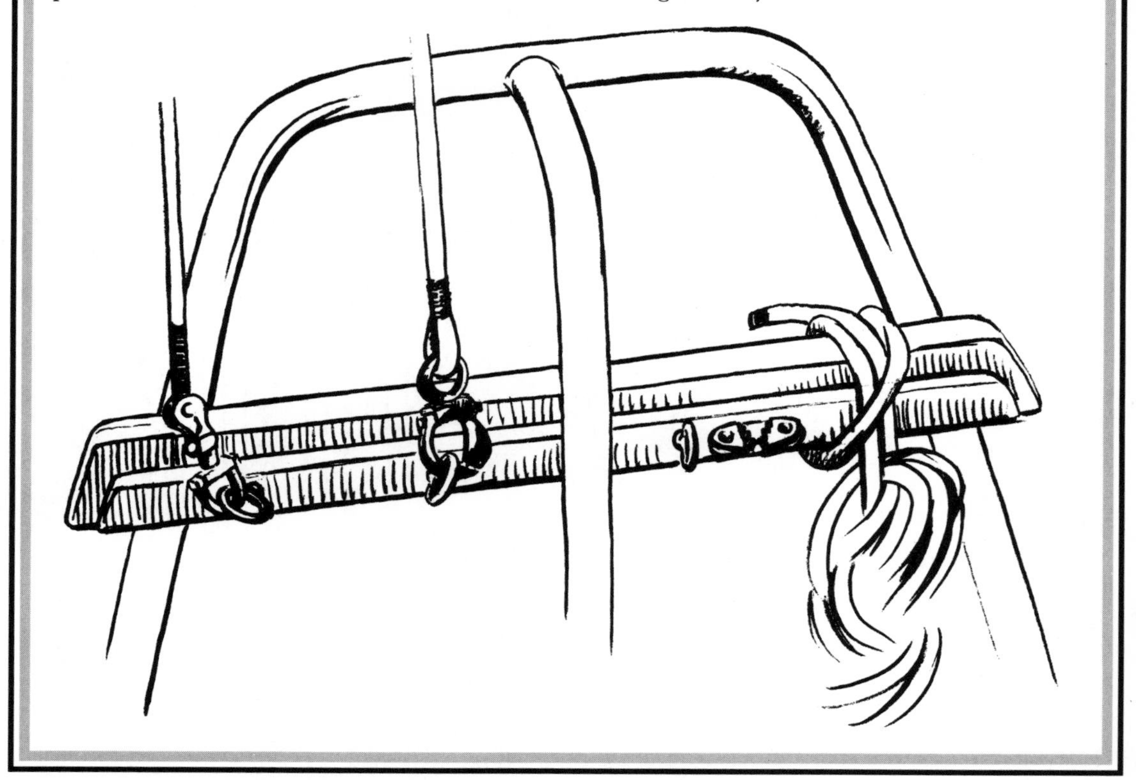

If our sample boat has a one-half beam of 5 feet 6 inches (1.67 m) the formula would read:

$$\text{Transverse Load} = \frac{RM_{30} \times 1.5}{\tfrac{1}{2}\text{ beam}}$$

and in our case that's:

$$\frac{50{,}000 \text{ lbs.} \times 1.5}{5.5} \quad \left(\frac{6{,}925 \text{ kgs} \times 1.5}{1.67}\right)$$

which equals 13,636 pounds (6,220 kgs).

At last! We now know how many actual pounds of force will be pulling on the weather shrouds when the boat is heeled to its maximum righting moment angle.

SELECTING WIRE

But you don't just go out and get a piece of wire with a 13,636-pound (6,220 kg) breaking strength, since unless the boat is very small or the mast very short you'll want at least two wires, attached at different points, to spread the load on the mast. And it is prudent to make these wires stronger than they absolutely have to be—two or three times stronger—as a safety factor for shock loads that could exceed the maximum R.M., to compensate for the eventual degradation of the wire, and because the wire will be permanently damaged if it is stressed much beyond 50 percent of its breaking strength.

In theory, you can share the load among any number of wires. In practice, you'd probably choose among a few tried-and-true configurations, depending on the type of hull, the type of sailing intended, and your own pet theories. Figure 7-3 shows two of these configurations, with the percentage of the load each wire will bear. Note that the numbers add up to more than 100 percent; different combinations of sail and sea condition will put varying loads on each wire.

For our example boat, a cutter with a length on the waterline of 35 feet, we'll use the second configuration shown, a double-spreader rig with a pair of lower shrouds on each side. The lowers share 50 percent of the load, and the uppers and intermediates take 30 percent each. Because the lowers share that 50 percent, many designers make them smaller than the other shrouds, assuming they'll take only about 25 percent each. But because the mast can shift fore and aft significantly under sail, each lower may take a great deal more than its supposed share at times. It is prudent to make the lowers as

Figure 7-3. Percentages of the maximum shroud load allotted to each shroud in a single-spreader and a double-spreader rig. (From Sea Sense, 3rd Edition, by Richard Henderson. International Marine, 1991)

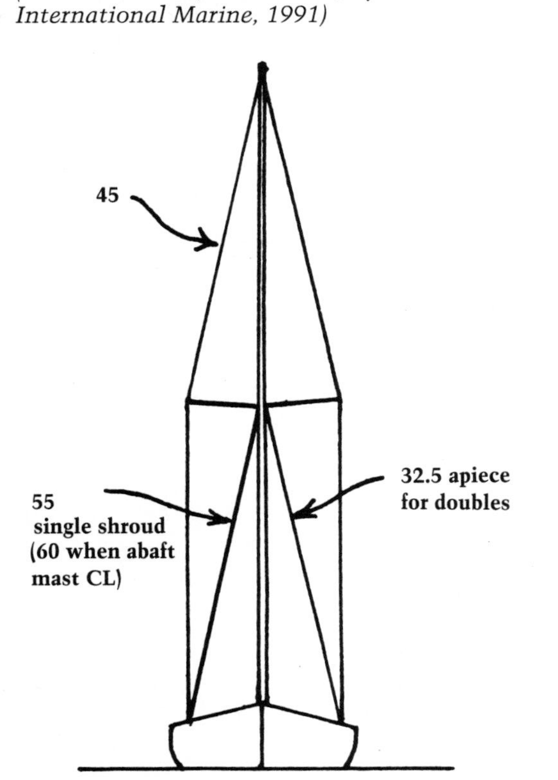

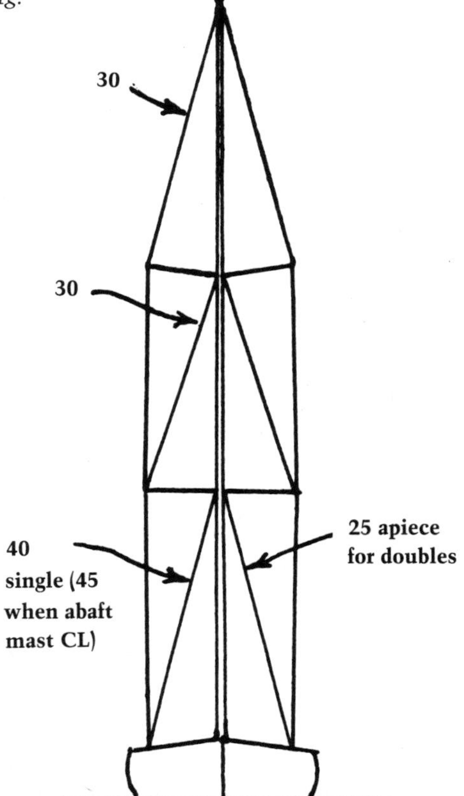

SHAG SPREADER TIPS

Synthetic carpet with short-to-medium-length shag makes great spreader-tip chafe guards, and won't soak up water and rot like Manila baggywrinkle.

heavy or heavier than the intermediates and uppers.

The jibstay is usually made at least as heavy as the heaviest shroud, to take the big loads from the genoa. When in doubt, make this wire heavier; it will fatigue more slowly and stretch less than a lighter wire.

The backstay can be as heavy as the jibstay or a little lighter—it usually has a leverage advantage from a wider angle on the mast and an attachment point farther from the mast.

For our sample, the minimum wire sizes run thus:

	% load	safety factor	load (lbs)	1 x 19 wire diam.
Lowers	25	2.5	3,400	¼"
Intermediates	30	2.5	4,090	9/32"
Uppers	30	2.5	4,090	9/32"
Jibstay	30	2.5	4,090	9/32"
Backstay	25	2.5	3,400	¼"
Forestay	25	2.5	3,400	¼"

Again, it would be better to use stronger wire than noted here for the lowers. Note also that the last-entered wire, the forestay, is a little lighter than the jibstay. Although it will be the sail you'll reef down to, its ultimate loads will be less than that exerted by a closely trimmed #2 genoa on the jibstay.

Bear in mind that none of these figures is cast in bronze. An extraordinarily tall rig, for example, would have a very steep angle on its jibstay, which means higher loads and the necessity for bigger wire. Also, cruisers often make all their shrouds and stays out of the same size wire so that they only have to carry one size of spare wire, turnbuckle, clevis, etc. An exception is sometimes made for running backstay pendants, which should be

RIGGING AND TUNING FOR DETACHABLE FORESTAY AND RUNNING BACKSTAYS

The cutter is a versatile rig, but in light airs, when you want to fly a big drifter, the forestay can be a serious impediment to tacking. It makes sense to be able to move the forestay out of the way when the wind is light, and secure it aft (made off to a chainplate with a bungy cord is ideal). Racers detach and reattach forestays and babystays frequently, using a compound lever quick-release device. Efficient, but prohibitively expensive; a positive-lock Fas-pin, the kind with a pushbutton and lanyard ring, is much preferable for cruisers. You loosen the turnbuckle, pull the pin, and you're clear. Takes a little longer, but it saves about $700. Sta-Lok turnbuckles are ideal for this application, since they are secured with low-tension nuts instead of cotter pins.

When the breeze comes up or you begin hitting chop, re-attaching the forestay helps stabilize the mast. And of course the forestaysail can be used in concert with the jib. In medium airs, tighten the forestay only moderately; too much tension and you'll slack the jib and flatten the forestaysail and main excessively. If breeze or chop builds more, it's time to set up the running backstays. These further stabilize the mast and tension the forestay, resulting in the flat forestaysail you want in heavy weather without causing jibstay sag.

SERVING LAZYJACKS ONTO TWIN TOPPING LIFTS

You can't splice rope lazyjacks into wire topping lifts, and hitching them on just leaves a bulky knot to chafe the sail. So fan out 3 to 5 inches of the lazyjack end, arrange the yarns around the wire, Constrictor them firmly on, then serve over the works. We show three constrictors; add a couple in the middle.

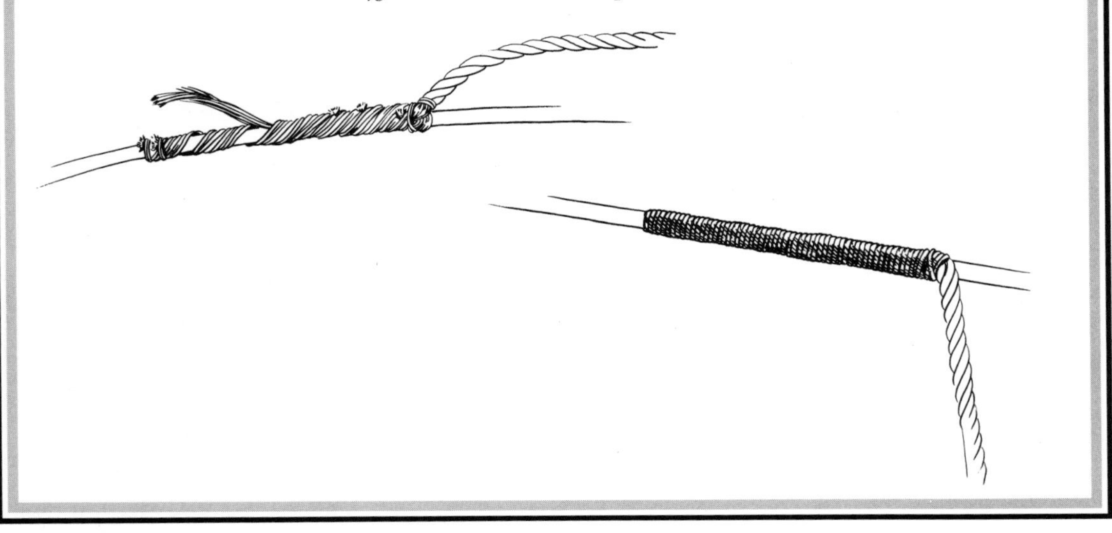

made of the more flexible construction of 7 x 7 (shrouds are usually the stiff but inelastic 1 x 19), preferably vinyl-coated to prevent chafe. Running backs can also be lighter than the forestay, since the mast itself takes up some of the load from that stay.

These and other considerations will affect what wire you hang on a mast, but it all traces back to that Righting Moment curve. As long as your decisions are based on that, you'll at least have an idea of how close to the edge a racing boat will be, or how reassuringly stout a cruising boat will be.

DOWN THE MAST

Getting to wire size takes time, but it's only a matter of matching expected loads to wire strength. Mast design, however, is a more complex, slippery design challenge. The object is to come up with a column that is stiff enough to take, without buckling, the standing rigging's compression loads, and yet light enough that the boat won't be top-heavy, and small enough in diameter that it won't offer unnecessary drag from wind. Unfortunately, these qualifications call for mutually exclusive responses. Fortunately, it's an old, old challenge; others have faced it in the past, and have left us formulas to plug into. Again, these formulas are meant to be patterns, not straightjackets; to use them creatively, it helps to understand some of the thinking that went into them. Later in this chapter, you'll see how designers have come up with some innovative answers to the mast design challenge. For now, let's return to our 35-foot L.W.L. cutter.

Extra Load

The shrouds put some compression load on the mast, but the fore-and-aft stays impose an additional load—boats also have a fore-and-aft righting moment, and some of that gets transferred to the mast via staysails and stays. Also, stays add compression to the mast simply by trying to keep it in place against boat acceleration and deceleration and the pull of various sails. Henry and Miller determined that stays added another 85 percent to the transverse load on the mast. So, revising our previous formula, we get

$$\frac{RM_{30} \times 1.5 \times 1.85}{\frac{1}{2} \text{ beam}} = \text{total mast compression.}$$

Condensed, the formula reads

$$\frac{RM_{30} \times 2.78}{\frac{1}{2} \text{ beam}}$$

In the case of our sample boat that comes to

$$\frac{50{,}000 \times 2.78}{5.5} = 25{,}272.7 \text{ lbs}$$

$$\left(\frac{6{,}925 \times 2.78}{1.67} = 11{,}528 \text{ kgs}\right).$$

Set that number aside for the moment; before we can plug it into one more formula, we need to consider some strength and stiffness variables.

Unsupported Length

The "Mizzens" chapter touches briefly on the significance of unsupported length—how a long column is more likely to buckle under a load than a shorter but otherwise identical column under the same load. Specifically, stiffness varies inversely with the square of the unsupported length; double the length of a mast without adding the intervening support of spreaders, and you have to make it four times as stiff to handle the same load. The "Mizzens" chapter includes a graphic demonstration of this using uncooked spaghetti.

Spreaders are commonly viewed as a means to widen the angle of shrouds to the mast, but they also serve to shorten unsup-

MASTHEAD DETAILS

This sturdy masthead has two big spinnaker halyard "wings" for the halyard blocks. The wings extend well forward and to the side to prevent chafe and fouling on the jibstay. Note the holes drilled in the wing gussets; these save some weight, and they can also be used to attach gantlines and safety tethers when aloft. The bail over the jib halyard mortise prevents those halyards from jumping off the sheave and jamming between the sheave and mast mortise.

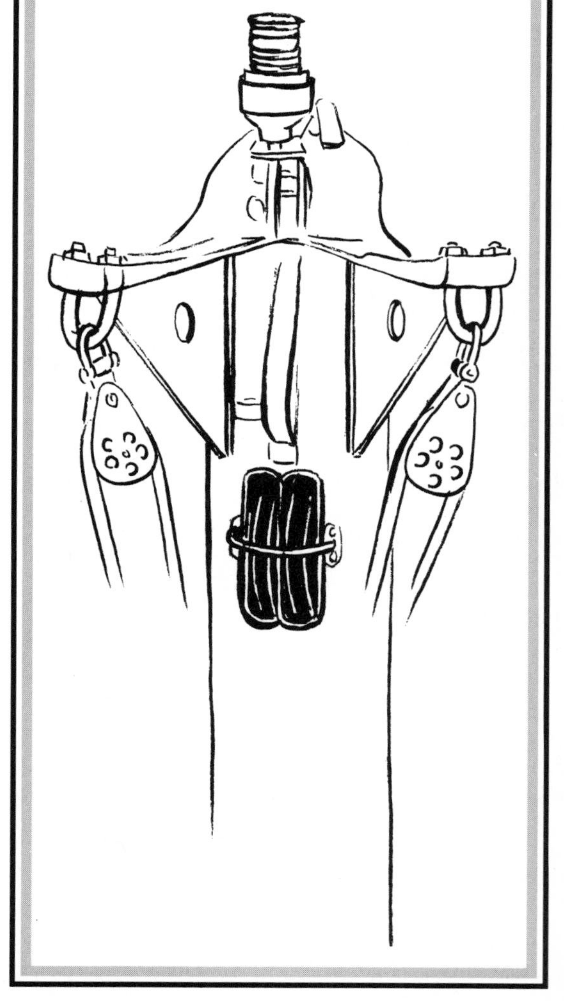

ported length, allowing designers to make masts adequately buckle-proof without making them massively heavy. Intermediate fore-and-aft rigging (running backstays, forestay) also help with unsupported length, but they, as well as spreaders, add complexity and thus vulnerability to a mast; more things to break. That's why you usually see more than two sets of spreaders only on racing boats, who expect to lose the occasional stick.

End Fixity

But there are other ways to increase stiffness without adding mass. Recall this other noodle demo from Chapter 6: Take two long pieces of spaghetti. Break a couple of inches off one. Press down on this shorter one and notice how much pressure it takes to bend it. Notice also that it bends in the middle. Now take the longer piece and pinch it a couple of inches from the bottom. Press. Even though the unsupported length above the pinch point is the same as the short piece, this piece is stiffer. And instead of bending in the middle of its unsupported length, it bends closer to the upper hand, where spreaders would be on a mast.

What you've just seen is a demonstration of the effects of "End Fixity." The short piece is said to have two "pin ends," while the long piece has "one pin end and one fixed end." A boat's partners do the pinching, so a keel-stepped mast will be stiffer than a deck-stepped mast of the same exposed length (see *Sojourner Truth* in "Mizzens"). A very stout tabernacle can add some degree of end fixity, but if you're interested in a light mast, keel-stepped is the only way to go.

Why, then, are so many masts deck-stepped? A classic reason is that they're easier to put up and down for traversing canals, passing under bridges, or trailering. The convenience compensates for added weight.

DECK-STEPPED SLOOPS: ECONOMICS VS. ENGINEERING

It's simpler, easier, and cheaper to deck-step a mast than to keel-step it; no hole to cut in the house, no complex thrust reinforcements, no hole to cut in the sole, no step to build on the keel. And of course it's simpler, easier, and cheaper to make a sloop rig instead of a cutter rig; no forestay, no running backstays, no tangs, no extra halyard, no extra deck fittings. The mast for a deck-stepped sloop is big, heavy, and doesn't have the versatility or redundant safety of a cutter, but you can understand why a manufacturer would find it attractive. No harm done, as long as this sense of economy doesn't extend to things like safety factors and deck bracing. . . .

But economics are also an issue, since it's cheaper to step a mast on deck than to cut a hole in the deck, and the sole, and install a mast step, mast collar, etc.

And even aesthetics can be an issue. I once heard of a couple who hired an interior designer to redo their boat. The designer came below, took one look at the base of the mast and said, "First of all, that has to go." And often it does go, to make possible an arrangement of staterooms, sofas, and entertainment centers that otherwise wouldn't be possible.

In sum, while pinning one end of a mast stiffens it without adding mass, there can be other factors to consider, just as with spreaders.

Radius

There's one other "weightless" means to mast stiffness: making the mast fatter. Stiffness varies with the square of the distance from the neutral axis of a mast to the mast wall. With a round or oval mast this neutral axis, around which all the forces are balanced, is in the center of the mast. It can take careful calculation to find this axis with exotic mast cross-section, but the square-of-the-radius formula holds.

This brings up the reason why most masts are oval instead of round: Because shrouds attach at more points than stays, masts are generally supported better laterally than they are fore-and-aft. You tend to get short, buckle- resistant lengths on the sides, and long, buckle-prone lengths on the front and back. Even on masts with multiple headstays and running backs, induced mast bend can put huge, let's-see-if-we-can-fold-this-thing loads on a stick. Therefore masts need to be stiffer fore-and-aft than they are laterally. The simplest way to accomplish this is to make a mast with the most heavily stressed sides farthest from the neutral axis. This makes the mast stiffer in one plane than the other. A rectangular cross-section works well, but an oval offers less wind-resistance and less weight, so that is the most common mast shape.

Wall Thickness

Mast stiffness also varies directly with the thickness of the mast walls. No free ride here; if you want to stiffen by thickening, you pay a price in weight. Some boats have masts that are "sleeved" with double wall sections over heavily stressed areas. This saves making the entire mast thicker-walled. But more often thick walls are resorted to when other stiffening methods have drawbacks.

For instance, racing sailors will usually select a very narrow mast for low wind resistance. They'll compensate for the loss of square-of-radius stiffness by making the mast very thick and by shortening the unsupported lengths with three, four, or even five sets of spreaders. And they'll accept a low safety factor.

Cruising sailors, though they'll be interested in good performance, will want to minimize the expense, vulnerability, and intricacy of many sets of spreaders, and will be less concerned about an absolute minimum of wind-resistance. A cruising mast section will have moderate radius, moderate wall thickness, and one or two sets of spreaders. And it will be built with a high safety factor.

In sum, mast design involves juggling various stiffness-inducing factors along with cost, performance, reliability, and even interior design. To get the mast you want, you just have to be able to express those factors with numbers.

The Formulas, Part II

All mast design formulas are variations and refinements on "Euler's Formula," an engineering cornerstone which predicts the behavior of columns under compression, with allowances made for all the significant variables. A predigested, easy-to-plug-into form appears in Skene's Elements of Yacht Design. I like this version because it's simple and conservative. For another, more race-oriented approach, see the USYRU formula in the accompanying "Miscella."

Meanwhile, let's start with the formula for the lateral, or transverse, plane. We're looking for a specific "transverse moment of inertia"—essentially stiffness—which will be expressed, due to multiple squarings hidden in the calculations, in inches to the fourth (in.^{4}). I_{tt} (in.^{4}) is our symbol.

So:

$$I_{tt}\ (\text{in.}^4) = C_{transv} \times \frac{L_t^2\ (\text{in.}^2)}{10{,}000} \times \frac{\text{Load}}{10{,}000}$$

Where C = a transverse constant

LT = the length from deck to lower spreader

Load = RM_{30} compression load.

USYRU SCANTLING FORMULAS

The United States Yacht Racing Union (USYRU) developed an alternative to Skene's mast scantling formulas for offshore racing yachts. The two sets of formulas are based on the same elements; the one from the USYRU combines them in a different form, uses a lower safety factor, and works with the righting moment at 40 degrees of heel instead of 30 degrees. Resulting masts are considerably lighter than ones designed by the Skene's method, but the USYRU formulas do prevent the worst excesses of hold-your-breath spindliness in offshore racing masts.

The formula for longitudinal inertia is

(Longitudinal Safety Factor x 40 x Righting Moment (ft-lbs) at 1° heel x [Mast Height (inches)]^2) ÷ (End Fixity Factor x ½ Beam (at chainplates) x π^2 x Modules of Elasticity).

More concisely, that's

$$\frac{FS_L \times 40 \times RMC \times P_L^2}{F \times CP \times \pi^2 \times E}$$

With an inner forestay, the longitudinal safety factor is 1.5, if the forestay is attached to the mast between .651 and .701 of the mast height above the sheer and is backed up with running backstays. Without a forestay, the safety factor is 2.

End fixity is 2 if the mast is keel-stepped, 1.5 if deck-stepped.

For minimum transverse inertia, the formula is

$$\frac{FS_T \times 40 \times RMC \times P_T^2}{F \times CP \times \pi^2 \times E}$$

Where P_T is the height of the mast to the lower spreaders.

The safety factor is 1.7 for a single-spreader rig, 2 for a double-spreader, assuming the single spreaders are more than halfway up the mast and at least four-fifths of the boat's one-half beam in length, and the lower set of double spreaders is at least .36 of the way up the mast and at least three-fifths of the boat's one-half beam. This spreader placement proviso has the effect of keeping the "P_T^2" measurement high, which results in a heavier, safer mast section.

Stiffness varies inversely with the square of unsupported length, so multiplying the load times the longest unsupported length—deck to spreaders—takes care of that relationship. The constant takes care of end fixity, a safety factor, and the properties (modulus of elasticity) of the material the mast is made of. The matters of wall thickness and mast radius will be dealt with later, on mast section charts which take these factors into account without further calculation.

Let's go shopping for a mast section. Assume the stick on this boat is 42 feet above deck, and that the lower spreaders are 17 feet 7 inches above deck. That's about 42 percent of the total exposed length, a number that lends itself to wholesome proportions, so that no part of the rig will take a disproportionate load. If the lower spreaders were higher up, the lowest section of the mast would have to be much stiffer—longer unsupported length. Since the entire mast is scaled to this section, that would make the

C. Transverse			C. Fore and Aft with Double Lowers	
Mast Material	Single Spreaders	2 or more Sets Spreaders	Masthead	Fractional
Spruce	6.78	8.11	4.0	3.74
Aluminum	.94	1.13	.54	.52

Figure 7-4. *Values for the constant "C" in the formulas for moments of inertia about the transverse and longitudinal axis of a mast. These values assume a keel-stepped mast. Deck- stepped masts require values for the constant perhaps 50 percent greater, which can be reduced to 20 to 30% if a big, stout tabernacle is present to provide partner-like support. (Adapted from Skene's Elements of Yacht Design, 8th Edition, by Francis Kinney. Dodd, Mead, 1981)*

higher, shorter sections far too stiff and heavy. Conversely, if the lower spreaders were a lot lower, we wouldn't need as heavy a section down low, but the upper and now longer sections would not be heavy enough to support their loads. So somewhere around 40 percent of exposed length is a good location for lower spreaders. For single spreaders, about 55 percent of the way up is a good location, for the same reasons.

In any event, 17 feet 7 inches is 211 inches. Square that and we have an L_T^2 of 44,521. Our load from the rigging formula at RM_{30} is

$$\frac{RM_{30} \times 2.78}{\frac{1}{2}\text{ beam}} = 25,272.7 \text{ pounds.}$$

Our constant, from the chart in Figure 7-4 is 1.13 for an aluminum mast, keel-stepped.

Plugging into our formula, then,

$$I_{tt}\ (\text{in.}^4) = 1.13 \times \frac{44,521}{10,000} \times \frac{25,272.7}{10,000} = 12.71 \text{ in.}^4$$

What the heck is 12.71 in.^{4}? It's just 127,100 reduced to a more compact form by those "10,000" divisors in the formula. In either form, it's our transverse moment of inertia.

The formula for the fore-and-aft or longitudinal moment of inertia is much the same, except for a different, smaller constant. It's smaller because the unsupported length is assumed to be the entire exposed length of the mast; less of a safety factor is needed to help compensate for mistuned shrouds, failing spreader fittings, or the like. I_{LL} is our symbol for longitudinal inertia. So:

$$I_{ll}\ (\text{in.}^4) = C_{f.a.} \times \frac{L_L^{2}\ (\text{in.}^2)}{10,000} \times \frac{\text{Load}}{10,000}$$

Where C = fore-and-aft constant

L_L = length, deck to jibstay

Load = RM_{30} compression load.

Accordingly,

$$I_l\ (\text{in.}^4) = .54 \times \frac{254,016}{10,000} \times \frac{25,272.7}{10,000} = 34.67 \text{ in.}^4$$

TOPPING LIFT/OUTHAUL

Here's an excellent small-boat innovation that reduces end-of-boom clutter: the outhaul sheave is let into the end of the boom, and its axle serves as a bolt for the bail that the topping lift attaches to.

As you'll note in the constants table (Figure 7-4), there are different figures for wood and aluminum masts, taking into account their different properties. It should not be thought that because the constants for wood are so much higher the masts will be that much heavier; the constants are scaled to stiffness in a given material, which is not necessarily the same thing as weight or even size. Aluminum masts should work out a little lighter nonetheless, but in these litigious times it is not unusual to find them at least as heavy as wooden ones, as an anti-lawsuit safety factor. Also notice that the constants assume a keel-stepped mast and two lower shrouds per side; if a mast you deal with is deck- stepped or single-shrouded, adjust accordingly.

MAST CHARTS

Figure 7-5 shows the final step in mast design: choosing a mast cross-section of sufficient stiffness to stand up to our calculated moments of inertia. The cross-hatched graph is a "mast chart," which precalculates for us the effects of radius and wall thickness in determining the desired section. We need only read up from the bottom, which has the transverse scale, to intersect a line drawn across from the side, which has the longitudinal scale. Where the lines meet, we can read off our dimensions—roughly 5 inches on the transverse axis and 8 inches on the longitudinal axis in our example.

A leafing-through-a-catalog-ish alternative is to consult a list of the dimensions and moments of available extrusions, and pick out the one that comes closest to our requirements (Figure 7-6). Either way, we'll usually be faced with two variables:

(1) Since calculated moments rarely coincide precisely with those of available mast sections, a choice must often be made between sections that are somewhat stiffer and somewhat more supple than our ideal.

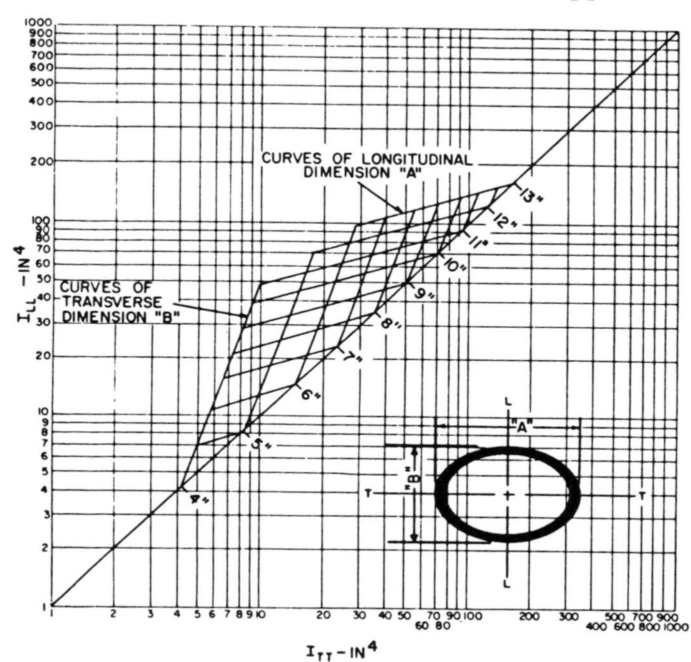

Figure 7-5. A mast chart for oval sections. (From Understanding Rigs and Rigging *by Richard Henderson, International Marine, 1991)*

SPAR SECTIONS

6061-T6 ALUMINUM

SECTION	SHAPE	DEPTH	WIDTH	WALL	LBS/FT	MOMENTS OF INERTIA I_{xx}	I_{yy}
077	B	2.84 in.	2.125 in.	.085 in.	.917	.60	.34
095	B	3.75	2.75	.085	1.35	1.92	1.15
100	C	3.75	2.25	.100	1.30	1.50	.63
110	C	4.75	2.75	.100	1.50	2.40	1.10
125	B	4.90	3.09	.091	1.48	3.50	1.48
128	B	5.00	3.00	VAR.	1.70	4.80	1.90
130	D	5.00	3.50	.120	2.08	4.22	2.43
150	D	6.00	4.00	.140	2.75	8.91	4.40
152	B	6.00	4.00	.130	2.80	11.50	5.60
165	A	6.52	4.10	.147	3.33	14.41	6.78
166	B	6.50	4.00	VAR.	3.30	16.00	5.70
180	B	7.15	4.50	.148	3.57	18.50	7.44
181	A	7.15	4.50	.148	3.57	18.50	7.44
185	B	7.15	4.50	.170	4.42	19.50	10.00
200	B	7.69	4.86	.180	4.54	27.40	11.45
202	B	8.00	4.25	VAR.	4.46	34.60	8.60
→205	B	8.06	4.88	.180	4.76	32.20	12.00
220	B	8.55	5.40	.188	5.20	39.00	16.60
221	A	8.55	5.40	.188	5.20	39.00	16.60
231	A	9.24	5.82	.188	5.60	49.60	22.20
232	B	9.19	4.86	VAR.	6.19	63.70	15.60
240	A	9.91	6.20	.188	6.30	68.40	27.20
280	B	11.00	5.88	VAR.	10.04	132.91	34.57
305	A	12.00	7.48	.204	8.32	125.90	53.70
061	E	2.25	.61	.070	.38	.13	.01
081	E	3.25	.88	.100	.79	.54	.06
121	E	4.75	1.25	.130	1.50	2.18	.24
171	E	6.75	1.80	.130	2.19	6.51	.77

Figure 7-6. *A page from a Yachttech catalog. For our example boat we choose mast shape B, section 205.*

(2) Since masts come in varying wall thickness, we have a range of choices, with a very small-diameter, thick-walled mast at one extreme, and a very fat, thin-walled mast at the other.

In practice, these two decisions are settled according to the nature of sailing the boat is intended for. Cruisers will naturally be inclined to go a size up when in doubt, and to opt for a moderately large section, if only because it is less expensive than a very thick one. And likewise, racers are likely to shave away scantlings in the interest of lightness, even as they minimize diameter to minimize windage, and hang the expense.

In the case of our cruising cutter, the catalog mast section that most closely matches our requirements has dimensions of 8.06 inches (205 mm) by 4.88 inches (124 mm) by .180 inch (4.6 mm) thick.

Options

Design is a search for the appropriate. The above bulletproof cruising cutter rig is not the only bulletproof cruising cutter rig; it and any number of other configurations fit the job description. We could have made a single-spreader, deck-stepped, single-shroud design as well. Or we could have made the mast out of wood, which can be custom-made and tapered to any needed dimension by a backyard builder. And we could have tapered the top of any of these rigs to reduce weight and windage aloft. The options are endless, and they all make at least some sense. But the time comes when you have to sit down and say, "It's going to look like this." When the gestalt of everything you know or think you know leads you to what you feel is most appropriate.

You've just been privy to the design of a "classic" rig, one whose form and particulars are the result of long evolution and conservative engineering, and whose physical components are readily available. Most rigs, cruisers and racers alike, come out of this heritage, but designers are always pushing at the envelope. They do this for their own satisfaction or at the urging of clients for whom "normal" is not enough. In the following pages, you'll get a perspective on what's possible. You'll see what can happen when a talented, innovative, and prudent designer goes to work. In this collection the emphasis is on "prudent;" anyone can come up with a novel rig, but only care, skill, and a realistic application of basic design principles will produce a rig that will perform as planned and stay in the boat.

TROUBADOR

Veteran yacht designer Ted Brewer is a direct heir to the people who developed the foregoing rig scantling formulas. But his designs, for everything from gaff-rigged schooners to BOC racers, have always been characterized as much by freshness and adaptability as by classical conservatism. So when client Dr. Paul Bubak asked for a very easily handled but very efficient fractional rig, Brewer came up with the artful sailplan shown in Figure 7-7.

Traditionally, cruising rigs have minimized the effort expended on staysails by breaking up the single large one of a sloop into two or three smaller ones, to make a cutter. But another method is to shift staysail area into the less truculent mainsail. This results in such a small staysail area that you no longer need to run the jibstay all the way to the masthead in order to have room for the sail. A "fractional rig" results, so named because the stay only comes ¾ or ⅞ or 15⁄16 of the way up the mast.

Less staysail-wrestling also means less jibstay tension, and thus less mast compression, so fractional rigs can have slightly lighter masts for the same total sail area (see table in Figure 7-4).

It's not all plusses, however. You might not have the chore of changing headsails, but you do have to reef the main sooner and more often. And savings in headsail costs are offset by increased mainsail costs, particularly if it has full battens, lazyjacks, and other options to make the sail more efficient and easier to handle.

Troubador's rig has a very small (non-overlapping) jib and a very large, very tall main.

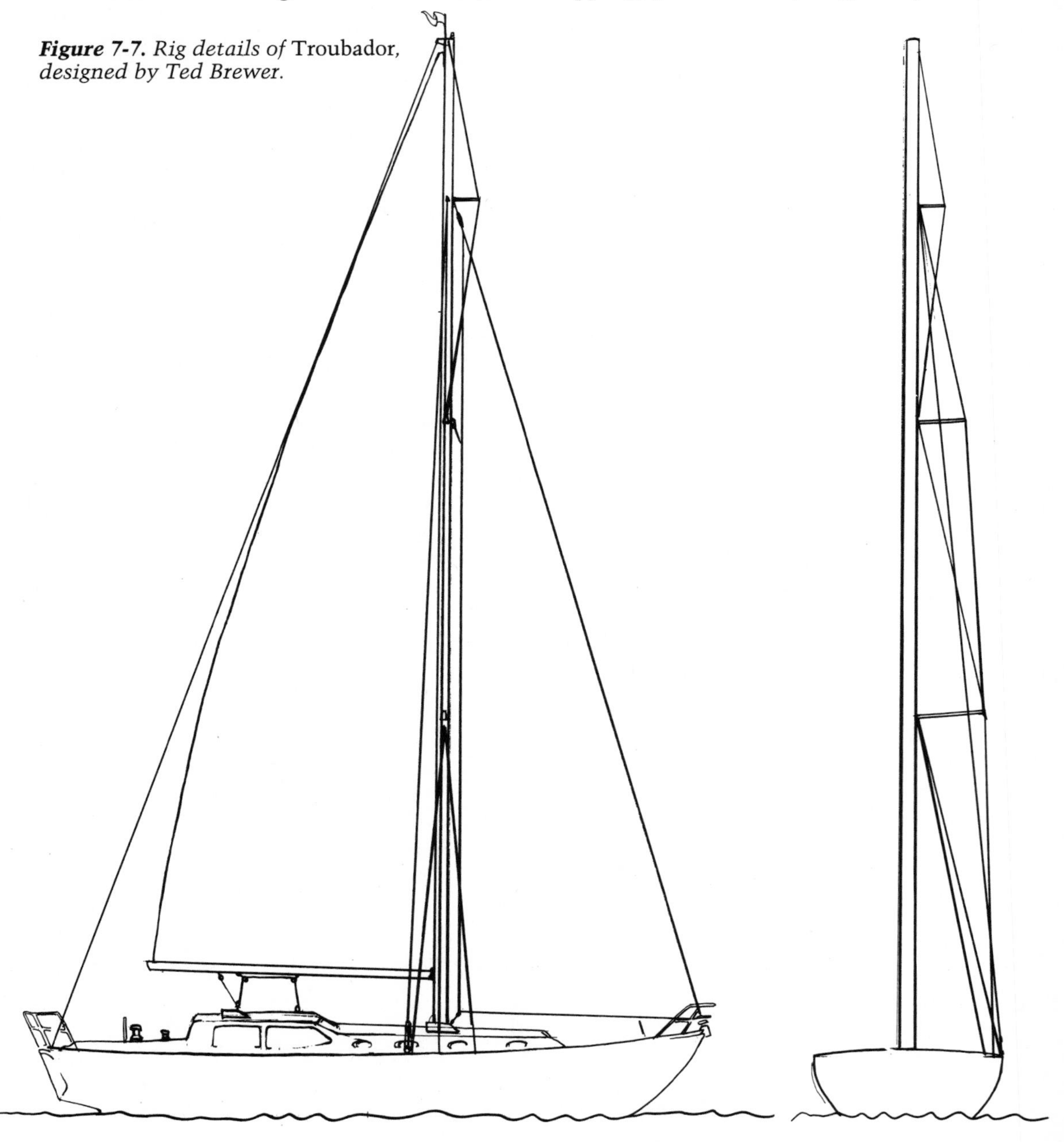

Figure 7-7. *Rig details of* Troubador, *designed by Ted Brewer.*

It's a bit radical-looking for a cruising boat, but Brewer simultaneously pushed the envelope and kept his scantlings conservative. The result is a safe rig and low work load for Bubak and crew, without correspondingly low performance.

Rig Details

With so little sail area above the jibstay, the masthead needs little lateral staying. Diagonal jumpers are sufficient, and they also act to brace the mast against the forward pull of the forestay. Running backstays act as backups for the jumpers for fore-and-aft pull; they can be set up in high winds or choppy water or when the mast is heavily bent by the backstay. Many fractionally rigged race boats dispense with the jumpers and rely solely on running backs. This saves weight and windage but means the runners must be set up promptly with each tack or risk losing the stick. Combining jumpers, runners, and a conservative mast section makes for low labor (you only need the runners about 20 percent of the time) and high safety, and they're still light enough not to cut too much into performance.

A final detail is that *Troubador* also has a forestay, something rare in a fractional rig. It helps brace this high-aspect mast, leaves enough space for a good-sized working staysail, and is a good wire to hang a storm staysail on—low and well aft. In light airs this stay can be disconnected and made off aft and to one side, for easy tacking of a big jib.

OTTER

Pump *Troubador*'s mainsail up even further, shrink the staysails down to zero, and move the mast right up into the bows of the boat and you have the rig for *Otter*, a 32-foot Mark Ellis–designed catboat created for Ed Scheu of Hanover, New Hampshire. It's another fractional rig (Figure 7-8), and a requirement of ultimate ease of handling resulted in the absence of staysails. Ordinarily, this would have resulted in reduced efficiency to weather. Many cruisers would have accepted this as the price you pay for ease of handling (and Ellis has designed the line of Nonsuch unstayed catboats for them). But Scheu is a racer. No longer interested in the physical challenge of running the Stars and Concordia yawls of his younger days, he was by no means eager to forgo the joys of passing other boats.

So just as Bubak had wanted an efficient but low-sweat, heavy-duty cruising boat, Scheu wanted a languorously attended racing catboat. Accordingly, this ultimately simple rig was fitted with a whole raft of sail controls: adjustable jumpers and backstay (the mast can be deflected fore-and-aft 20 inches (508 mm)); high-powered vang and sheet traveler systems; a Cunningham; and single-line reefing. All of this makes Otter more work than a Nonsuch but much less work than any comparably efficient staysail-equipped boat.

Two Design Challenges

Otter's main has to compensate, in size and aerodynamic quality, for the missing staysails. But the size of the main is limited by the presence of the standing backstay, a piece of wire crucial not only to rig integrity but also to mast- bending sailing efficiency. The obvious response is to make the mainsail very, very big, and Ellis has done this—note the high aspect ratio and the generous boomkin increasing the backstay angle as far as practicable. But even with a heavily roached, full-batten mainsail filling every possible square inch of the resulting space, there's just not enough area to move anything but a notably light hull. And because of the hefty righting moment of such a tall rig, that hull also needs to be stiff. So Otter is almost three tons lighter than a cruise-ori-

Figure 7-8. *Otter's* *sail and rigging plans. The diagonal jumper stays stabilize the upper end of the mast fore-and-aft; the forestay and strut stabilize the lower section. The lower ends of the jumpers lead internally through the mast to deck level for adjustment, which is necessary when the backstay is adjusted. The mast's position in the bow of the boat creates a very narrow staying angle. Dotted lines show the initial upper shroud angle, which proved insufficient to prevent lateral bend. This problem was solved by adding a third, upper set of spreaders to take the upper shrouds to the masthead, and lengthening the intermediate spreaders (which had been the uppers). The running backstays are not shown here.*

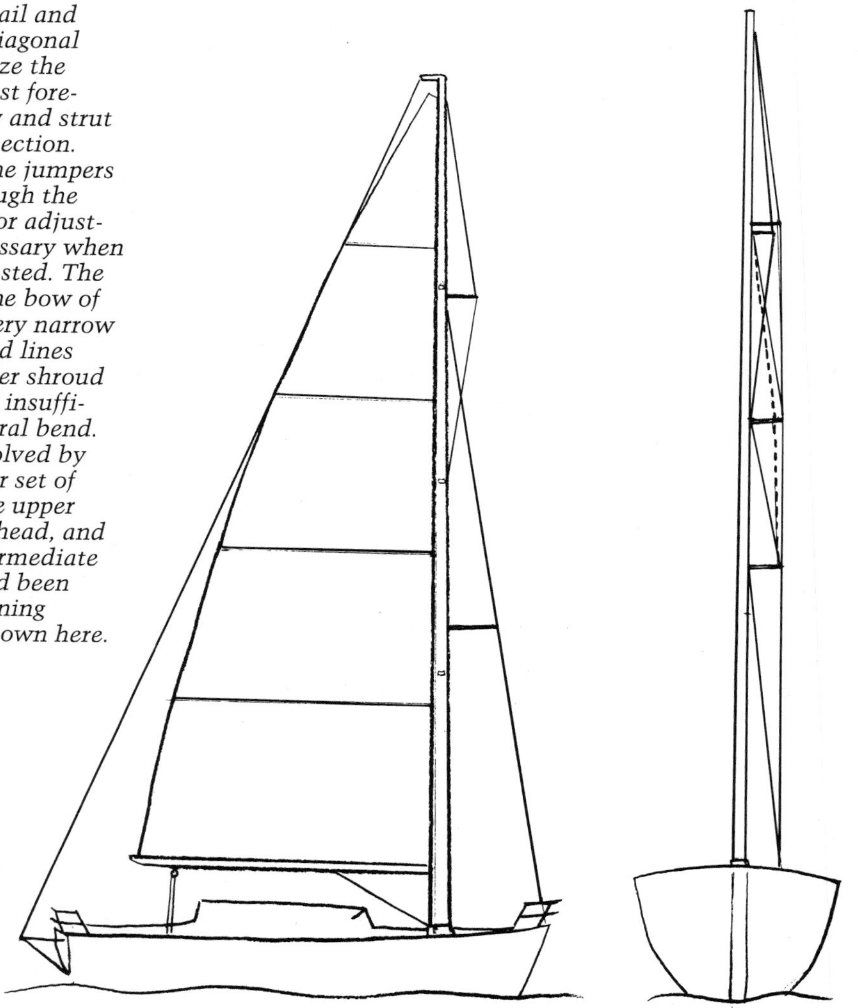

mast section

ented Nonsuch 33, and has a deep keel with all the ballast in a bulb at the end.

Good designers seek to mesh rig and hull qualities, but they usually tailor the rig to a given hull or racing rule. In this case, though, the hull was largely molded by a single piece of standing rigging.

The first challenge was to modify the hull to accommodate limitations imposed by the rig. The second challenge is the hull's revenge: The mast sitting so far forward, where the hull narrows, means adequate staying angle for the shrouds is hard to come by. Moving the mast aft would help this problem, but that would just mean more area cut out of the mainsail by the backstay. The stick's final location was a compromise between staying angle and sail area.

Rig Details

The rig is ¾ fractional, with an extra bit of rigging in the area you'd usually expect to find a forestaysail: a removable forestay strut controls mast wobble below the forestay. If a forestaysail were ever installed, this strut would have to be permanently removed and replaced with running backstays. In other words, removing a labor-intensive sail made possible the elimination of a set of labor-intensive running backs. This leaves only one set of runners, at the top of the forestay, to reinforce the diagonal jumper stays.

As originally designed, *Otter* had two sets of shroud spreaders, with the upper shrouds reaching to the height of the forestay, and diamond jumpers for lateral and fore-and-aft staying from there to the masthead. But do you see how long the mast section above the forestay is? It's a lot of mast for skinny-angle jumpers to control (compare with *Troubador*). Too much, in fact, since in initial sailing tests the masthead sagged to leeward.

Returning briefly to the drawing board, the Ellis office moved the upper spreaders to the level of the forestay, added a third set of slightly longer spreaders where the upper spreaders had been, ran the upper shrouds to the masthead for improved lateral staying, and beefed up a bit on the strength of the standing rigging. In other words, having pushed the design envelope as far as they could with this novel configuration, they grudgingly added weight and windage rather than compromise sailing efficiency or rig integrity. A recipe for evolution.

THE FREEDOMS

Because of its standing rigging, Otter makes the most efficient use of its one big sail—at the price of a considerable design challenge, focusing on the relationship between standing rigging and hull. A free-standing mast, as on Tillotson-Pearson's Freedom line of sailboats (Figure 7-9), removes that design challenge; the hull and sailplan are no longer subject to backstay- and shroud-imposed limitations. But, as is always the case with technology, an entirely different challenge arises from this sweeping solution: how to make an unstayed mast small, light, and inflexible enough to approximate the performance of a well-designed stayed mast.

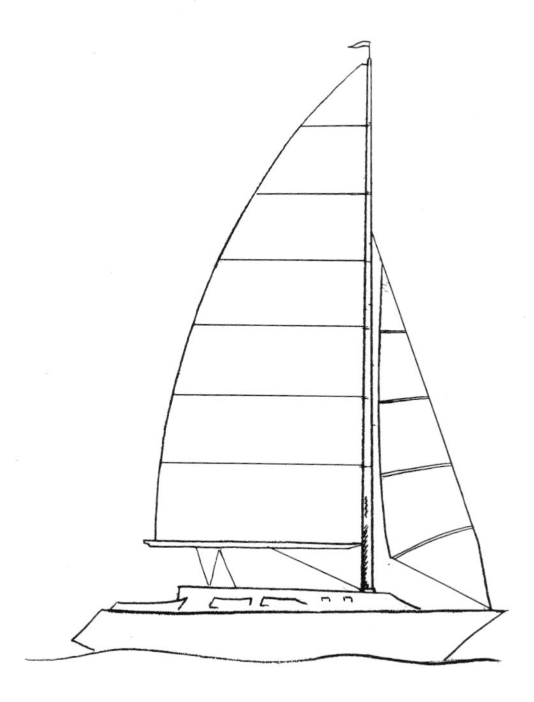

Figure 7-9. *Sail plan of the Freedom 32.*

Unstayed masts are by no means a new idea; they've been characteristic of many traditional craft all around the world since sailing's beginnings. It's just that until relatively recently they had to be made of solid wood, and there was no way to keep them from bending excessively under load, thus compromising sailing efficiency, without making them extremely large and heavy, thus compromising sailing efficiency.

In this century, hollow aluminum and wooden masts have been developed that are sufficiently light, durable, and deflection-resistant to be useful for sailboats, as long as you don't care a great deal about efficiency to windward and have a taste for the fairly tubby hulls you need to hold up a big, heavy cat rig. Perhaps because sailors tend to like sleek, weatherly boats, unstayed aluminum and wooden rigs have not taken over the market, even though they are usually far simpler and physically easier to sail than their stayed brethren.

Carbon fiber, an even more recent mast material development, is far less elastic than wood or aluminum, so it can be used to make masts, unstayed, that are nearly as deflection-resistant as ones that are stayed. Carbon fiber is spun from, take a deep breath, thermally decomposed polyacrilonitrile, known more familiarly as "pan." Filaments of pan are pre-stretched, then stabilized with a relatively mild heat (428 degrees F). Then the filaments are placed in a nitrogen atmosphere and really cooked (at about 2,700 degrees F), a process which essentially burns away everything but tough little strings of carbon.

Sound complicated? Nah, that's just the chemistry. The real chore comes in trying to build a mast out of this stuff. First you have to make it into a kind of fabric, just so it can be handled. Then you have to lay it onto a form—usually either an aluminum mandrel or a bladder-filled clamshell mold—under maximally nitpicky levels of heat, pressure, and tension, in a pattern which has been precisely engineered to produce optimal stiffness and durability with minimal weight or size.

And when all that's done, you have to convince sailors—a notoriously conservative bunch—that this isn't just some loony engineer's Bright Idea. It doesn't help that other manufacturers, some of whom are not as skilled or careful as the people of Tillotson-Pearson, produce masts which perform poorly and/or break. It doesn't help that the process is very expensive. And it doesn't help that carbon fiber masts still deflect under load more than a comparably well-designed stayed mast (though see accompanying Miscella for a positive side to mast deflection). In particular, unstayed masts cannot endure compression loads from big staysails, the ones you trim in to go to weather.

With these drawbacks, carbon-fiber unstayed masts might have faded into obscurity but for the evangelical zeal of one Garry Hoyt. Early on, Hoyt pitched the advantages of the rig with such force that the public began to focus less on the disadvantages. Hoyt founded Freedom Yachts, convinced that he could do not just a better job, but a revolutionary one. In addition to shrouds, he did away with stays, staysails, vangs, and conventional booms and luffs (he used wishbone booms, which were supposed to be self-vanging, and used mast-encircling luffs).

To make the sail hold an aerodynamic

MAST BEND, STAYED AND UNSTAYED

A stayed aluminum mast can be bowed in the fore-and-aft plane by tightening the backstay and/or forestay, usually with hydraulic adjusters. This action flattens the jib by tensioning the jibstay, and flattens the main by pulling the sail's draft forward. This results in improved efficiency when going to windward, and it delays the need to reef because moving the draft of the sails forward lessens heel and weather helm.

Unstayed rigs do not provide the same sail control options, or at least not in the same way. Full-length battens control sail shape (along with the vang), preventing the draft from moving too far aft. Reefing is delayed not by flattening the sail, but by engineering the mast so that the head sags well to leeward in gusts, effectively depowering the big main.

shape on a relatively noodly mast, he installed full-length battens. To make the sail easier to handle he installed lazyjacks and a single-line reefing system he invented. Then he led these and all other control lines to the cockpit. This last detail is important; if you go forward on an unstayed boat, there's nothing to hold onto but the mast.

Hoyt unleashed his early Freedoms on the world of yachting, demonstrated that they were efficient, easily handled, and durable, and waited for sailors to drop stayed rigs like last year's software. He made a dent, but not a big one. Why? For one thing, he'd made a lot of noise about how horrible stayed rigs were supposed to be. To hear Hoyt tell it, going out on a stayed-rig daysail involved an exhausting physical workout, along with a strong chance of suffering a dismasting from "one missing cotter pin or one broken wire." People looked around and saw that it just wasn't so. They also saw that non-racing boats didn't require the "deck apes" that Hoyt said were necessary to handle staysails. And they saw that some of his rigging ideas—the wishbone boom, wraparound luff, and lack of staysails—were not only unfamiliar, but also offered no efficiency advantage. It didn't help that his hull designs were less than graceful.

So over the years, Freedoms have edged closer to the mainstream in terms of overall appearance as well as rig details. This process was accelerated when Tillotson-Pearson bought Freedom Yachts and hired naval architect Gary Mull to redesign the boats. The new Freedoms are sleeker, and sport a low-mounted stay with self-tending staysail, along with an ingenious Hoyt-patented "Gun Mount" spinnaker (again, you don't have to go forward to tend it) and a conventional mainsail luff groove, boom, and vang.

The masts themselves are still marvels of engineering (they're guaranteed for the life of the boat), arguably the best production carbon fiber spars made. They are built to extremely high safety factors, partly because that's how Tillotson-Pearson does things and partly because carbon fiber masts are very, very difficult to repair once they've been damaged. An impact shock, for example, can break fibers on the inside of the mast. Carbon fiber can also degrade from reaction with stainless fasteners, so fasteners must be isolated from the fibers and installed only at places that are specially reinforced during construction.

Mull-era Freedoms have made an impression on the designers of conventional rigs. Seeking to make sailing easier, they've adapted Hoyt-ish details: full-battened mains; smaller, more easily handled running sails; and aft- led control lines.

With the two approaches to mast design moving closer in terms of secondary detail, and roughly equivalent in terms of perfor-

CARBON FIBER AND STAYED RACERS

Because carbon fiber is vulnerable to impact loads and corrosion from improperly mounted fasteners, and because it is extremely tricky to repair, responsibly made unstayed masts are quite heavy for a generous, lifetime-guaranteed safety factor.

But carbon fiber is also extremely light for its strength and extremely inelastic. So if safety factors aren't a big issue, fiber masts can be much lighter. And if they have standing rigging to support them, they can be far lighter still. This is why, rules allowing, the very hottest race boats have spindly, stayed carbon fiber masts. Expensive like you wouldn't believe, and so, so fragile, but just a hair faster.

mance, how is a sailor to decide on which is the better rig? As a rigger who works primarily with stayed masts, I might have a heavily biased view, but here it is anyway: Well-designed and made unstayed rigs are for people who are disinclined to be involved with their boats. Someone else has done all the work; you just step in and sail. If something goes wrong, you get someone else to fix it. Meanwhile, you just sail.

A well-designed and built stayed rig practically invites involvement on the part of the sailor. Sure, you can leave all the work to others, and many sailors do. But a stayed rig allows you to see to the health of your rig, allows you to make repairs and modifications. The rig is built of a series of simple, durable components; if you take care of them, they will take care of you.

Diversity is wealth. Sailors tend to cluster around styles of sailing, and tend to bad-mouth other styles, but diversity in rig and hull design has little to do with superiority and inferiority. It is a reflection of human diversity. Unstayed rigs have had a hard time gaining a toehold in the market, but they're now sufficiently developed—and accepted—that they form a strong part of the rigging spectrum.

TUMBLEHOME

Of course, evolution is a tricky thing; lots of rigs less radical-seeming than Freedom's have appeared over the years, been tried briefly or at length, and vanished. It's hard to tell what will be an advance and what will be a dead end. One concept that, after years of halting development, is on its way to the mainstream is the Rotating Wing Mast.

Ordinary masts create a lot of windage and turbulence in front of the mainsail, decreasing its efficiency. But if the mast is made in an airfoil or "wing" shape it provides turbulence-free lift, augmenting or even completely replacing the sail. Of course, wing masts need to be able to rotate, so that the leading edge is always facing into the wind, and they need to be very narrow to function as efficient airfoils. These two factors have created the greatest challenges for

WING MASTS AND UNINTENDED SAILING

A wing mast represents sail area which cannot be lowered. Under sail it is the ultimate reef; in extreme conditions there's no such thing as "running under bare poles." You're stuck with that sail area, like it or not.

And at anchor, a wing-masted boat will sail all over the place all by itself, unless you remember to turn it sideways to the wind, which will effectively stall it. If a boat has two masts, you can turn them in opposite directions to keep things quiet.

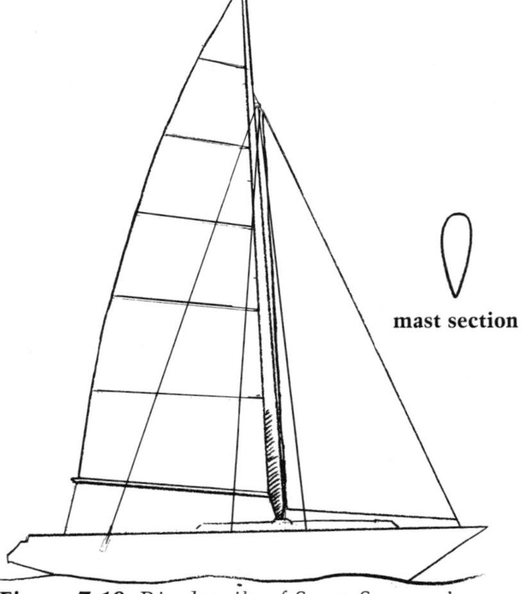

Figure 7-10. *Rig details of Scott Sprague's* Tumblehome.

designers. If the mast rotates, how can you attach standing rigging to it? And if it's so skinny, how can it keep from buckling under standing rigging-induced compression?

Of all the responses tried so far, the most workable has involved a sort of glorified gooseneck fitting attached to the face of the mast three-quarters or so of the way up. Shrouds, forestay, and running backstays attach to this fitting, which remains stationary as the mast pivots behind it. The mast is further reinforced laterally with diamond stays which, since they do not attach to the deck, also do not inhibit rotation.

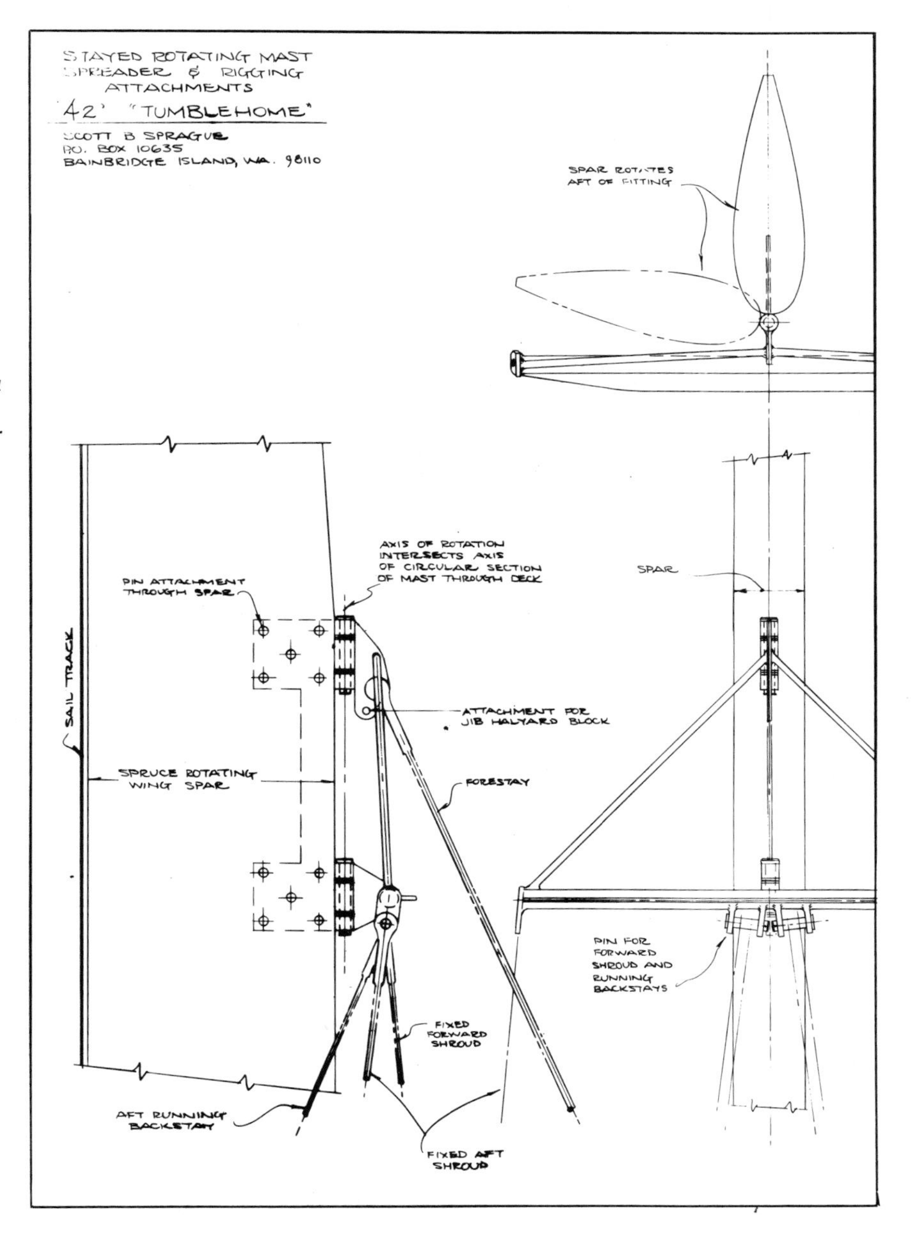

Figure 7-11. *Mast goosenecks are a necessity on stayed wing masts; if shrouds were attached to the mast's sides, the mast could not rotate. Scott Sprague's ingenious gooseneck for* Tumblehome *goes things one better, with the addition of a lower shroud spreader bar. This bar allows him to use aft-leading lower shrouds, which ordinary wing masts cannot carry. The spreader bar is slightly wider than the mast is deep (see top view), so that even when the spar is rotated for running, the wires have a fair lead aft to the chainplates. The running backstays are set up at the base of the spreader bar, along with the forward lower shrouds. When sailing, the leeward runner is slacked, so does not interfere with mast rotation.*

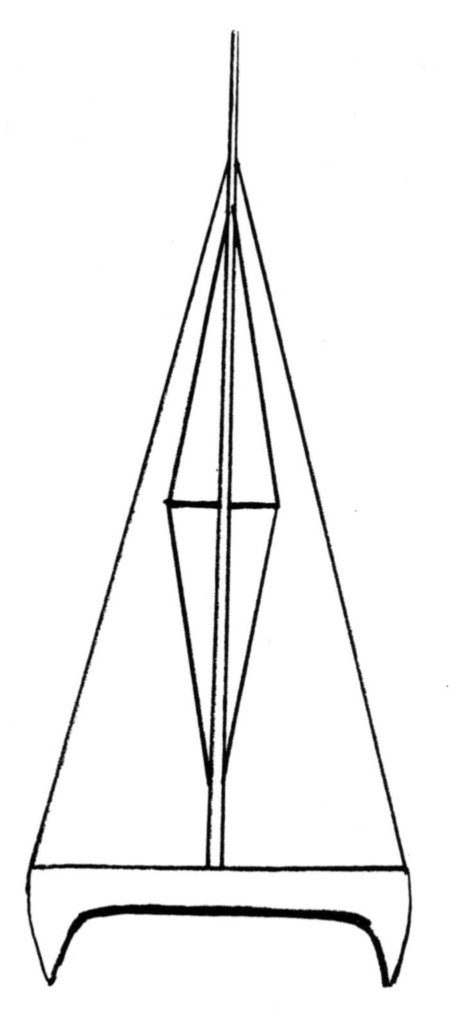

Figure 7-12. *Diamond stays, shown here on a multihull, provide lateral support for a mast, preventing buckling without being attached to the deck. This is ideal for wing masts, whose small athwartships dimensions make them weak laterally.*

This was the configuration chosen for *Tumblehome*, designer Scott Sprague's sleek, fractional-rigged sloop (Figure 7-10). (The diamond stays are very small and do not show on the drawing.) Sprague is best known as a designer of heavy deep-sea boats like the Hans Christian line, but when he went to design his own boat, he had Puget Sound sailing in mind. He wanted to try something "a little different—lighter, more challenging, and a lot of fun."

Tumblehome's wing mast (4 inches by 14 inches at its largest, with an average wall thickness of ¾ inch) is a clear vertically grained Sitka spruce work of art. Here's Sprague on its construction:

"Originally I was going to cold mold the mast, but after studying the shapes and wall thicknesses, I concluded that for this spar solid spruce made the most sense. The mast is really made a lot like a traditional spar. Gluing was a bit of a problem since there was no way to clamp the spar as you would a box mast, so I made a series of female molds out of plywood and set them up on a 50-foot workbench with a transit. The mast was then glued up in two halves and flopped together like a giant submarine sandwich. Blocking, halyard runs, wiring, and internal epoxy sealing were all done prior to putting the two halves together. Lettuce and mayonnaise are optional."

Stayed wing masts are more popular on multihulls than monohulls, since the wide shroud angle of the former makes for less mast compression. Sprague widened his shroud angles by mounting the hinge fitting well down on the mast. This limits the size of the staysail, but with no backstay to deal with, there's room for a big, efficient, full-batten main. Note that the section of the mast above the hinge is unstayed; like an unstayed mast, it can deflect to leeward in gusts to depower the main. Call it a semi-stayed mast.

LYLE HESS CUTTER

To many people, the gaff cutter rig is not simply old, it's anti-modern. And yet, for eyes weary of balancing out the high-stress details of the Bermudian rig, the gaff rig is a marvel of engineering.

Just think of it: no standing backstay, so no limitation on mainsail area. Sail shape controlled not by bending the entire mast, but just by the height and angle of the little spar at the top of the sail.

You want light-air performance? You get room for a conventional drifter and a gaff topsail, way up there where the light breezes blow. Some gaffers even carry a wardrobe of topsails to suit varying wind conditions. True, you can put a great big main on a Bermudian rig, but when the wind comes up you have to reef that big main. With a gaffer, your first reef consists of dropping the topsail. Much easier.

If the wind continues to rise, reefing a gaff main results in a dramatically lower center of effort compared with a triangular sail of similar area. And since you bring the gaff down at the same time, you also lower the center of gravity appreciably.

A nifty design note relative to reefing: In

GAFF STRIPS

Gaff jaws abrade the mast where they bear when the sail is hoisted. The traditional fix is to tack sheet copper or bronze around the mast. But even when well-bedded, this invites moisture to gather underneath. Rot follows. Though the sheeting stops chafe, it won't keep the mast from being dented.

As an alternative, screw half-round bronze or hardwood strips vertically onto the mast. The strips take the load better, drain more readily, and are easily inspected and repaired.

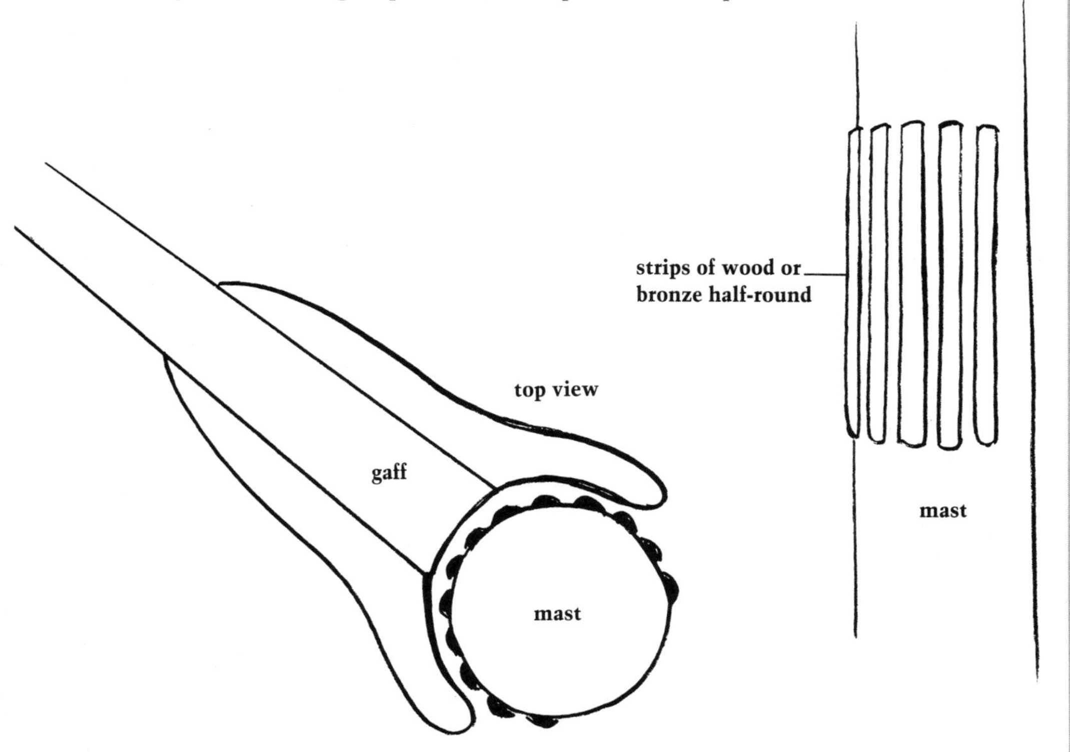

fully hoisted position the gaff, loaded by the sail and peak halyard, thrusts diagonally down against the mast. Aft lower shrouds take up this thrust, preventing the mast from bowing. But when reefed, the gaff lies against an unsupported section; you might think the thrust load would burden the mast unfairly. But as you can see in Figure 7-13, the peak halyard is more nearly vertical when the sail is reefed, so that more of the sail load goes into compression on the mast, and less goes into compression on the gaff. The deeper the reef, the less the gaff's thrust.

Stay tension is a bit different on gaffers, too. Aft-swept swinging spreaders are backed up by, of all things, the mainsheet—via the weather quarter lift and the peak halyard. This won't give you the profound tension you get from a Bermudian backstay, but it's

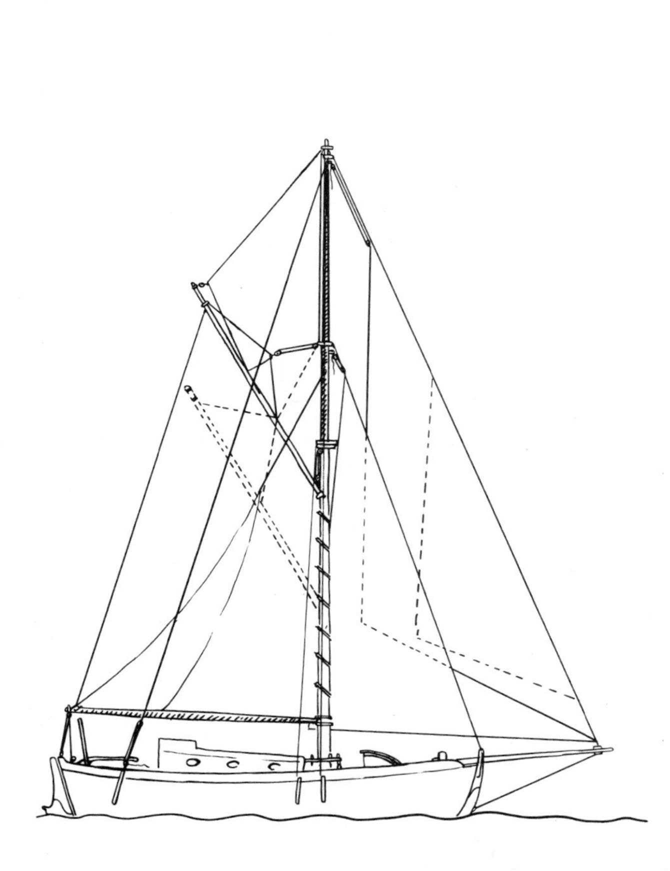

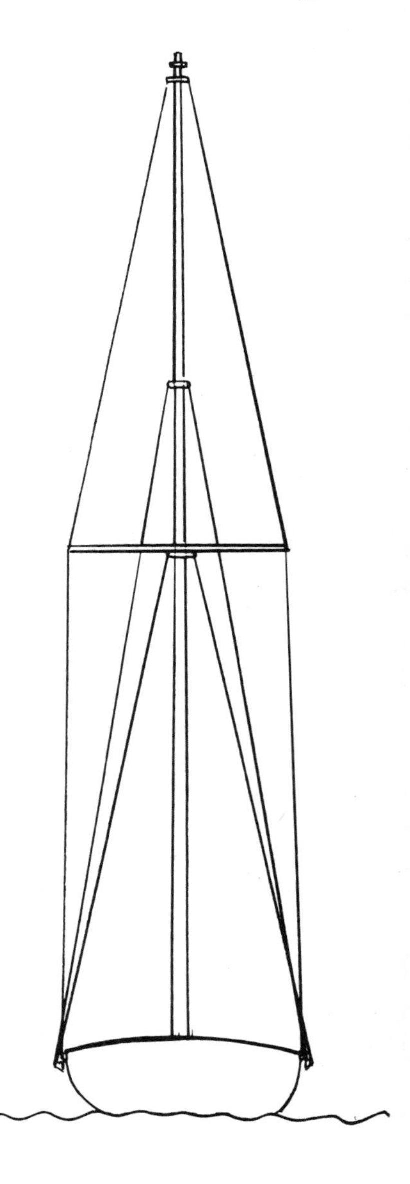

Figure 7-13. *Lyle Hess's Falmouth cutter, length on deck 29 feet 9 inches. The reefed gaff position is also shown.*

plenty for the light- to medium- air sails you'd hang from a gaffer's jibstay. The forestaysail can be set in combination with the jib or by itself, the latter in storm conditions or for shorthanded short tacking. Slightly aft-leading intermediate shrouds can oppose its pull, and in some cases they are backed up by running backstays.

Gaff running rigging traditionally relies on block and tackle instead of winches for

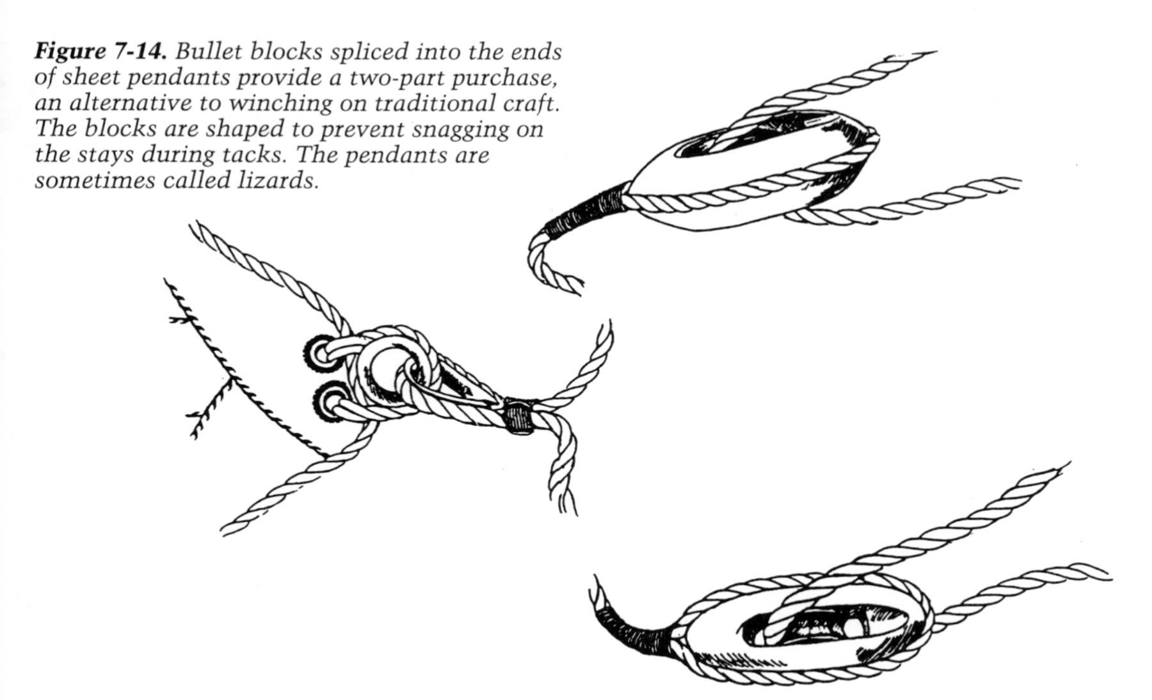

Figure 7-14. *Bullet blocks spliced into the ends of sheet pendants provide a two-part purchase, an alternative to winching on traditional craft. The blocks are shaped to prevent snagging on the stays during tacks. The pendants are sometimes called lizards.*

JIGGERS

Load on a halyard varies over its hoist; at first you're only picking up the weight of the sail, and perhaps a gaff or square yard, and the going is relatively easy. But at the end, when you want to tighten the sail up for an efficient shape, the going is decidedly difficult.

With most contemporary boats, it's the winch that applies the final, heavy load. But on gaffers and square riggers, it's done with a jigger: The halyard is double-ended; one end leads down to deck on one side of the mast and is pulled on by hand as far as possible, then belayed; the other end leads down to deck on the other side of the mast and has a three-, four-, or five-part block and tackle hanging from it. This purchase compounds the primary purchase aloft. That is, if your throat halyard has a four-part purchase, a jigger of three parts will give you a 3 x 4 = 12-part purchase at the end of the hoist for easy luff-tensioning.

You could, of course, make up a 12-part primary purchase, but then you'd need a halyard three times as long, and there's already enough string to deal with on a gaffer.

LEECH TWIST (THE MOTHER OF ALL WINDS IS APPARENT)

The direction of apparent wind is determined by vessel velocity and direction relative to wind velocity and direction; but apparent wind always moves forward as vessel speed increases, and aft as wind speed increases. Wind at deck level is slowed by friction with the surface of the water, hence windspeed is higher at the masthead, and apparent wind is farther aft at the masthead than at deck level.

Sail efficiency requires that the sail "attack" the wind at a specific angle relative to the boat's fore-and-aft centerline; since apparent wind moves aft as you go up, the sail angle should also change, becoming wider with height. This progressive angle change produces "leech twist."

The sail assumes a shallow spiral shape as the leech gradually twists to leeward. A little twist is a good thing, but too much will leave the sail too tight at the bottom and too loose at the top. This is a particularly vexing problem with gaffers, because the gaff sticks out so far, and thus can "fall" so far to leeward. This action can be controlled with vangs on the gaff, but practically speaking only on schooner foresails, with the vang leading from the mainmast. Leech twist can also be minimized, as noted in the accompanying text, by raising the angle of the gaff, and widening the mainsheet traveler, to allow the boom to go to leeward and thus keep under the gaff even when reaching.

mechanical advantage. This makes sense for the big heavy mainsail at least; an advantage of two or three is enough for a fit, medium-sized person to raise a yacht-size gaff hand-over-hand. Final tension can be obtained either with a jigger (see "Miscella") or with a small winch. It's worth noting here that fully battened Bermudian sails can be as heavy as a gaff sail of comparable area; with no block and tackle they can be slower and harder to hoist than their archaic cousin.

As for staysail sheets, you have a choice: either lead the sheets to conventional winches, or hang blocks on short "lizards" attached to the clew for a 2:1 advantage (see Figure 7-14). The latter is suited to small staysails, but watch out for flogging blocks. If you go with winches, you'll find they needn't be huge, since the staysails are relatively small.

So many advantages. And wait, there's more! Modern sailcloth is far more stable, lightweight, and strong than the canvas of yesteryear, so the gaff rig just about can't help but perform better than it did in the days of cotton. But it helps to have a sailmaker who knows the peculiarities of cutting gaff sails if you want to make the most of the improvement.

Sail design has also evolved. Note for instance the high "peak angle" of our example boat: The higher this angle, the smaller the arc the gaff can swing through, and thus the less leech twist you'll get. Combine this with an extra-wide mainsheet traveler and twist can be minimized to near-Bermudian levels (and see accompanying "Leech Twist" Miscella).

For even more efficiency, you can give the gaff its own gooseneck and run it up a heavy-duty sail track on an aerodynamically-shaped wood or aluminum mast. And how about making the gaff out of aluminum instead of wood, to save weight? With these and other simple, inexpensive details you can have a gaff rig that goes very well to weather, with none of the high-tension, money-intensive, vulnerable tweaking gear of an only-slightly-higher-pointing Bermudian rig. And when you bear off the wind, you'll find that a gaff rig has infinitely adjustable sail shape for high-powered, low-effort reaches and runs.

So how come this rig is so rare? Partly because pointing, at any price, is what counts in racing. Partly because handling a big gaff sail can take some skill and, in the absence of winches, physical effort. Partly a lot of other reasons, perhaps the biggest one being that it is no longer available "off the shelf"—you pretty much have to build a gaff rig yourself these days, since the supply and design infrastructure is almost entirely Bermudian-oriented.

Bermudian was the new kid on the block once, too. It grew to dominance because sailors grew to prize its strength—windward ability—above all others. Indeed it might be said that they became blinded to all others, to the point that I can only half-jokingly reintroduce the gaff rig as radical. This chapter began with some nice, safe, linear formulas. But those formulas are put to work in the service of wildly varying human preferences, based on stupefyingly diverse human assumptions of what will work. The only reason that chaos isn't the rule is that all these formulas and preferences are formed with regard to the ocean, which doesn't care what we like and which will smash us if we don't take it fully, earnestly into account.

"Taking the ocean into account" is as close a definition as I can come to for the term "Seaworthy." It's not a term as readily reducible to formula as mast design, but like a lot of other things, you can know it when you see it. In the next chapter you'll find an utterly subjective view of seaworthiness, with the emphasis on matters affecting rig integrity. Think of it as right-brain design.

Chapter 8
Seaworthiness

"There are four kinds of seafarers under sail: dead; retired; novices; and pessimists."
– Tristan Jones

A client once came to me wanting to know how to run his exterior-mounted chainplates around or through the unusually tall bulwarks he had built onto his boat. He was a thoughtful, meticulous man, and I was sure he had a specific reason for his bulwarks' height. So I asked him, and he pointed to his feet and said, "My sneakers."

It turns out what he meant was that if his boat ever approached knockdown, the bulwarks were tall enough that he could stand in them, in relative security, while dealing with getting the boat back up. If I were to give his attitude a name, I would say it was seaworthy.

Monohull sailors commonly deride provisions, in seagoing multihulls, for living in the hull with the hull upside down. "Why," the complaint goes, "don't they just get a boat that won't capsize?" But well-designed multihulls are very, very difficult to capsize (see later in this chapter); these provisions are just "sneaker bulwark" prudence. They are seaworthy.

Contrast that with the experience of a friend who'd been foredeck crew in a racing monohull one day. His boat won, and he began jumping up and down for joy. "Stop!" shouted the afterguard, "You'll go through the deck!" Obsessive paring away of weight is a hallmark of racing craft, and this vessel's foredeck, where people had to work in often strenuous circumstances, was deemed strong enough if it could just support a cautious man. That is not seaworthy.

Seaworthiness is like pornography; you know it when you see it, but it's very difficult to define. Seaworthy vessels are strong, but how strong? Easy to steer, but how easy? Have a gentle motion, but how to qualify gentleness? To compound things, the popular perception is that the more "seaworthy" a vessel is, the slower and more sluggish it is. Is this true? To what extent? And why?

In terms of rigging, there's essentially just one question: Is it strong enough? That is, regardless of the seaworthiness of the hull, is the rigging likely to hold together under the loads the hull will put on it? But an appropriate rig is intimately wedded to its hull; in order to answer our single question, we must look at many others, like the ones in the preceding paragraph. The formulas in "Mains" gave us basic integrity, but now we will be looking for subtler, more elusive harmony.

EFFECTS OF ATTITUDE

A boat grows, bit by bit, out of the attitude of the designer and builder. The client with the sneaker bulwarks is also likely to pay similar attention to handholds, engine access, navigation—he'll build himself a liveable world. And the multihuller who makes careful provision for being upside down will devote even more attention to the far, far likelier state of being upright. And it also follows that the skipper of the thin-decked boat will be edgy not just about the deck, but about the mast, the stability of the vessel, and the stamina of the crew.

EFFECTS OF EDUCATION

But there's more to seaworthiness than attitude; very careful, well-intentioned people can go very wrong from simple technical ignorance. Take, for example the couple whose interior decorator thought their keel-stepped mast took up too much room below (see "Mains"). Here's a scenario, based on actual events, of what could happen:

The couple pay a yard to convert to a deck-stepped rig. No one involved in the project realizes that deck-stepped masts have to be stiffer than keel-stepped ones of the same length. So they just cut the bottom off the old mast and set it on a reinforced house top. First time out sailing, the mast jumps around alarmingly, frighteningly close to folding. Back to the yard to have a "sleeve" inserted in the mast to stiffen it. Now the mast is okay, but the decorator has wreaked other havoc. In changing around the accommodations, he's eliminated a bulkhead on the port side, amidships, a structural bulkhead where the inboard chainplates used to attach. The starboard shroud chainplates still attach to a corresponding bulkhead, but the port ones now go to a big tie rod which pierces the cabin sole, attaching to a lug on the hull. Tie rod and lug are both massively strong, as is the clevis pin that connects them. But the worker who does the installation doesn't bring along the right size of cotter pin. The ones he has are a little too big, but he jams one in about halfway, and decides that it is good enough. The sole covers up this arrangement, and it is never noticed by even the very maintenance-conscious owners. Over several years the cotter works its way out, then the clevis gradually wiggles free, then one day in a fine breeze the port shroud chainplate comes up through the deck and the top of the mast goes over the side.

There was nothing wrong with wanting a comfortable, workable interior. And nothing wrong with properly engineered deck-stepped masts or chainplate tie rods, or

SPARE CLEW

If you must insist on cruising with a roller-furling stays'l, have your sailmaker install a "spare clew" directly above the regular clew. Then if the regular clew pulls out you can move the sheet to the other clew. Part of the sail will be flapping in the breeze, but at least you'll be able to sail. This procedure is a good idea for hanked sails, too, but the second clew can be where a reef would fall, so you have both a reef clew and an emergency clew.

proper-sized cotter pins. What was wrong was the assumption that any of these items could be considered separately, that all of the details of construction would be naturally and inevitably comprehensible. One of the most profound and challenging joys of sailing lies in mastering the intricacies of a little floating world. It's just dangerous to anticipate mastery.

EFFECTS OF CULTURE

It is difficult to see through the assumptions and norms of one's own society—they're so reflexively there—and yet they shape our craft at least as much as the ocean does.

Consider, for example, the great racing yachts of the turn of the century. Here was the last gasp of a feudal society, complete with a hierarchy of indolent aristocrats, knight-helmspeople, and forelock-tugging menials. The boats were the size of castles, and the entire ocean was a moat.

Or consider the recently vanished sailing junk of China. Here was a trading vessel which contained a complete merchant community/family. These easily handled boats were designed around a vastly different social order than that of the European yachts, and the result was a correspondingly different design, construction, and handling.

Now look at a typical IOR racer of the recent decades. What are we to make of this incredibly expensive, fragile, ill-handling craft? What are we to make of the human ballast that lines its rail, of the vestigial mainsail, of the ludicrously large headsails? If we see these craft as "normal" or "desirable," it might just be because of some unfortunate conditioning.

HARMONIC TOWING

When towing a dinghy—or anything else—in a swell, extend the painter so that both vessels are at the top of different swells simultaneously. This prevents shock loads on the towline.

DRAGGING WITH A HOOK

When dragging an item with a hook, make sure the hook goes in with the bill pointing down, so it won't fall out if the load is released.

EFFECTS OF HYPE

Shortcomings in attitude, education, and culture lead to shortcomings in boats. And that, in our era, has led to vastly profitable "solutions." These are generally based on sound principles, but are hyped to mask the nature of the flaws they are supposed to solve.

For example, we've all seen boats trumpeted as having "all control lines led aft, so you never have to leave the comfort and safety of the cockpit." But the question arises: When was it that the area outside the cockpit became uncomfortable and dangerous? Similar hype exists for roller-furling headsails, which save us from "venturing onto a heaving, wet foredeck."

It is true that aft-led control lines are a boon for singlehanders and racers, for speed and convenience. And it's true that roller-furling has opened sailing up to a lot of non-acrobats. But that's not the same thing as danger and discomfort being vanquished by these devices. I believe that it's no coincidence that this "comfort and safety" concern arose with the dominance of the IOR. If you have a wet, skittish boat, you'll leap at any gizmo that promises to make life a little less alarming.

And in any event, the "solutions" themselves have downsides: When the roller-furling mechanism breaks, jams, or otherwise goes into a snit, usually in high winds and seas, you have to go onto that foredeck anyway, armed with a crescent wrench and a lifejacket; and aft-led lines can overflow a cockpit in a hurry, making it a crowded, confusing place. It can be enough to make you want to escape to the comfort and safety of the foredeck.

To compound things, hype has obscured a host of simpler, more dependable, and invariably cheaper alternatives. One dramatic example is the "trapdoor" for staysails aboard Linda and Steve Dashew's *Sundeer*. Instead of being fixed on deck, the fore and jibstays run right through trapdoors to the keel. Lowering the staysails is a matter of opening the doors and slacking away the halyards. The entire sail inventory is stowed in the room below the doors, where they can be hanked on or off in a cockpit-surpassing level of comfort and security.

The *Sundeer* approach would be hard to retrofit to most boats, but how about this one: bear off. That is, when it's time to deal with headsails, change your heading to bring the wind aft. The boat's motion will be much gentler, the mains'l will blanket the heads'l and the apparent wind will be much less. It's amazing how few sailors think of this, but then we've all been overly influenced by the press-on-regardless acrobatics of racing.

There are many other ways to respond to boathandling challenges, rig-related and otherwise. But no matter how effective these responses may be they are only bandaids, things we do after the damage has been done, unless we employ them as part of a conscious design sensibility. I do not want to, and in any event am not equipped to, write a comprehensive treatise on yacht design. What I do want, as a rigger, is to understand how details of rig and hull can be optimally interrelated, and I have found that a few design considerations can help illuminate this question.

MOUSING A HOOK

Mousing a hook prevents the load from hopping out, and a stout, tight mousing also reinforces the hook, preventing it from stretching open. But make certain before you apply the load that the load is on the hook, not on the mousing.

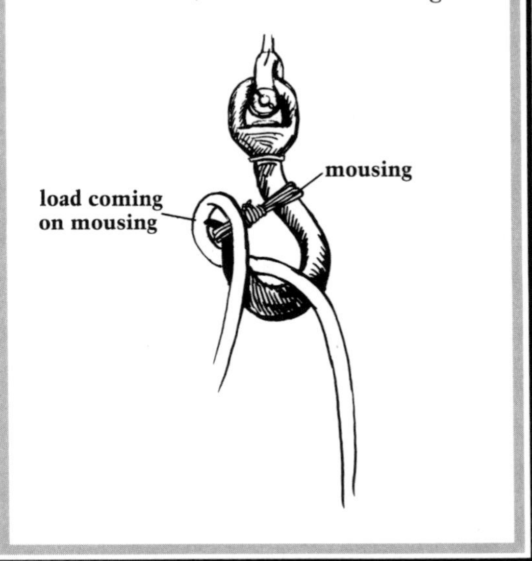

THE TRANSVERSE RIGHTING MOMENT CURVE

In "Mains," the transverse Righting Moment (R.M.) curve revealed the maximum load that the rig would have to bear. But this curve can also reveal a lot about how a given hull will behave in varying conditions, and thus what kind of rig design details are appropriate.

Figure 8-1 will look familiar from Chap-

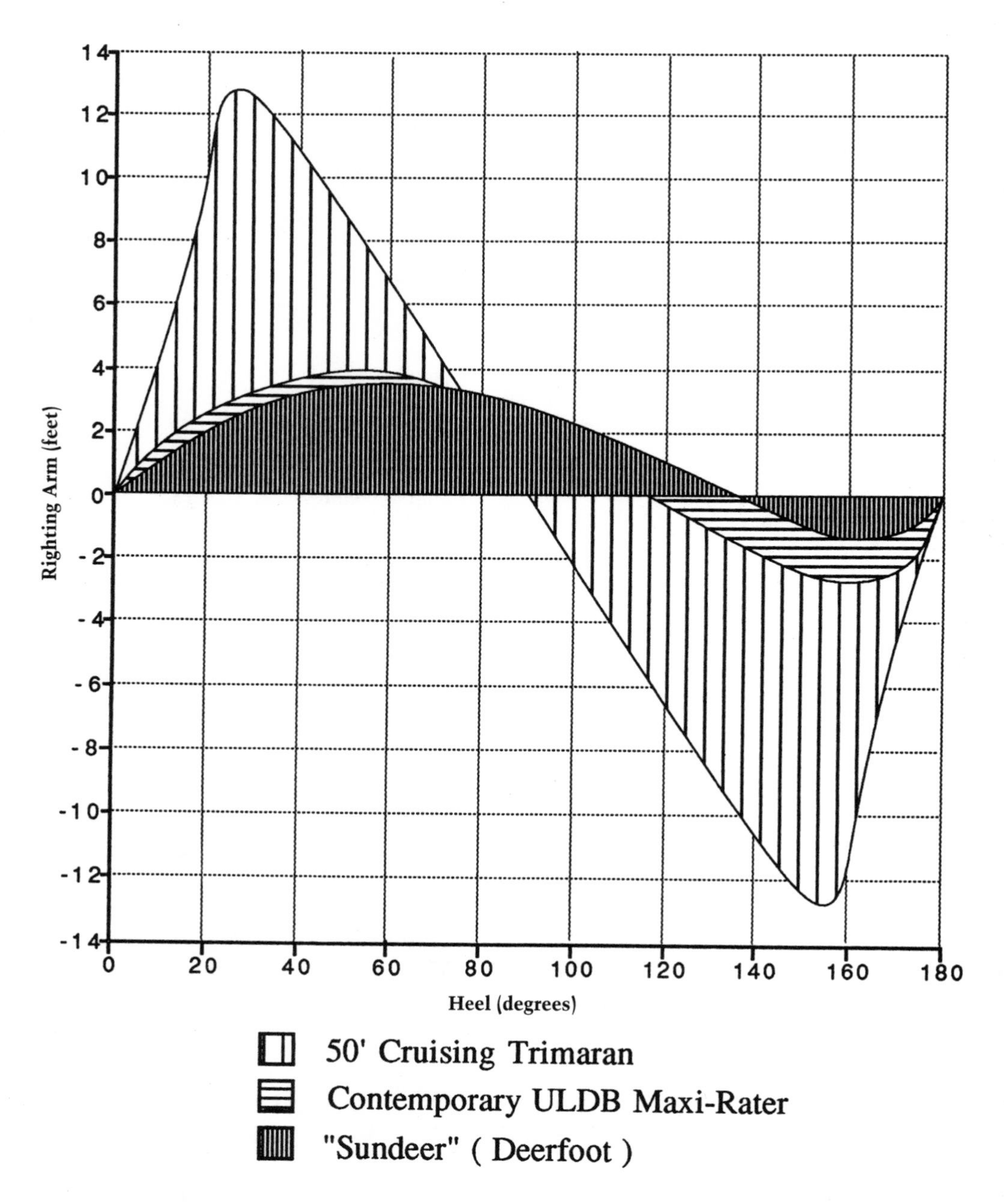

Figure 8-1. *The righting moment curves for three vessels, as seen in Figure 7-2. (Righting moment is the righting arm multiplied by vessel displacement.) The ultralight maxi-rated racer has high initial stability but a capsize angle of 115 degrees—pretty much the minimum for an offshore monohull. The exceptional stiffness of the multihull renders its low capsize angle a distinctly secondary consideration to any right-thinking multihuller. The capsize angle of 135 degrees for Sundeer means that the boat would roll back upright almost instantly were it ever capsized (barring structural damage or loss of keel). (Courtesy Jeff Van Peski)*

STABILITY ATTRITION

Yacht design since the 1960s has followed a steady downward trend in terms of stability. According to C.A. Marchaj (*Seaworthiness: The Forgotten Factor*, Adlard Coles/Nautical), "The positive area under the righting moment curve has been reduced to less than half, and at the same time the negative area has increased dramatically."

Marchaj also notes that the angle of vanishing stability has been reduced from about 180 degrees to about 120 degrees, and that the righting moment maximum is now "about half of that characteristic of the traditional yacht form."

ter 7. It shows the R.M. curves of three vessels: a cruising trimaran, a racing monohull, and the Dashews' Deerfoot design.

The tri's curve rises very steeply until about 25 degrees, then plummets abruptly. From 90 degrees to 180 degrees the hull is in "negative stability;" that is, it wants to remain upside down. The racing monohull's curve rises less sharply, peaking at about 55 degrees, then drops—also less sharply—entering negative stability at about 115 degrees. Note that the monohull's maximum negative stability is considerably less than that of the trimaran; the monohull will be easier to right. The stability curve for this boat *as sailed* will have a much steeper initial section, since such boats invariably carry large crews, most of whose members spend much of their time perched on the weather rail to provide ballast. Thus the boat can stand up to breezes without being slowed down by extra-in-keel lead ballast.

The curve of the Deerfoot, also a monohull, rises most gradually of all, but enjoys a long plateau, peaking at about 60 degrees. Then it drops slowly, but look! It maintains positive stability right through 135 degrees, and the maximum negative stability is very small. Should this boat ever be capsized, she'll be boosted back on her feet almost immediately by any modest passing wave.

None of these curves is "bad" or "good," although they are often interpreted as such. Monohullers, for example, never tire of deriding the spectacular range and intensity of multihull negative stability, but fair consideration is hardly ever given to the equally spectacular range of positive stability.

Multihullers are just as nasty about putting down monohulls. "Lead mines," they call them, in which you are always "sailing on your ear." But this is just a way of saying that, with a combination of pendulum stability (the "lead mine" in the keel) and form stability (the buoyancy of the hull), a monohull bends before the wind, spilling gusts instead of taking them full on. For a given displacement, a multihull's sails, rigging, and hulls must be much more strictly and heavily engineered, since they literally have to stand up to every gust. It is good practice to make multihull rigging at least 50 percent heavier than rigging for a monohull of the same displacement.

Multihullers and racing monohullers do agree on one subject: that cruising monohulls are too slow and too heavy. But although the cruiser won't sail as upright, it has a higher ultimate resistance to knockdown or capsize. And if it ever is capsized, it will always come back to upright. For some people, this can be a compellingly attractive feature.

So there are plusses and minuses to each design approach, the details of which I've barely touched upon. And there are plenty of boats in each category whose vices far outweigh their virtues. By studying transverse stability curves, we can get an idea of how to maximize the virtues and minimize the vices of each type of craft.

To build a better trimaran, for instance, we'd work to make it still harder to tip over, whether from wave action, "tripping" on the outer hulls, or any other reason. And we'd probably try to make un-capsizing easier, just in case. And this is what good multihull designers are trying to do.

The cruising monohull can be improved by making the stability curve steeper in the first 30 degrees, if this can be done without sacrificing ultimate stability. This generally involves lowering the vessel's center of gravity and fussing with the amount and location of the ballast. Comparing Figures 7-1 and 7-2, you can see that while the righting arm of the Ohlson 38A—a conservative, wholesome cruising monohull—is 1.5 feet at 30 degrees of heel, the 30-degree righting arm for Steve Dashew's *Sundeer* is over 2 feet. *Sundeer* has almost no form stability, but Steve sweated to get all the weight down low. This low center of gravity, combined with a perfectly balanced hull, made for a very stiff ocean cruiser. It can be done.

Racing monohulls are not so susceptible to improvements in seaworthiness. They could be, but their sailors are inclined to sacrifice everything to going well to weather. That and rating rules (see below) pretty much hamper virtue. But even here, responsible designers and rule-making bodies can prevent the worst excesses of unseaworthiness.

> **INGENIOUS SELF-STEERING**
>
> When his Dove lost her rudder at sea, Winston Bushnell coiled and stoppered 660 feet of spare lines into a 6-foot–long bundle, then trailed the bundle about 30 feet behind his boat on two lines, one leading to each quarter. By adjusting the lines, he was able to steer his boat.
>
> Source: *Capable Cruiser*
> Lin and Larry Pardey

FORE-AND-AFT STABILITY

The transverse righting moment curve is only the simplest, most accessible of many seaworthiness-indicating factors, and it is by no means definitive. C.A. Marchaj beautifully addresses these other factors in his book *Seaworthiness: The Forgotten Factor*, far better than I'd ever know how to. But many of those factors are outside the scope of rigging per se. So I'll touch briefly on just one, an item of gratifyingly far-reaching significance, and one we can study without having to decode any of Mr. Marchaj's graphs and formulas.

A hull also describes a curve of fore-and-aft stability, again dependent on hull form and ballast amount and location. A heavy vessel, fat at bow and stern on the waterline, will be very stiff fore-and-aft, but with so much hull dragging through the water it will also be very slow. By making the hull—and particularly the bow—relatively skinny at the waterline, but flaring above it, you get minimum resistance to forward motion in light conditions, but "reserve buoyancy" in heavier conditions, when the hull will tend to dip at the ends. The amount and degree of flare is crucial. Too little too high will not keep the bow from diving, or might result in a "hobbyhorse" motion. Too much down low, and the hull will slap and pound in a sea.

Ultimate fore-and-aft stability must be sufficiently high to discourage the somersaulting action known as "pitchpoling." As with transverse loads, sometimes wind and waves can gang up on the most stable vessel, but good design can minimize a flip in either direction.

END-OF-BOWSPRIT ANCHORING

For boats with bowsprits, rig a stout nylon "snubber" line from deck, out to the end of the bowsprit, through a turning block, and back to deck after dipping underneath the bowsprit shrouds. After the anchor has been lowered and set, hook or tie the end of the nylon to the rode, then veer more scope so the nylon takes the strain.

The pendant absorbs shock loads, keeps the bow down and controls yawing and pitching, prevents the rode from chafing on the bobstay or hull, and can be quickly and easily slacked or retrieved from deck. Altogether superior to the practice of rigging a snubber from a fitting on the stem, at the bottom end of the bobstay.

Source: *Capable Cruiser*
Lin and Larry Pardey

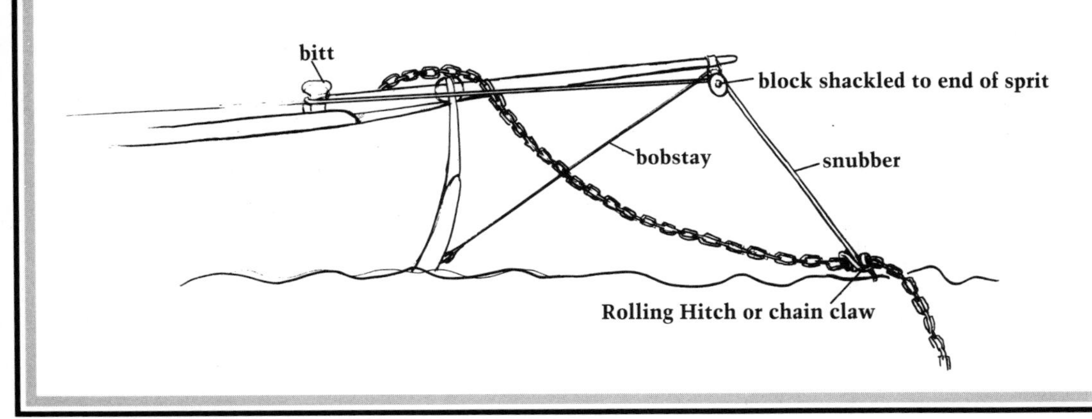

To sum up, we've worked out stability in two planes, and seen that this stability must be played off against sailing performance. Different hull types and sailing styles, of course, demand different stability compromises. And if this were all there were to the question of stability, rigging sailboats would be a much simpler subject. But aside from matters of preference like rig configuration and materials, there's one big monkey wrench deeply lodged in the design works.

RATING RULES

The noble, sensible, sporting idea behind most rating rules, which address hull and rig designs, is to provide a basis for comparison among vessels, so that they might be assigned appropriate handicaps for even competition. In yacht racing, handicaps take the form of time allowances; a favorably rated (ostensibly "slow") yacht can finish a race well behind its competitors and still win if its handicap allowance is great enough. This seems fair enough, but rules have historically had unfortunate effects on the integrity of the vessels built to suit those rules. Designers and sailors naturally want to see their boats finish first, and creative exploitation of rules can give them ways to do this without necessarily making the boats go any faster, let alone be more comfortable, stable, or stout.

So for example, rules often "tax" the waterline length, since in theory a vessel's maximum non-planing speed is approximately 1.1 times the square root of the waterline length. An easy way to exploit such a rule is to design a vessel very short on the waterline, but with great long overhangs at bow and stern. When the vessel heels under sail, more hull length is submerged. As a result, the boat will sail faster than the rule predicts.

The history of sailing rule-making is essentially an endless exercise in loophole-plugging. So rules have grown ever more Byzantine, and hull shapes have grown ever more bizarre in attempts to exploit the endlessly generated loopholes.

The International Offshore Rule (IOR), which predominated from the 1970s until recently, was originally intended to be an all-time loophole-plugger. But its creators underestimated the resourcefulness of—and perhaps overestimated the principles of—yacht designers. In a nutshell, boats that most successfully exploited the rule were very fine forward, very beamy aft, high-sided, lightly built (rig weights in the 1980s were typically a third lighter than in the '60s), and relied heavily on human ballast for something like transverse stability. Fore-and- aft stability was severely compromised, making for very wet (read: submerged) foredecks, and the area and degree of negative stability was very high. Many IOR boats became negatively stable at angles less than 120 degrees. Knockdowns and dismastings became commonplace, and crews suffered from the discomfort of riding in sharp-motioned, skittish, wet boats.

Sounds like pure hell, doesn't it? But on the other hand, these boats were also incredibly expensive and had huge, hard-to-handle staysails. I'm speaking of them in the past tense, but in fact IOR-style boats continue to have a strong presence in the marketplace. I mention them here partly as a diatribe against their continued existence, but mostly because they make a great Bad Example, illuminating the relationship between fore-and-aft and transverse stability.

HEELING AND DIPPING

When a vessel is at rest and upright, form and ballast stability makes it float on its lines. If it doesn't, and if it's not too far off, moving ballast or gear around will level it. This is extremely important because the boat will only go where you point it to the extent that it pushes through the water in a balanced fashion. If there's too much hull in the water forward, the force of the water will push the bow to windward, resulting in weather helm. A common response to this is to rake the mast forward, since this moves the sail area forward, reducing weather helm. But it's usually much simpler and more effective to leave the rigging out of it and just trim the hull down aft. Conversely, a hull trimmed too much aft will generate lee helm.

But fore-and-aft trim is a tricky thing; most vessels trim differently when heeled than they do when upright, because the shape of the hull in the water is no longer symmetrical, as it is when the boat is upright. This is very hard to avoid, since you want a boat to be fine forward, to move easily, yet have a beamy midsection, to resist heeling. But when you do heel, that beamy midsection gets submerged, and is a lot more buoyant than the corresponding area on that skinny bow. So most boats nose down as they heel and their weather helm increases, no matter how beautifully balanced they are when upright. To counteract heel-induced weather helm, sailors will (a) reef the main, which moves the center of effort down and forward, or (b) move the draft of the sails forward using mast bend or luff tension, or (c) both.

Here is a place the rigger can shine, using the running and standing rigging not just to deliver power but to balance the helm. That is, unless the hull is so sensitive to heel that no amount of rig adjustment can make it right. A worst-case scenario for

EFFICIENCY

Cost and structural vulnerability increase exponentially with sailing efficiency. So a well-built, moderately efficient boat will cost far less and be far more durable than a well-built highly efficient boat.

riggers was the nose-diving, broach-prone IOR fleet. About all a rigger can do for these boats is to help keep people in the cockpit. And as noted earlier in this chapter, that's what roller-furling, aft-led control lines, and other conveniences have done. But ameliorating the effects of bad design does not constitute seaworthiness. For many boats, adding sail-control conveniences is like putting a muzzle on a rabid dog.

SIDE EFFECTS

There's an old saying that "if you take care of the molehills, the mountains will take care of themselves." A seaworthy rig is one in which the molehills are taken care of. A vessel shaped by concern for bulwarks, handholds, human comfort, longevity, simplicity, and a zillion other details can be engineered to acceptable standards of Speed, Weatherliness, and Maneuverability. But it is an oft- demonstrated fact that vessels dedicated only to Speed, Weatherliness, and Maneuverability are possessed of precious little else. Take care of the mountains, and the molehills seek more hospitable quarters.

Another effect is lowered cost. Gear for racing rigging is ruinously expensive, with price rising exponentially for every tenth of a knot gained. Cruising gear may be heavy, but you don't pay for frantic engineering.

A third side effect is peace. Once you get out of obsessive race mode, you enter a world in which dismastings are rare, which means you don't have to sweat bullets worrying about a dismasting all the time. Seaworthy rigs sometimes do go overboard, but when they do, it is usually from gear failure, not scantling failure. That is, while a racing mast will collapse because it's just too flimsy to withstand a little extra load, a sensibly scaled mast will do so because something—a clevis pin, a tang, a bolt—has worn out, and the sailors have just not attended to it.

This brings up one last side effect: Molehill care makes you part of the boat, brings you into its life. There are precious few sane reasons to go sailing; no need to throw out the elegant relationship possible between humans and their vessels.

Chapter 9
Tricks and Puzzles

"There ain't no such animal."
– Farmer, on seeing his first giraffe

A good magic trick is about deception on the surface and a sense of wonder at heart. When, after patter and prestidigitation, the magician presents you with an absolute impossibility, you are taken right out of your everyday, everything-has-an-explanation world, if only for a moment.

If you are the magician, and if you're paying any attention to your audience at all, you will find that simply being around delighted, mystified people is far more gratifying than any ego boost you might get from being able to "trick" people. You'll find that that delight is addictive, so you'll practice to make your illusions effortlessly convincing, just to help produce that delight.

Some audiences are kinder than others. If someone is intent on interrupting, trying to expose your technique, they're missing the point; try to see it as their problem. No matter how kind and appreciative an audience is, resist the temptation to repeat these tricks more than once; illusion is a fragile thing.

Yes, I am revealing "how it's done" here, but no, you must not; by some accident of fate you are about to have Great Secrets re-

BAG CARRY

If you're moored at the end of a long dock, you have to carry your garbage farther than most people to dispose of it. But put it in a bag and hold it as shown, and you won't end up with aching, paralyzed fingers.

vealed to you, but I urge you to use them to generate wonder, not deflate it with explanation. If someone really wants to know "how it's done," they'll bug you about it for weeks until you give in. But then they'll have earned it, and will be receptive to your admonition "never to share this with another soul."

All of the accompanying string tricks can be performed anytime, anywhere, with no preparation. People expect magic tricks to involve gimmicks, so when you just up and make a miracle, they're doubly delighted. "The Professor's Nightmare" is the only one of the bunch that requires serious study, but that's only to be expected for a routine that many professional magicians call "the world's greatest string trick." Take your time with it; all by itself, it's worth the price of this book. But only if you do it right.

THREADING A RING

But before we get into the Nightmare, let's start with a simpler impossibility: threading a solid ring onto a string without passing either end through the ring.

"Impossible? Of course it's impossible," you tell your audience, as you drape about 6 feet (1.8m) of ⅛-inch (3 mm) or so diameter string over your hand.

"The trouble," you continue, picking up

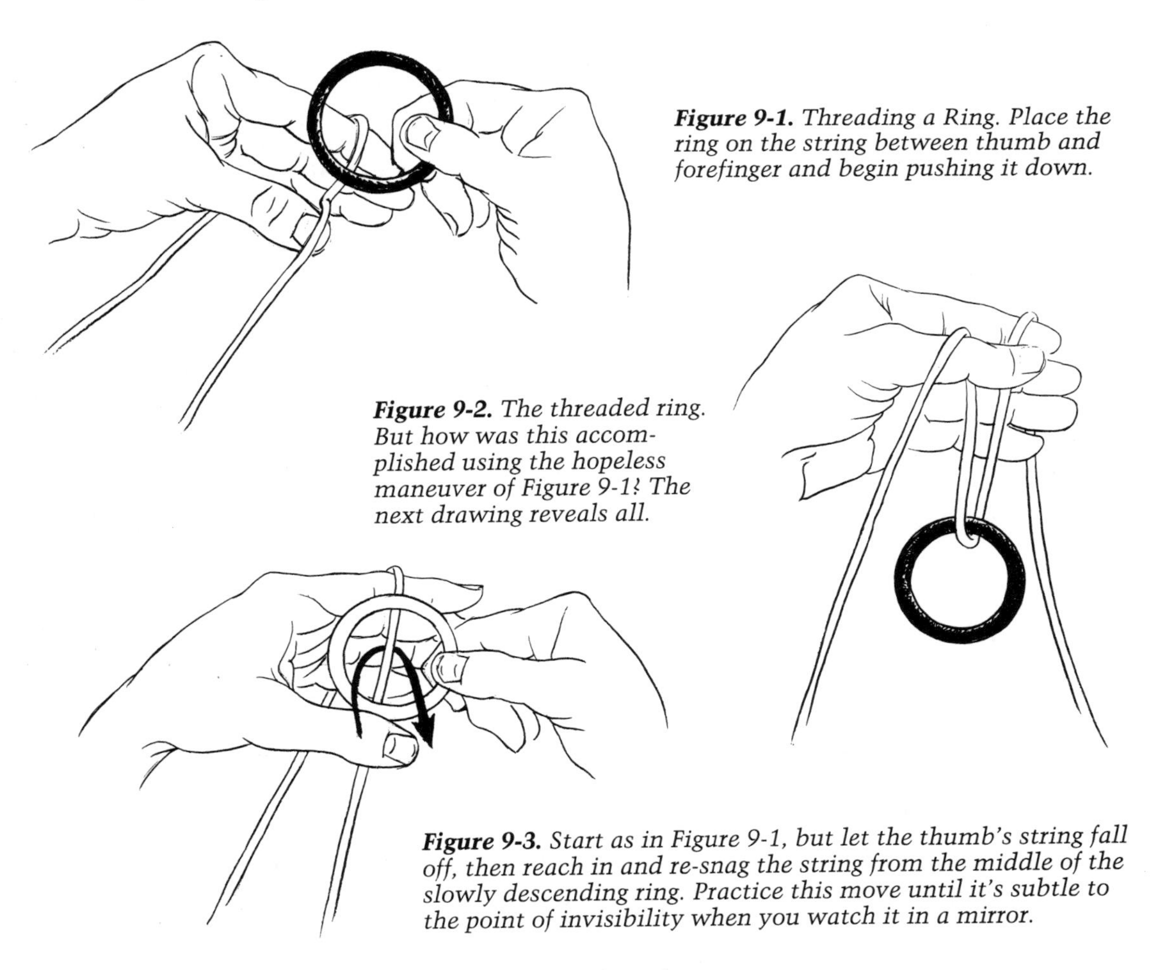

Figure 9-1. *Threading a Ring. Place the ring on the string between thumb and forefinger and begin pushing it down.*

Figure 9-2. *The threaded ring. But how was this accomplished using the hopeless maneuver of Figure 9-1? The next drawing reveals all.*

Figure 9-3. *Start as in Figure 9-1, but let the thumb's string fall off, then reach in and re-snag the string from the middle of the slowly descending ring. Practice this move until it's subtle to the point of invisibility when you watch it in a mirror.*

a wide, thin ring or bracelet, "is that when you try to put the ring on in the middle, the middle is at the top and the ends are at the bottom. And if I push the middle down to the bottom, the ends just go to the top."

You demonstrate this laying the ring on the string, in the gap between thumb and forefinger, and slowly pushing the ring down (Figure 9-1). Sure enough, the middle goes down and the ends come up. Repeat this self-evident procedure a couple more times, slowly, with lots of inane, self-evident commentary; you are accustoming your audience to reality.

"So it looks," you say as you begin to press the ring down once more, "as though sometimes the impossible really is impossible, and you'll never get a ring threaded onto the middle of a string. Like this." And with that you take your hand away, and the ring is threaded on to the string (Figure 9-2).

As you spoke, it turns out, you let the string slip off your thumb, which was hidden by the rest of your hand, which was up near eye level and tilted slightly toward you (Figure 9-3). Practice in the mirror to get just the right non- suspicion-inducing minimal attitude and amount of tilt. As you passed the ring down between thumb and index finger you slipped the tip of the thumb back under the string. A little dexterity is required here so that the ring can continue downward without hesitation. The ends will come up as they have before, but this time one of them will be pulled through the ring on the way up.

THE JUMPING RING

"But some impossible things are possible," you continue. And as you launch into an idiotic monologue on modern physics—Schrodingers Cat works well here—you tie three overhand knots in your 6-foot (1.8 m) string. The first knot is in the middle, the next one a handsbreadth away, and the third a handsbreadth beyond that. Slip the ring on and tie it to the string with the third knot.

"The ring," you announce with exaggerated confidence, "is on the right." Then you pause, appear to consider that statement, then continue with somewhat less confidence. "That is, it's on my right, which is to say it's on your left, right? I mean, well, I'm going to make the ring jump magically to my left, which is stage left, assuming you are the stage. Unless we were both facing the other way, in which case, um. . . . Confidence is no longer in evidence, but you plunge ahead:

"I'll just put the string behind my back," which you proceed to do, letting go with the hand holding the long end, then reaching behind your back with both hands.

"Now I'll bring the string back out and ______" you stop, hands still behind your back. You fight back panic. You move your arms first one way, then the other, painfully indecisive. "The ring will come out on my, uh, left, which is your right, right?" And you finally bring the string back out and sure enough, the ring is on your left now because, of course, you switched it while it was hidden, either by turning it over or by simply changing hands.

Before anyone can comment on this obvious, obvious move, you say, "I know, it's amazing, but there's more." You put it behind your back again and say, "Now I'll make it return to your right. No, I mean to

STEVEDORE GRIP

When handling slings, get in the habit of holding them with your fingertips, the "Stevedore Grip," instead of wrapping your fingers around the line. You'll avoid getting your fingers caught between the sling and a load.

UNIVERSAL MARLINGSPIKE PROPORTIONS

The late Nick Benton, master rigger of extraordinary talent, turned his mind to every aspect of rigging. Disgusted by the blunt clubs sold these days as marlingspikes, Nick analyzed the proportions of classic Drew spikes and came up with the accompanying diagram, in which "X" refers to the diameter of wire you'll be working with. A spike for ⅜-inch (9.5 mm) wire, for example, would be 9 inches (229 mm) long, with a diameter of 9/64 inch (3.6 mm) at the widest part of the "duck taper" at the tip.

This design is suited to more than splicing; it will let you get the point into shackles of a size you're likely to use with ⅜-inch (9.5 mm) wire. There'll be enough meat that you won't have to worry about bending or breaking the spike. For most yachts, a spike scaled to 5/16-inch (8 mm) wire—7½ inches (191 mm) long—is appropriate .

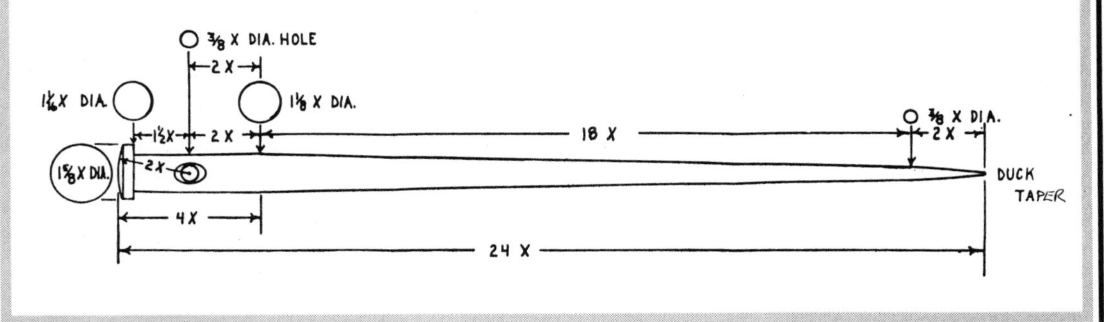

my, I'll make, uh . . ." And while panic returns, you appear to be moving your arms indecisively again and the audience is thinking, "This is funny—is he really that dyslexic?" But while they're laughing, you are quickly and smoothly tying a fourth overhand knot, a handsbreadth past the third one. It takes only a little practice to be able to do this in the right place without fumbling. The string should need to be out of sight for only four or five seconds.

When you bring the string out again, having once more end-for-ended it, the ring has indeed jumped back to the original side. Your short-end hand covers the new knot—thumb and index finger is all you need. You appear to be very pleased with yourself, and more than a little relieved.

At this point, particularly if your audi-

SNAP-ON SPIKES

Snap-On, makers of seriously high-quality tools, is under the delusion that they produce three sizes of scratch awls. What they actually make are three sizes of small marlingspikes with a taper very similar to the classic Drew pattern. The steel in these alleged awls is very hard and highly finished; all a sailor has to do is file the point into an acceptable, rounded, not-too-sharp "duck taper" and drill a lanyard hole in the plastic handle. Not as classy as having the real thing machined up, but an excellent tool nonetheless. The middle size is about right for the average yacht.

ence contains small children or other snide know-it-alls, someone will proclaim that it's rilly, y'know, a stupid trick, that you just turned the string around when it was hidden.

You are shocked. You are insulted. You are dumbfounded. "How could you think I would do anything so simplistic, so obvious, so easy, so unmagical? You think I just turned it around? Okay then, this time I'll make the ring jump to the middle knot." And with that you whip the string behind your back, grab it by the first-tied knot, and whip it right back out, with no pause at all. This time when you hold it up, the new knot is exposed, and the ring is hanging in the middle.

You have just gone from buffoon to miracle worker. Before your audience can recover, give a wicked smile, stuff string and ring into a pocket, pick up another 6-foot (1.8 m) piece of string, and prepare to perform the World's Greatest String Trick.

THE PROFESSOR'S NIGHTMARE

"Sometimes, even when a thing's possible, it's impractical," you begin. "For instance, if I were a very logical, educated person—say, a math professor—and I had to cut this string into three equal pieces, I'd probably measure the overall length, divide by three, make my marks at the appropriate spots, and cut." You pantomime all this with great solemnity.

"But that's so time-consuming, when all you have to do is hold the ends and fold the string in thirds." And here, you hold the string with about 2 inches of end projecting between thumbs and forefingers (Figure 9-4). Then you reach across with each ring and little finger, snag the line under the opposite hand, and move your hands apart.

"Quick and easy, right? Now you just cut the line where it is bent." And here you let go with your right hand and appear to bring "the line where it's bent" up alongside the

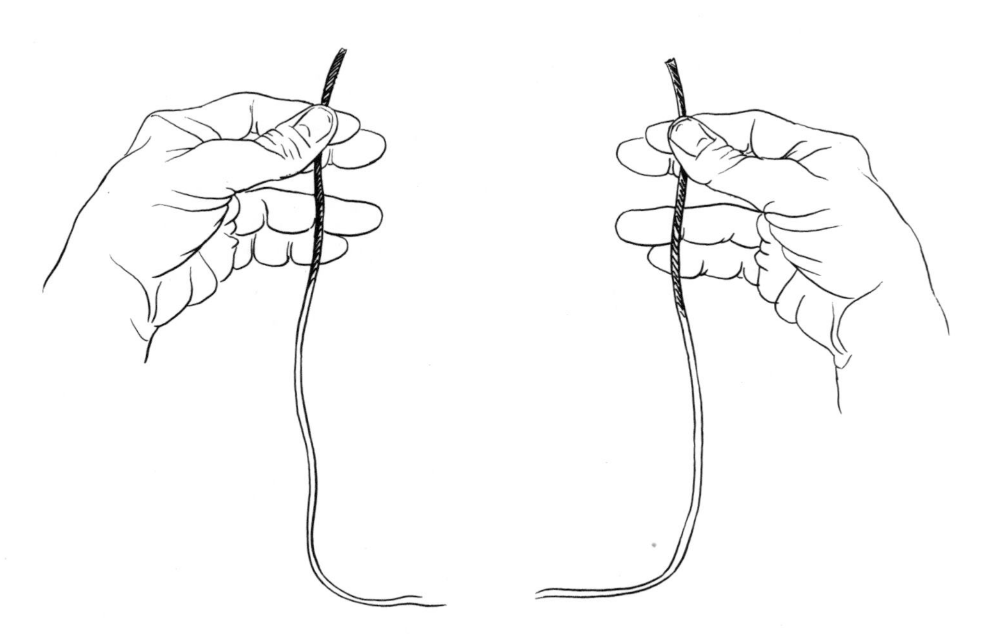

Figure 9-4. *The Professor's Nightmare. Start with the ends of a 6-foot piece of string held lightly in fingertips, ends projecting.*

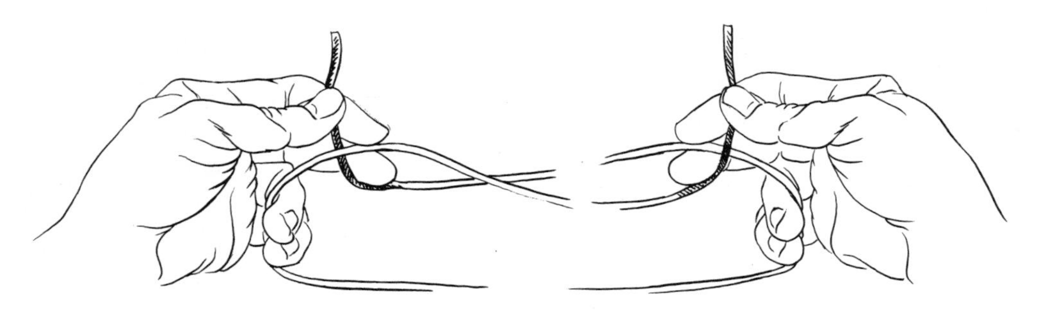

Figure 9-5. *Bring hands together and snag opposite standing parts with ring and little fingers. Move hands apart. As string goes taut, insert middle fingers as shown.*

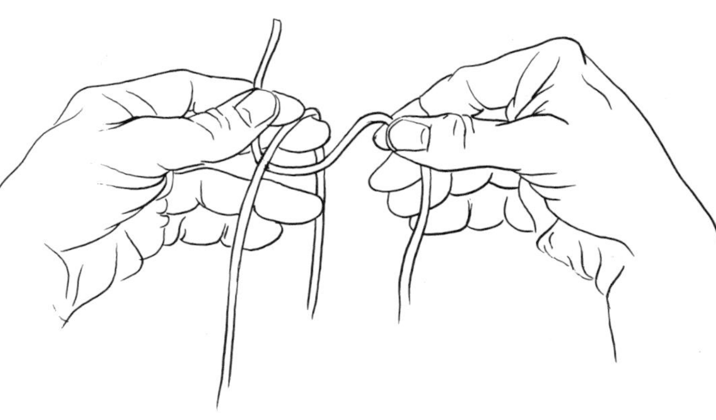

Figure 9-6. *Let go with one hand and use it to bring up the part that leads from the end on the other hand. Here we show the hands open for clarity, but in performance this move is concealed by the fingers of the left hand.*

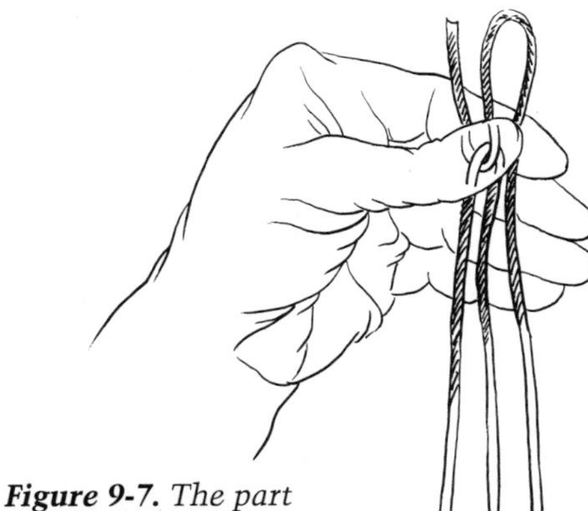

Figure 9-7. *The part brought up is laid alongside the end and pinched between fingertips. To viewers, it looks like the bight one-third along the length of the string, which is actually hidden.*

end in your left hand. Then you cut it with a sharp pair of scissors. But there's a deception here. What you actually do, with a smoothness born of long practice, is to bring up a bight of line from near the end. The real "bend" is hidden behind your hand. So when you cut, you're only snipping a short piece off the end. There's a bend at the bottom, and you reach down and cut that for real. The original bottom end might be hanging down a little, but you trim it to match the others, saying, "Hey, so what, college professors just don't know how to be expedient."

You now appear to have three equal-length pieces, but you're not through with pedagogues yet: "A professor would probably say something like, 'A' is the same as 'B', and 'B' is the same as 'C', so 'C' is the same as 'A'. Or something ponderous like

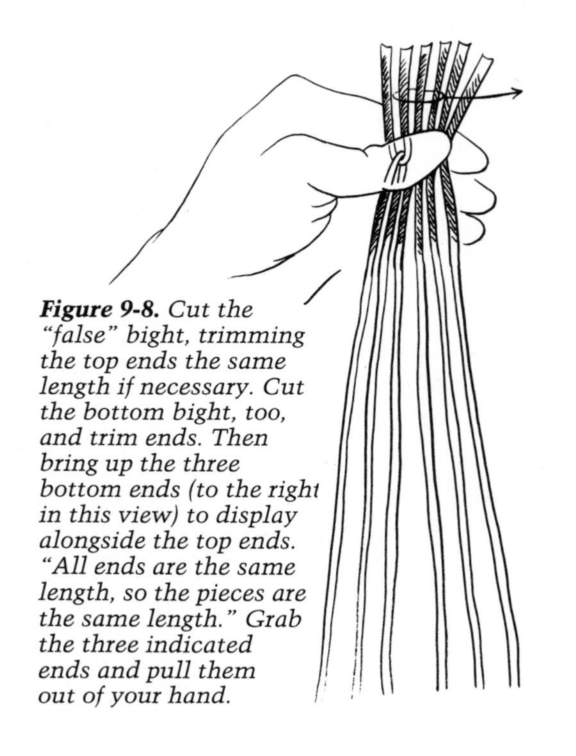

Figure 9-8. Cut the "false" bight, trimming the top ends the same length if necessary. Cut the bottom bight, too, and trim ends. Then bring up the three bottom ends (to the right in this view) to display alongside the top ends. "All ends are the same length, so the pieces are the same length." Grab the three indicated ends and pull them out of your hand.

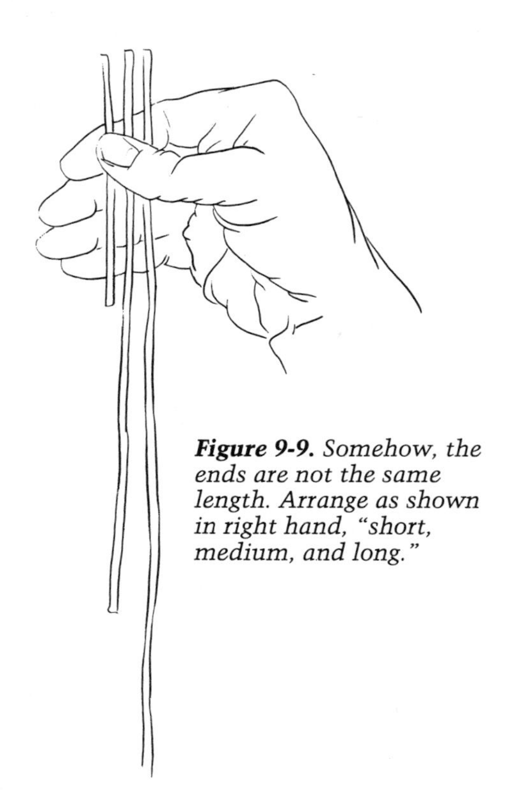

Figure 9-9. Somehow, the ends are not the same length. Arrange as shown in right hand, "short, medium, and long."

Figure 9-10. After thoroughly displaying the three different lengths, hold them in your left hand in the order shown. Bring up the bottom end of the long piece and lay it between the short and medium pieces. Practice diligently to make this a smoothly deceptive move. Then bring up the other two ends and lay them alongside the short piece.

that. But in the real world, all you gotta know is that all the ends are the same length. The top three here are even, and so are the bottom three. If the ropes are the same length, they stop at the same place. Simple."

Just to emphasize you bring up the lower three ends, one at a time, and lay them alongside the others in your fingertips.

"See? All even. So we don't need any formulas. And with that, you grab the second, third, and fourth ends from the left with your right hand, yank them out, and present three straightened pieces with a flourish.

"Yessir, three equal-len—wha?"

Panic returns. You are holding a long, a medium, and a short piece in your hand. You are as amazed as your audience is.

"But the ends were the same length—I saw them, they were—did you see them? Well, just wait a minute, let's back up here."

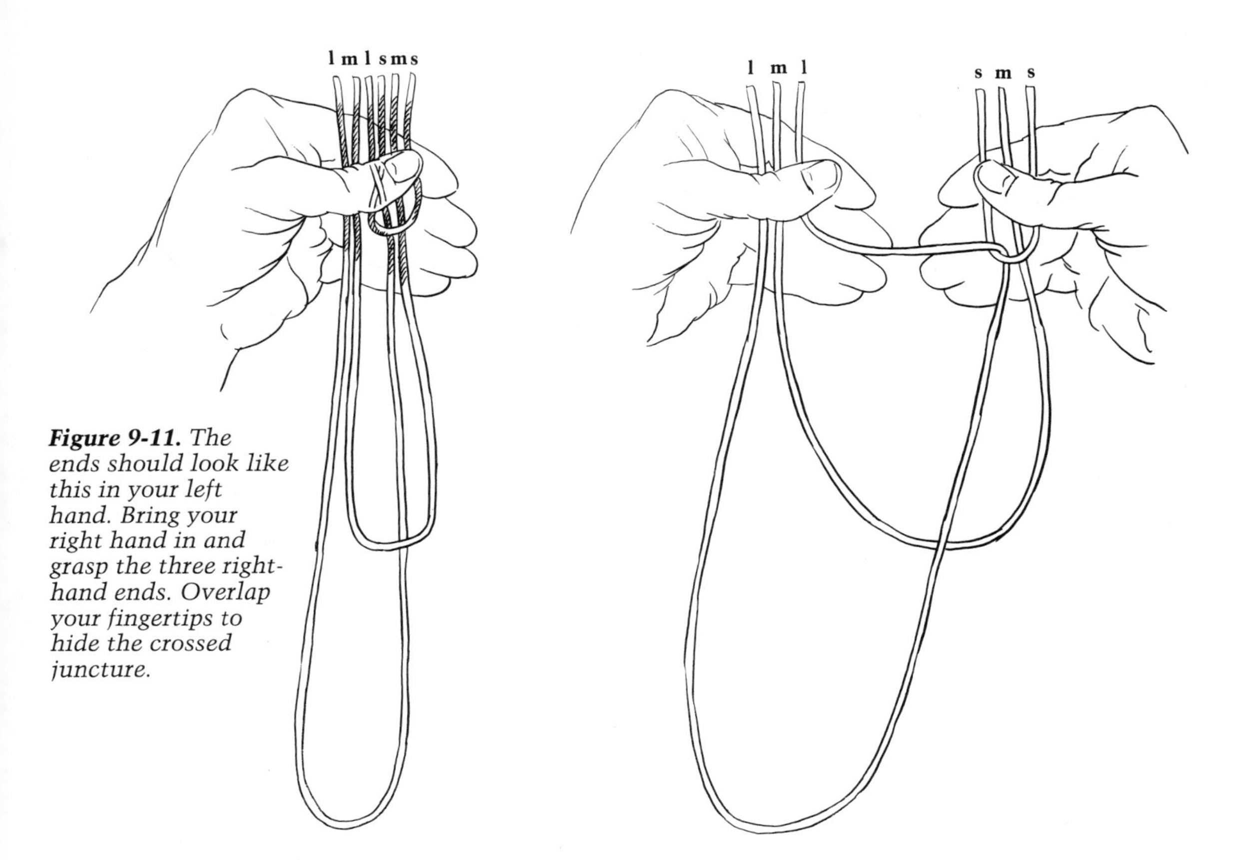

Figure 9-11. *The ends should look like this in your left hand. Bring your right hand in and grasp the three right-hand ends. Overlap your fingertips to hide the crossed juncture.*

Figure 9-12. *Move your hands slightly apart, showing the "short," "medium," and "long" pieces. Then move your hands completely apart, apparently transforming the strings into three medium-size pieces.*

With your left hand you take one piece at a time from your right, with slow, hypnotic movements, "A long, a medium, and a short." With your right hand you take them back again, "short, medium, long." And again.

You point to the pieces now lined up in your left hand. "A professor would say that 'A' is longer than 'B', which is longer than 'C'. It's all very logical. But all the ends were the same length." As you speak, you pick up the bottom end of the longest piece, which should be farthest from your fingertips. You bring it up behind the medium piece and appear to set it in place outside the short piece. But you actually set it to the left of the short piece. It's an easy sleight. It helps to look up and catch your audience's eye as you do it. It also helps to practice the move until you can fool yourself in a mirror.

Now you bring up the medium end and lay it outside the others, then the short end. Note that the short piece passes behind the folded long piece.

Time for another crucial move. You bring your right hand up so that it overlaps your left hand, grasp the three rightmost ends, and move your hands slightly apart. It will look exactly as though the long, medium, and short pieces were draped between your hands. In reality, you'll have both ends of the short piece and one end of the medium piece in your right hand, and

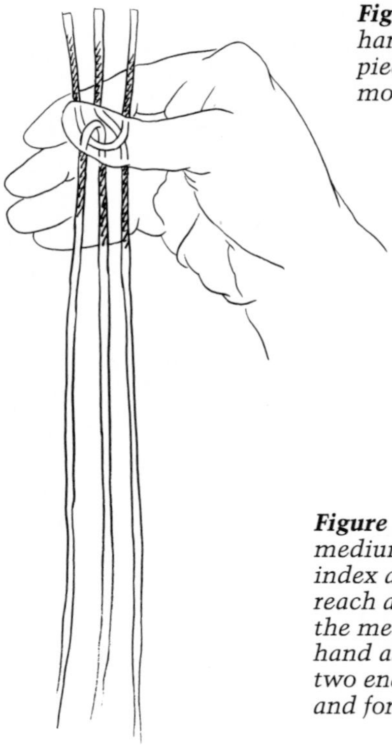

Figure 9-13. *Hold the ends now as shown in your right hand. With your left, reach across, grasp the middle piece, and draw it out. This should echo your previous moves of showing "long, medium, and short."*

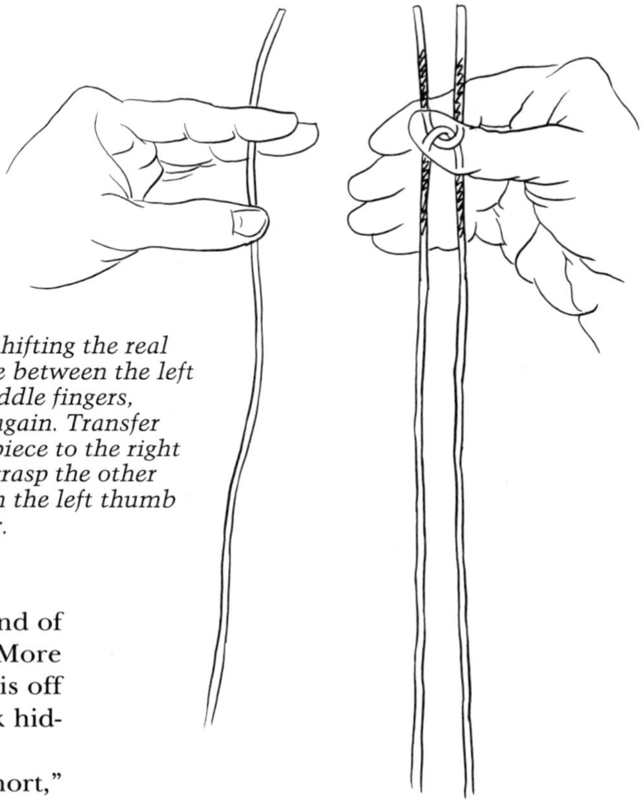

Figure 9-14. *Shifting the real medium piece between the left index and middle fingers, reach across again. Transfer the medium piece to the right hand as you grasp the other two ends with the left thumb and forefinger.*

both ends of the long piece and one end of the medium piece in your left hand. More mirror time here, so you can bring this off casually without exposing the interlock hidden by your right hand.

"We have a long, a medium, and a short," you continue, "but common sense tells us that if the ends are the same length, the pieces are the same length."

And with that you move your hands smoothly apart, snapping the strings between them smartly as they fetch up. Then you let go with your left hand and hold up the miraculously restored three medium-length pieces.

Resuming your hypnotic examination movements, you reach across with your left hand and take the middle end, the actual medium-length piece, and draw it out. Move just fast enough, and the audience will not notice that it isn't one of the end pieces. As

GEORGE PITKIN'S CRAZY IDEA

When doing fancywork with small cordage, the ends ravel easily, making it difficult to tuck them through tight spaces. Taped ends are too bulky, as are Constrictors in all but the finest, most-difficult-to-manipulate twine. But if you dip the ends in Krazy Glue and wait a few seconds, your ravelling days are over.

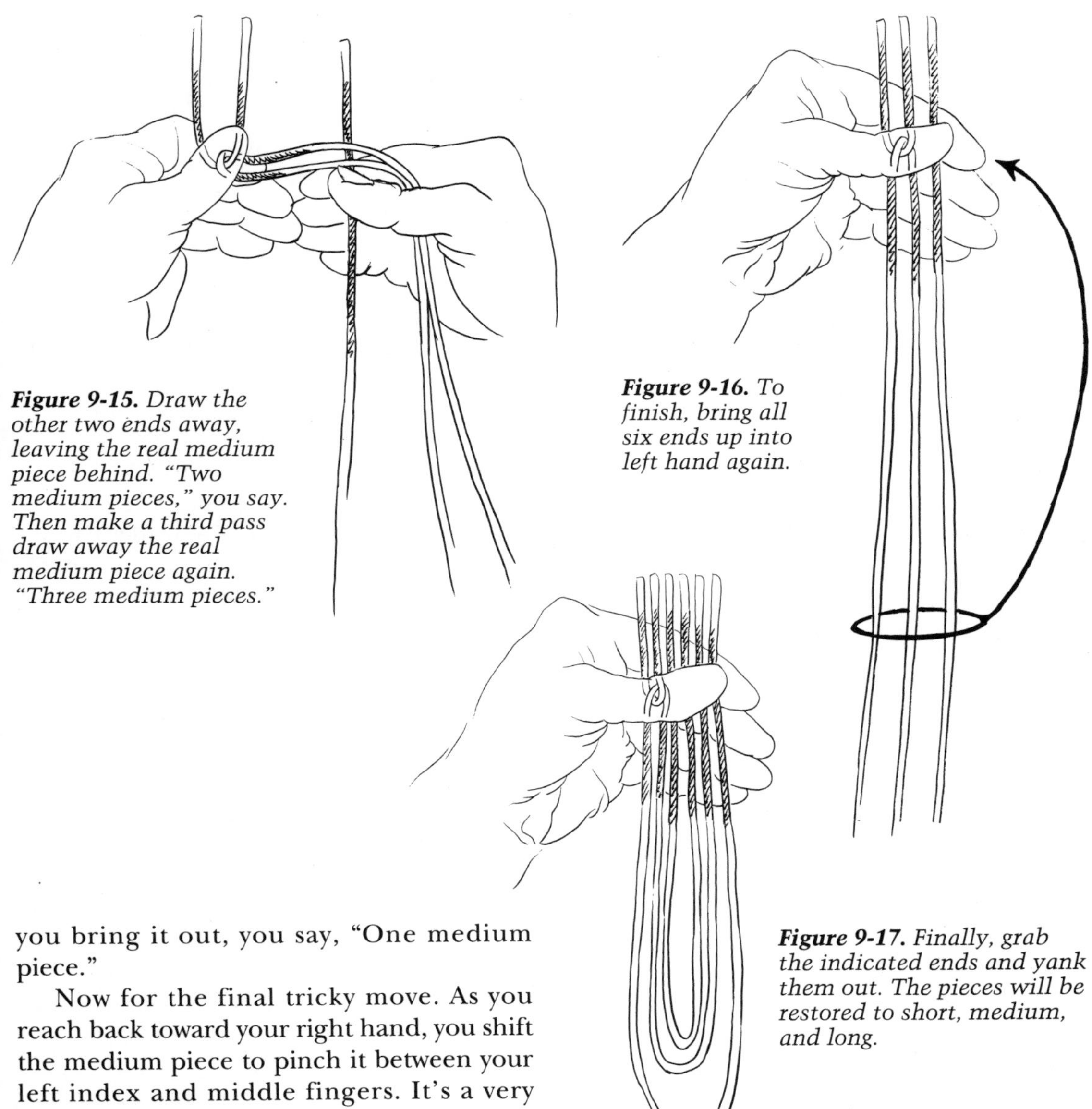

Figure 9-15. Draw the other two ends away, leaving the real medium piece behind. "Two medium pieces," you say. Then make a third pass draw away the real medium piece again. "Three medium pieces."

Figure 9-16. To finish, bring all six ends up into left hand again.

Figure 9-17. Finally, grab the indicated ends and yank them out. The pieces will be restored to short, medium, and long.

you bring it out, you say, "One medium piece."

Now for the final tricky move. As you reach back toward your right hand, you shift the medium piece to pinch it between your left index and middle fingers. It's a very subtle move. When you bring your hands together, your left thumb and index fingers grasp the other two ends and your right thumb and index fingers grasp the medium piece. You draw out the two ends, the interlock hidden behind your left hand, and leave the medium piece behind.

"Two medium pieces," you say, and it looks indeed as though you'd simply pulled out a second medium piece. It's impossible, but the audience sees it happen. Now you reach across, again, and draw away the real medium piece once more, saying, "And three medium pieces."

To finish, you once again bring the bottom ends up one at a time, and lay them alongside the upper ends.

"But because we live in a world where professors have to make a living, too"—you grab the second, third, and fourth ends from the left and yank them out—"We also have"—once more with the hypnotic moves—"a long, a medium, and a short."

Applause, applause.

CIRCLING THE WORLD

And now a brief interlude while you rest your fingers. Regard your bit of string contemplatively, and relate to your audience the following:

Imagine a string tied around the earth at the equator. Call the Earth's diameter at this point an even 8,000 miles. How much longer would the string need to be in order to be one foot above the earth's surface all the way around?

The amazing answer is: 6 feet 3½ inches!

Here's the math: A circle's circumference is its diameter multiplied by π (3.14159...). 8,000 miles x π = approximately 25,000 miles. To get the string a foot off the surface all the way around, you just add 2 feet to the diameter (a foot at either end). The diameter is now 8,000 miles plus 2 feet, and π x 2 feet is 6.283 feet or 6 feet 3⅜ inches. If you want to work it out in big numbers, 8,000 miles is 42,240,000 feet (8,000 x 5,280). Eight thousand miles plus 2 feet is 42,240,002 feet. The circumference of the second is 132,700,873.7 feet. The difference between the two, somehow, against all intuitive math, is 6.283 feet, or 6 feet 3½ inches. I don't know about you, but no matter how painstakingly and exhaustively I prove it, my brain says, "It just can't be so." Go figure.

THREADING THE NEEDLE

Magicians are fond of saying that the hand is quicker than the eye, but they usually don't explain how to train a hand for such speed. But you can let your audience in on a little exercise called "Threading the Needle," using the long piece from the "Professor's Nightmare."

Wrap the string around your thumb, from the base out, making the turns come toward you over the top of the thumb (Figure 9-18). 4 feet (1.2m or so) of the end hangs down below your hand. After you put on 8 to 10 turns, bring up a bight, with the end on the inside (away from your fingertips), and pinch it between your thumb and index finger at their very tips.

"Now this is just a beginner's exercise," you explain. "When you get really good, you can thread a sewing machine while it's running. But this is a good start. See, I'm going to pick up this end here"—and you pick up the end you left hanging down at the beginning of the turns—"and thread it through this little-bitty eye, so it'll look just like this"—and here you slowly thread it through the eye, which is just barely big enough for the line—"only I'm going to do it faster than you can see, so fast that I need all these extra turns here to keep the string from flying out of my hands. When I'm through, it'll look just like this"—and after threading it through, you grab the end and pull it taut, so it looks like Figure 9-19.

The scene set, you take the end back out of the eye, grasp it once more near its end, and commence a squinty-eyed, ludicrously intense warmup-and-taking-aim ritual. Then in a Bruce Lee–like blur of motion (Bruce Lee–like sound effects are a nice touch, too) you shoot your right hand forward. And sure enough, you've threaded the eye at impossible speed.

Or so it appears. What has actually happened is that as the line came taut, the first turn automatically undid itself and was yanked up into the eye. Play with it and you'll see what I mean. No digital dexterity

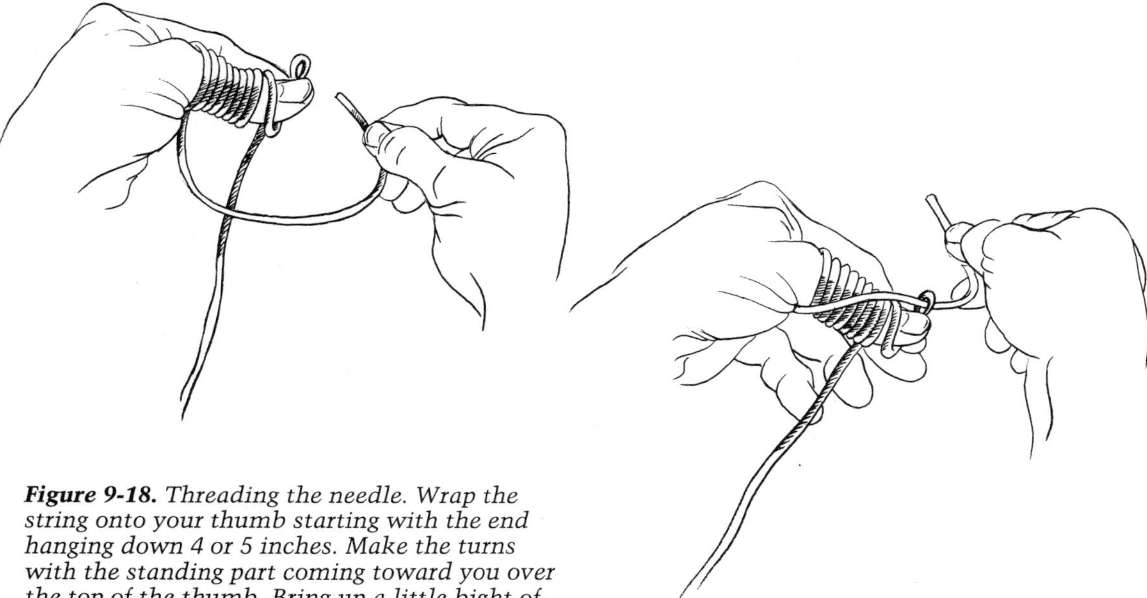

Figure 9-18. *Threading the needle. Wrap the string onto your thumb starting with the end hanging down 4 or 5 inches. Make the turns with the standing part coming toward you over the top of the thumb. Bring up a little bight of the standing part after you have enough turns, give it a half-turn counterclockwise, and pinch it with the long end innermost from your thumb and forefinger tips. Pick up the short end and make as though to thread it through the small bight with lightning speed . . .*

Figure 9-19. *. . . but instead simply pull the end taut. This will cause one of the turns to fly off the base of your thumb and emerge in the little bight. It looks exactly as though you've threaded a needle at impossible speed.*

required. But you don't tell that to your audience. And you don't do this trick anywhere near a sewing machine.

THE MOBIUS BOWLINE

The Bowline is such a friendly, familiar knot, but do we really know it? If you hitch your "Nightmare" long piece to a chair, rail, or such, then tie a Bowline, leaving a long end, and then tie that long end around the loop of the knot, can the Bowline then be untied? Potential puzzle-solvers can do anything they want except untie either end. Be prepared to step in, untangle, and retie; people can get carried away.

The solution, as you can see (Figure 9-20), is quite simple: Loosen the turn at the top of the knot, pull it down, pass the body of the knot through it, and you have a slipknot which can easily be straightened out.

When someone does manage to untie the knot, they will have displayed rare topological ability; be sure to congratulate them. Then invite them to retie it.

By the by, this geneticaly altered third cousin to a Bowline-on-the-Bight is good for more than giggles; if, as the drawings show, the small edge at the end is spliced, you can make a loop knot that will always stay tied, even in the most slick, springy, jerked-upon line, and yet is easily tied, untied, and adjusted in size just like a regular Bowline.

THE GOOD OLD ACTUAL INDIAN ROPE TRICK

"But enough of puzzles and deceptions," you say, "it's time for some real magic. During one of my trips to India," (pause to glare

CHINESE GOOD LUCK KNOT

I have a friend named Joseph Roberts, whose work as a geologist has taken him into the outback, upland, and darkest depths of every continent. And wherever he has gone, he has gained immediate entree to local society by the simple act of approaching the village shaman or equivalent and asking to learn how to tie a knot that brings good luck. And there always is such a knot, and Joseph always picks up another for his collection, and he is always warmly received because he's asking for something simple and nice instead of the shaman's usual diet of requests for love potions, curses on enemies, and relief from colds.

It turns out that just about every basic knot, and quite a few complex ones, are considered good luck by someone, somewhere. Only the Granny and the Hanging Noose stand out as ligaturistically malevolent. It might be argued that most knots are metaphors for luck; that every time you tie a Bowline, or a Butterfly, or a splice, you are engaging in a ritual that speaks to connection, unity, trust.

If that's so, then this particular knot has earned its preeminence as a Good Luck Knot, for it combines an elegant tying procedure, graceful proportions, some utility (it's very secure, but will jam), and a vivid (if pagan) *picture* of the nature of Good Luck. You'll find front and back views of this knot on this book's frontispiece and endpiece. One side forms a cross, which even in pre-Christian China represented Heaven: linear, yang, to the point. The other side is a square, ancient symbol of Earth: great, broad, enduring. It is in the harmonious interplay of these two profound ideas that good luck arises.

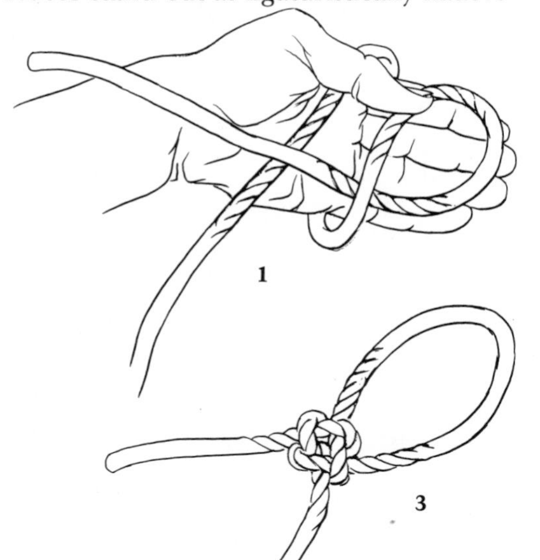

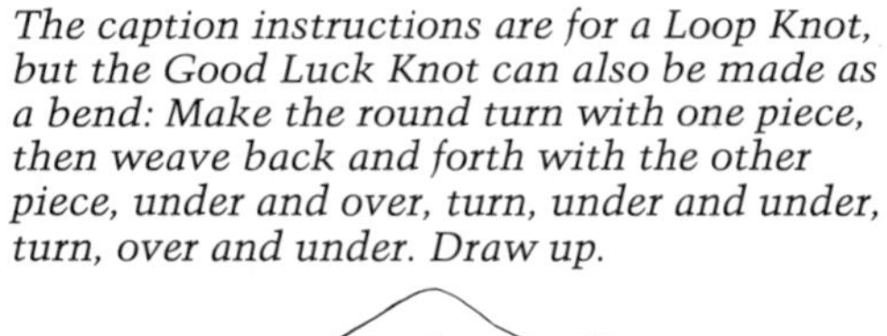

The caption instructions are for a Loop Knot, but the Good Luck Knot can also be made as a bend: Make the round turn with one piece, then weave back and forth with the other piece, under and over, turn, under and under, turn, over and under. Draw up.

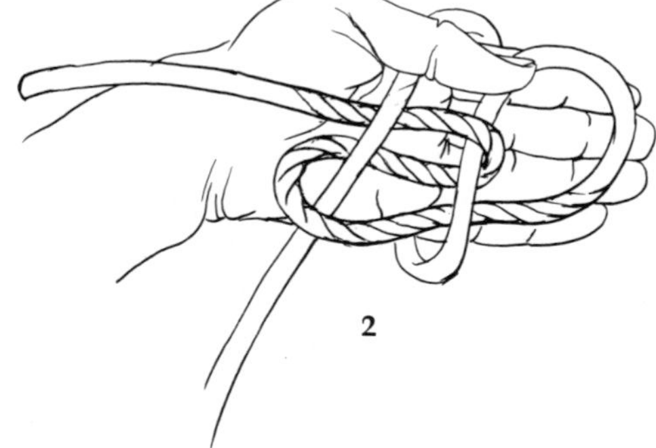

1. Make a round turn on your hand. Then pick up and form it into a loop on the palm, with the end under its own part and on top of the standing part.
2. Lead the end back under both parts on your palm. Then reverse direction again and lead it over its own part and under the standing part. Remove your hand and draw up carefully, working the square down smaller and smaller.

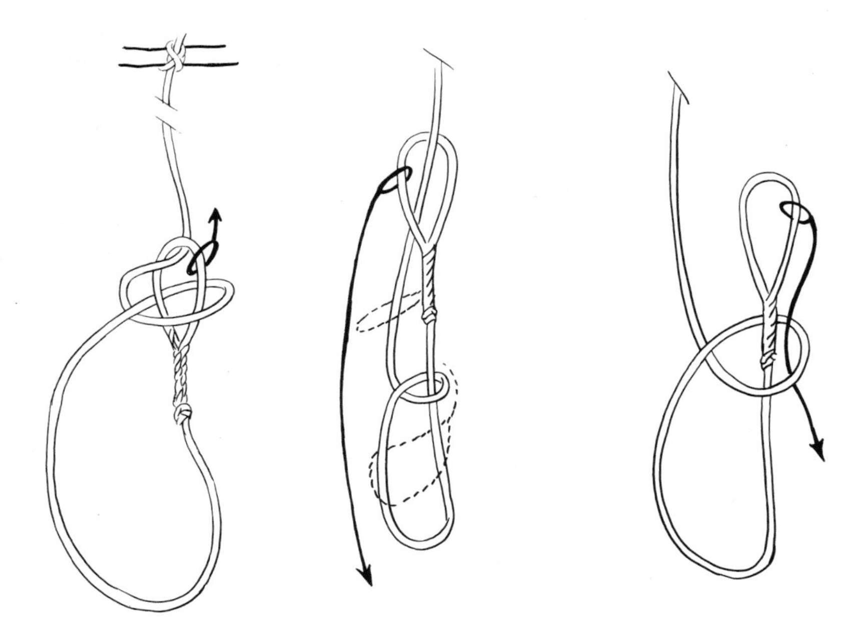

Figure 9-20. *Untying the Mobius Bowline, a topological adventure.*

at scoffers, unless you happen to have been to India, in which case they actually might believe what you're about to say), "high up in the Himalayas. . ." And you go on to some unlikely meeting with a swami who taught you the famous Indian Rope Trick, wherein "an ordinary piece of cordage can be made to stand vertically."

Pick up your much-used big piece of string, hold one end with one hand and slide your other hand up to the other end, stretching the string out vertically. You do this after having entered a deep trance, of course, and you accompany your moves with a semi-intelligible mantra.

You pause a moment in silence, string held taut, intense concentration etching your features. Then you let go with your upper hand, and the string collapses.

Hmmm. After a moment's puzzlement, you resume your chant, louder and even less intelligibly, and slowly stretch the string taut once more. Another pause, you let go and. . . it collapses again.

Consternation. Then your face brightens—you have remembered the secret. You repeat your actions and mantra again and then you let go again. With your bottom hand.

"Heh, heh, heh."

Index

I

J

K

L

M

N

P

R

S

T

W

The Good Luck Knot, back view. *See page 204.*